Boston Studies in the Philosophy and History of Science

Volume 333

The series *Boston Studies in the Philosophy and History of Science* was conceived in the broadest framework of interdisciplinary and international concerns. Natural scientists, mathematicians, social scientists and philosophers have contributed to the series, as have historians and sociologists of science, linguists, psychologists, physicians, and literary critics.

The series has been able to include works by authors from many other countries around the world.

The editors believe that the history and philosophy of science should itself be scientific, self-consciously critical, humane as well as rational, sceptical and undogmatic while also receptive to discussion of first principles. One of the aims of Boston Studies, therefore, is to develop collaboration among scientists, historians and philosophers.

Boston Studies in the Philosophy and History of Science looks into and reflects on interactions between epistemological and historical dimensions in an effort to understand the scientific enterprise from every viewpoint.

More information about this series at http://www.springer.com/series/5710

Florence Bretelle-Establet • Marie Gaille
Mehrnaz Katouzian-Safadi
Editors

Making Sense of Health, Disease, and the Environment in Cross-Cultural History: The Arabic-Islamic World, China, Europe, and North America

Springer

Editors
Florence Bretelle-Establet
Université de Paris
SPHERE, UMR 7219, CNRS-Université de Paris
Paris, France

Marie Gaille
Université de Paris
SPHERE, UMR 7219, CNRS-Université de Paris
Paris, France

Mehrnaz Katouzian-Safadi
Université de Paris
SPHERE, UMR 7219, CNRS-Université de Paris
Paris, France

ISSN 0068-0346 ISSN 2214-7942 (electronic)
Boston Studies in the Philosophy and History of Science
ISBN 978-3-030-19081-1 ISBN 978-3-030-19082-8 (eBook)
https://doi.org/10.1007/978-3-030-19082-8

This Springer imprint is published by the registered company Springer Nature Switzerland AG.
The registered company address is: Gewerbestrasse 11, 6330 Cham, Switzerland

“Dry, open situations meanly elevated, […] are above all others the healthiest; for such habitations have a free, pure, open air […] Low habitations, especially on stiff clay, rotten earth, or near a level with the sea, great rivers, marshes, lakes or putrid standing waters. These are worst of all; for their air is always moist, gross, and loaded with exhalations often putrid”. Short, 1750

“The invisible shapes of smells, rhythms of cricketsong, and the movement of shadows all, in a sense, provide the subtle body of our thoughts. Our own reflections, we might say, are a part of the play of light and its reflections. ‘The inner—what is it, if not intensified sky?’”. Abram, 1996

“In the wild mountain forests, a humid Qi accumulates. Near the sea, it is even worse and this Qi causes diseases. […] Also there can be a putrid Qi coming from rotten herbs and plants. We call it the toxic mountain miasma .山林曠野常有濕氣集乎其間。而瀕海尤甚其氣病人[…]或腐草污木淤泥之臭氣。名為山嵐瘴毒”.
Chen Zhen’ge陳珍閣, 1892

“In fact, winds are the main agents of all diseases Φαίνονται οὖν αἱ φῦσαι διὰ πάντων τῶν νοσημάτων μάλιστα πολυπραγμονοῦσαι ”. Hippocrates, 5th BCE

“ I corpi di pori assai apperti sono piu facili ad infettarsi ”. Soldi, 1431

“People from Yue, their pores are often open, their skin and flesh are not closed, this is because of an excessive sudation which lets evil from the outside (enter)粵人疏理元府常開毛腠不掩每因汗溢即致外邪”. Qu Dajun屈大均, 1678

“The wind is the origin of the one hundred diseases 故風者百病之始也”.
Huangdi neijing 黃帝內經, 3th BCE-2th CE

“La Nature a fait tous les êtres les uns pour les autres, et n’a point voulu qu’aucun d’eux fût isolé ou inutile”. Moreau, 1798

“Both dangers and benefits arise from our food”.
Mollison, 1988

“The harmful species are as important as the useful species… Their harmful side is of capital importance, for all these beings are complementary and indispensable to ensure general interest and global welfare

"فسبحان من جعل منافعها نعمة، ومضارها ترجع إلى أعظم المنافع ... وجعل في الجميع تمام المصلحة، وباجتماعهما تتم النعمة، وفي بطلان واحد منها بطلان الجميع، إنما هو واحد ضم إلى واحد وواحد ضم إليهما." الجاحظ Ǧāḥiẓ, 9th

“There are good scientific and historical reasons for regarding the state of our soils as the best measure of the future health and well-being of society”.
Holmgren, 2011

“The physician’s treatment should be adapted to the bodies in Egypt […]. The harmful effect of such drugs can be rapid in the body, especially in Egyptians, whose bodies are quick to suffer from injury. He should select from purgative drugs and others the gentlest in strength, so that there is no discomfort or harm to the bodies of Egyptians. He should not prescribe drugs found in the medical books of Greeks and Persians, for most of them are aimed at bodies with strong constitutions and with strong coarse humors

”وجب على الطبيب ان يختار من الأغذية ومن الأدوية [...]ما كان من الأدوية المسهلة وغيرها ألين قوة حتى لا تكون على طبيعة المصريين فيها كلفة ولا يلحق أبدانهم منها مضرة. ولا يقدم على الأدوية الموجودة في كتب الأطبا اليونانيين و الفرس فإن أكثرها عملت الأبدان قوية البنية غليظة الأخلاط.“ ابن رضوان Ibn Riḍwān, 11th

“in the North, [..] (People’s) constitution is robust, and people eat and drink with moderation. At home they are thrifty and plain and it is very rare that they contract a disease that injuries their original Qi. If they have a disease, one must use bitter and cold remedies to course and disinhibit. The disease disappears and immediately people feel comfortable. Conversely, Southeastern people’s constitution is delicate, their pores are not closed, they take too much food and drink, and have too much sex, they are very different from Northwestern people, if you give them bitter and cold remedies to purge them, wouldn’t it be the same as killing them with a blade?. 蓋北方[...]，稟賦雄壯，兼之飲食倍常。居室儉素，殊少強賊元氣之患。一有疾病，輒以苦寒疏利之。若夫東南之人，體質柔脆而飲食色欲之過侈，與西北之人迥異，概以苦寒之劑攻之目不幾有操刃而殺人乎”. Yu Bian 俞弁, 16th

Acknowledgments

This volume is one of the results of a collaborative project that started in 2012 between a small group of historians and philosophers—belonging to the SPHERE research team—working on the history of science and medicine in the Arabic-Islamic world, in China, in Europe, and in Latin America. This project began as a collective reflection on the issue of the relationships between health, disease, and the environment in a SPHERE seminar—*Environment, Disease, Therapy*. The financial support provided by SPHERE allowed us to organize a workshop in 2015, *Environment and Health between Observation, Imagination and Theories in Different Parts of the World*. This workshop provided the opportunity to transform our initial questions into a more elaborated reflection and to enlarge the context in which the relationships between health, disease, and the environment should be approached, as researchers from various parts of the world, notably from Morocco, China, and Tunisia, made presentations in this workshop. It also allowed us to determine the organization and the content of this book.

This volume owes its success to different people to whom we would like to express our gratitude. Our thanks first go to all the contributors of this volume, for their research, their cheerful input into this collective book, and their patience. This book includes many authors who are not English native speakers and needed to have their contribution translated or edited. Our thanks thus also go to Richard Kennedy for his careful translation or editing of the various contributions and to several institutions, namely, l'Institut Confucius Diderot, l'Institut Humanités, Sciences et Sociétés (IHSS), and SPHERE (Université de Paris, UMR 7219, CNRS & Université de Paris), for their financial support for this work.

Writing in a language which is not your mother tongue always carries something like an accent, which in spite of efforts to erase it, is like the local soil that sticks to your shoes, to paraphrase the poet Miguel Zamacoïs (1866–1955, "L'accent"). Each time you speak, your words while addressing something else, still speak of your region. "L'accent ? c'est un peu le pays qui vous suit… Avoir l'accent, enfin, c'est chaque fois qu'on cause, parler de son pays en parlant d'autre chose !". Let's hope that the contributors of this book, speaking from and about various geographical regions, will be understood and will also echo the cultural and linguistic diversity essential to this cross-cultural book. Book of this sort, written in a second language requires the combined effort of those who write and those who read… so, we would also like to offer our thanks to all our readers!

Contents

1 **Introduction** 1
Florence Bretelle-Establet, Marie Gaille, and Mehrnaz Katouzian-Safadi

Part I Observations, Definitions, and Theories About Environment, Disease, and the Body: How to Embrace the Whole World

2 **Creation, Generation, Force, Motion and Habit: Medieval Theoretical Definitions of Nature** 27
Isabelle Draelants, with the collaboration of Eduard Frunzeanu

3 **The Animal Environment and Human Health: The Approach Followed by the Medieval Zoologist Ǧāḥiẓ (Ninth Century)** 61
Ahmed Aarab, Kaouthar Lamouchi-Chebbi, and Mehrnaz Katouzian-Safadi

4 **Landscaped Environment and Health in Han China (208 BCE–220)** 79
Catherine Despeux

Part II Observations, Definitions, and Theories About Environment, Disease, and the Body: Questioning the Meaning of the Environment

5 **The Construction of Thinking on the Environment: The Words, Their Meanings, and Their Uses from 1790 to 1970** 105
Valérie Chansigaud

6 Environment in Relation to Health, Wellbeing and Human Flourishing: The Contribution of Twentieth-Century Continental Philosophy of Life and of the Subject 125
Marie Gaille

7 Environment and Chagas Disease: An Elusive and Diverse Relationship 145
Gabriel Gachelin

Part III Healthy or Unhealthy Environments: Sensorial Experiences of Lands, Spaces, and Milieus

8 The Worst Environment in Which to Live in China: A Question of Points of View. The Legendary Miasmatic Far South of China Challenged by Local Doctors in Late Imperial China 165
Florence Bretelle-Establet

9 Inhabited Lands and Temperaments: Observations and Therapeutic Solutions, the Views of Scientists and Medieval Physicians – Ǧāḥiẓ (9th), Rāzī (9th–10th), Ibn Riḍwān (11th) 209
Mehrnaz Katouzian-Safadi and Meyssa Ben Saad

10 Health and the Environment: Aldo Leopold, Land Health, and the First-Person Ecology Approach 239
Catherine Larrère

11 Urban Space of the Living and Dead: The Conception of Environment and Death in Beijing from the Eighteenth Century to the Middle Twentieth Century 255
Xi Ju

Part IV Healthy or Unhealthy Environments: Techniques, Tools, and Concepts to Reconsider the Relationship Between the Environment, Disease and Health

12 Urban Nature: (The) Good and (The) Bad 287
Nathalie Blanc

13 Health and the Environment in Ecological Transition: The Case of the Permaculture Movement 309
Laura Centemeri

14 "Affordances": A Notion that Reveals How Artificial It Is to Separate the Body and Its Environment 333
Anne-Lyse Chabert

15 Gestalt Therapy and Its Contribution to the Understanding of the Link Between Health and the Environment 347
Perrine Michon

Index 367

Contributors

Ahmed Aarab Faculty of Sciences and Techniques, University Abdelmalek Essadi, Tanger, Marocco

UMR 7219 – SPHERE, CNRS/Université Paris Diderot, Paris, France

Meyssa Ben Saad Université de Paris, SPHERE, UMR 7219, CNRS-Université de Paris, Paris, France

Nathalie Blanc Université de Paris, LADYSS, UMR 7533, CNRS-Université de Paris, Paris, France

Florence Bretelle-Establet Université de Paris, SPHERE, UMR 7219, CNRS-Université de Paris, Paris, France

Laura Centemeri CEMS, CNRS-EHESS-Inserm, Paris, France

Anne-Lyse Chabert Université de Paris, SPHERE, UMR 7219, CNRS-Université de Paris, Paris, France

Valérie Chansigaud Université de Paris, SPHERE, UMR 7219, CNRS-Université de Paris, Paris, France

Catherine Despeux INALCO, Sorbonne Paris Cité, Paris, France

Isabelle Draelants Institut de recherche et d'histoire des textes IRHT, UPR 841, CNRS, Paris, France

Eduard Frunzeanu Institut de recherche et d'histoire des textes IRHT, UPR 841, CNRS, Paris, France

Gabriel Gachelin Université de Paris, SPHERE, UMR 7219, CNRS-Université de Paris, Paris, France

Marie Gaille Université de Paris, SPHERE, UMR 7219, CNRS-Université de Paris, Paris, France

Mehrnaz Katouzian-Safadi Université de Paris, SPHERE, UMR 7219, CNRS-Université de Paris, Paris, France

Kaouthar Lamouchi-Chebbi Université de Paris, SPHERE, UMR 7219, CNRS-Université de Paris, Paris, France

Manouba University, Tunis, Tunisia

Catherine Larrère ISJPS, UMR 8103, Université Paris 1 - Panthéon-Sorbonne, Paris, France

Perrine Michon Lab'Urba EA 3482, CNRS-Université Paris Est Créteil UPEC, Créteil, France

Xi Ju School of Sociology, Beijing Normal University, Beijing, China

List of Figures

Fig. 9.1 Architectural construction: cooling system based on wind and water. The wind tower in the city of Yazd, in the center of Iran, at the edge of the desert. (Photo, François Caudron) 228

Fig. 14.1 The main stages of the project to set up a football 5-a-side team in Mali (2011–2013) 339

List of Maps

Map 8.1 Jiangnan and the far southern margins (Yunnan, Guangxi, Guangdong) 167
Map 8.2 Prevalence of *zhang* during the Ming and the Qing dynasties 181
Map 9.1 Map of the Arabic-Islamic world: Abbasid Caliphate (750 CE) 210

List of Pictures

Picture 11.1 Beijing City in the early eighteenth to the middle twentieth century 257
Picture 11.2 The distribution of the sites where people burned funerary objects in Inner City Beijing in 1930–1950 270
Picture 11.3 The distribution of temples for keeping coffin in Beijing in 1930–1950 275
Picture 11.4 The graveyards in Outer City in 1937 (the western part) 277
Picture 11.5 The graveyards in Outer City in 1937 (the eastern part) 278

List of Tables

Table 3.1 Presentation of the authors cited 76

Table 9.1 Transliterations of Arabic and Persian words into the Latin alphabet 233

Table 9.2 Summary of the names of books and authors 234

Chapter 1
Introduction

Florence Bretelle-Establet, Marie Gaille, and Mehrnaz Katouzian-Safadi

"Climate Change Is Bad for Your Health", New York Times, Oct. 30, 2017
"Environnement : les enfants naissent pré-pollués", Le Monde, 2 octobre 2015
"家里养这10种花, 身体越来越健康"
(Growing these 10 types of flowers in your home will improve your health), 人民日报 *2017-08-17*

Keywords Health · Disease · Environment · Cross-cultural investigation · World history · Historiography · Milieu · Interpendecy · Relationship · Sensorial experiences · Aesthetics

One need only look in any newspaper and observe the flourishing market in promises of good health through diets or ways of inhabiting the earth to find evidence that the "configurationist model," suggested by Charles Rosenberg in 1992 to characterize one archetypal response to explain epidemics in the history of the West, is still very much alive. For Rosenberg, in the long history of attempts to explain epidemic diseases, two fundamental approaches existed and often coexisted in the West: contamination and configuration.[1] While the former, reductionist and monocausal, consisted in the "idea of person-to-person contagion, of the transmission of some morbid material from one individual to another," the latter implies "a vision of health as a balanced, integrated, and value-imparting relationship between

[1] Rosenberg (1992: 293–305).

F. Bretelle-Establet (✉) · M. Gaille · M. Katouzian-Safadi
Université de Paris, SPHERE, UMR 7219, CNRS-Université de Paris, Paris, France
e-mail: f.bretelle@univ-paris-diderot.fr; marie.gaille@cnrs-dir.fr;
mehrnaz.katouzian-safadi@univ-paris-diderot.fr

F. Bretelle-Establet et al. (eds.), *Making Sense of Health, Disease, and the Environment in Cross-Cultural History: The Arabic-Islamic World, China, Europe, and North America*, Boston Studies in the Philosophy and History of Science 333, https://doi.org/10.1007/978-3-030-19082-8_1

humankind and its environment."[2] From the eighteenth through the mid-nineteenth century, configurationist arguments, refined notably in miasma theory, clearly dominated medical and nonmedical thinking in Europe and North America. For the adherents of this theory, epidemic diseases could be caused by a very wide range of environmental factors: topography, air, water, food, miasmas, or garbage. By the end of the nineteenth century, this approach, which focused more on factors from the social environment, like urbanization, working conditions, and poor education, was undermined by germ theory, which directed medical opinion toward contamination in "its modern, laboratory-oriented guise." However, as Rosenberg states, "an interest in environmental determinants of health and disease remained alive and in continued dialogue with the new bacteriological etiology."[3] The contributions in this volume dealing with views on health, the natural environment, and disease in nineteenth- and twentieth-century Europe confirm this last statement.[4] We argue that, in present times, the notion of "exposome," coined by Christopher Paul Wild (complementing the "genome")[5] and defined as a person's global exposure to the environmental factors potentially altering health and leading to our modern epidemics of chronic diseases, relates to this long history of Western medical thinking that focuses on the role of the environment to account for health and disease. As an answer to the successive pitfalls of bacteriological and biological determinism and to the globalization of the health market, the environment is again at the forefront in our struggle to make sense of health and disease. Alternately or simultaneously suspected, cherished, and feared, as in the past, for its potentially good or bad effects on human health, the environment, which in some parts of the world has a legal personality, is considered as playing an essential role in public health programs. It is scrutinized with new methods and new measures and protected nationally and internationally within public, scientific, and political institutions.[6]

Taking this long run of history into account, this book particularly addresses the issue of how humankind has observed, imagined, and conceptualized the links between the environment and health. We will comment on this trio of observation, imagination, and conceptualization later on. But let us first make some remarks on the scholarly knowledge gathered on this issue up to now. The history of the relationships observed and believed to exist between human health and the environment in the West has been addressed extensively in recent decades and has been an inspiring point of departure for our own project. This history has at times

---

[2] Rosenberg (1992 : 295).

[3] Rosenberg (1992: 299).

[4] While elaborated for describing the archetypal patterns of response to epidemics in Western history and to provide analytical tools for the historian, these models, themes, or views can be used to make sense of the larger issue of visions of health and disease.

[5] Christopher Paul Wild "Complementing the Genome with an "Exposome": The Outstanding Challenge of Environmental Exposure Measurement in Molecular Epidemiology". AACR, 2005, pp. 1847–1850. DOI: 10.1158/1055-9965.EPI-05-0456 Published August 2005. This concept was referred to in the reform of the French health care system in 2015, see: http://www.assemblee-nationale.fr/14/pdf/rapports/r3356.pdf. Consulted on September 4th 2019.

[6] Shelton (2015).

been addressed incidentally in wider cultural studies on the history of health practices. It is, for instance, through a cultural history of what was considered healthy and unhealthy in France since the Middle Ages that Georges Vigarello (1993) examined and interpreted the changing views on the relationships between environment and health, themselves defined in different terms too, according to time. For his part, Alain Corbin (1982), through a history of odors, analyzed how the various smells in the natural environment were thought to affect human health. However, the story of the relationships observed and thought to exist between the human health and the environment in the West has been addressed more directly by scholars interested in the emergence of numerous scientific and political movements which occurred in the eighteenth and nineteenth centuries in Europe and North America and led to a new vision of this relationship. For them, the emergence of medical geography, at the end of the eighteenth century, the beginnings of parasitology in the first decades of the nineteenth century, and the development of bacteriology in the last decades of the nineteenth century, together with the triumph in the nineteenth century of statistics including new medical statistics, were among the important developments which undoubtedly changed the ways of understanding the relationships between the environment, disease, and health. While medical geographies and cartographies had made it possible to visualize that local environments favored certain diseases and epidemic diseases, either facilitating their spread or being their cause, through the local soil, air, or climate,[7] parasitology and bacteriology revealed the complex mechanisms of the living world and, thereby, the interdependency of all its elements, with its risks for human health. With the refinement of microscopes and laboratory techniques, applied in Europe, the United States, Japan, as well as in their colonies, the enormous realm of the invisible and of the infinitesimal was progressively revealed to humankind during the nineteenth century. While an increasing number of bacteria (and later viruses) were discovered and while more and more insects were identified as the vectors of bacteriological or parasitic diseases in humans, the environmental miasmas together with the pathological foul odors or exhalations from stagnant waters left the scientific arena, at least for a while. "Le microbe, voilà l'ennemi (The microbe, this is the enemy)" proclaimed the French Academy of Medicine, in 1896, while the successes of bacteriology encouraged scientists of the day to find bacteriological causes for all diseases, with potential curative serums or protective vaccines.[8] The catchphrase of the day, attributed to the biologist Paul Bert (1833–1886), testifies to an immense confidence placed in the medical sciences of the time: "une maladie, un germe, un

[7] By the end of the eighteenth century, large investigations were carried out in different parts of Europe (in France, Vicq d'Azir in 1777; in Germany, Leonhard Ludwig Finke, in 1792) in order to highlight a possible causal relationship between incidences of disease and elements from the natural environment (Vigarello 1993, [1999], p. 183). These investigations were also carried out in order to understand, from the very end of the eighteenth century, the diffusion of epidemic diseases. In the second part of the nineteenth century, medical geographies, accompanied with disease maps, flourished in Europe, and they were also made for European colonies (Picheral 1982; Koch 2011; Hanson 2012; Monnais-Rousselot 1998).

[8] Vigarello (1993 [1999] p. 254).

vaccin (one disease, one germ, one vaccine)." In the twentieth century, the new disciplines linked to ecology would reveal another facet of this life that all beings share—this time underlining the risks for the health of the environment itself.[9]

The extensive scholarship on the relationships between health and the environment in the West, examined from the perspectives of scientific debates, political movements, and health practices, allows us to have a better understanding of what, in Western history, was understood to be a good or a bad natural environment for humankind. It also reveals how such distinctions were made and what types of practices were promoted, individually or collectively, to maximize the beneficial environmental influences or minimize the worst.

Considering the knowledge it brings together, this scholarship leaves us some room to address the issue from a slightly different perspective. First of all, as far as science is concerned, one has to be careful not to impose the norms of contemporary science on past scientific activities. This may lead to their disqualification as nonscientific. What do we actually mean by "science"? It is a question one should always bear in mind, taking into account the different types of work in time and space. On this matter, we agree with the perspective proposed by Scott L. Montgomery and Alok Kumar; according to them, in the past, human beings "created systems of principles about how nature worked. These systems were not rigorous or theoretical in the same sense as today, yet they acted to guide the development of new knowledge about how nature worked and also the place of human beings within the whole."[10] As a result, this volume, when dealing with scientific approaches to the relationship between health, disease, and the environment, does not ground itself on a distinction between what is scientific and what is not (yet) or is prescientific. It even rejects the labels "prescientific" or "proto-scientific," which, to draw on Lichtenthaeler's words, not only obfuscate "the merits of the ancients as initiators of the present but also the real meaning of science."[11] This volume favors the history of science, which is defined here as the exploration of nature, an approach that allows a diversity of processes and methods, in the making. We thus tackle this issue in the same spirit as Armand Marie Leroi when, as a contemporary biologist, he assesses the knowledge and process used by Aristotle in his exploration of nature, seeing the latter a scientist.[12] And we also address this issue in the same spirit as most historians of science in the non-Western world who not only assess that every attempt of the past

[9] One can, for instance, refer to the works carried out since the 1970s by Lovelock (1979) and more recently by Latour (2015, 2017).

[10] Montgomery and Kumar (2016, p. 6).

[11] Lichtenthaeler's discussion on what is science and being scientific is translated from German into English by Paul U. Unschuld (2003, p.322-323). For Lichtenthaeler, science and being scientific rely on a scientific tripod-taking nature as a research object, recognizing the repetitive character of natural phenomena under the same conditions, grasping phenomena mentally—which, he shows, is not unique to modern times but already existed in Greek classical antiquity. For Paul Unschuld (2003, p. 323), it is clear that this tripod also existed in ancient China. For a discussion on the question of what pertains to science or not, from the perspective of world history, one can also refer to Unschuld (2003, p. 319–325), Voguel (2010, p. 472) and Sivin and Lloyd (2002).

[12] Leroi (2014).

to observe and explain natural phenomenon can be considered to be science but also emphasize that each scientific milieu develops questions, methods, tools, and answers in its own way.[13]

However, another specific feature of this volume is that it does not focus only on a scientific approach to the relationship between the environment and health but also consider it from a moral and aesthetic point of view. Furthermore, we consider it as multisensory beings. It is also embedded in various practices such as eating customs, funerary rites, and so on. It may be part of a conception of how human beings should live on Earth and use its resources and may be assessed from a therapeutic point of view. It is for these reasons that one of the primary intentions of this volume is equally to stress the various faculties and dimensions with which human beings envision the relationship between health and the environment—imagination as much as the theoretical faculties and the body as much as the mind—and the diverse ends or concerns that lead them to become interested in this relationship.

Finally, the feeling, the conviction, and the observation that the natural environment can affect human health have been, and are still recurrent and common, throughout the world. The citations in the epigraph to this book bear testimony to this common and, most likely, universal belief. This led us to raise the issue of the links observed and believed to exist between human beings and the natural environment in a broad geographical framework. In this investigation, without looking for spatial or historical completeness, we bring the reader from ancient, late imperial, and modern China to the medieval Arabic-Islamic world up to medieval, modern, and contemporary Europe and contemporary South and North America. By multiplying the historical and geographical settings, this book aims to enrich our knowledge of how people, in different places and at different times, have thought of and still think of the relationships between environment and health. By its methods, it strives to avoid some of the pitfalls of comparatist studies as it does not examine these relationships through the prism of the knowledge of our modern contemporary European experience. Neither does it try to find germs of thinking in ancient sources that modern sciences or contemporary epidemiology could validate as accurate today and nor does it aim to be an account of successes and priorities. It takes another route: it examines protagonists who, in different times and in different places, have reflected, on their own terms, on the links between the environment,

[13] As Chemla and Fox Keller (2017) recall, several concepts have been elaborated to account for the diversity of doing science, in the past as today, from Crombie's "style of thought," Hacking's "style of reasoning," Knorr's "epistemic cultures," or Fox Keller's "epistemological cultures," among others. We agree with these scholars that all these concepts are useful for the history of science, and we also agree that none of them can be used to characterize, in a perspective of culturalism, the history of science in one specific area. While this volume aims to be cross-cultural, at no moment do we think of cultures as something homogeneous, fixed once and for all, but we follow Chemla and Fox Keller's idea, suggested in their title "Cultures without Culturalism: The Making of Science Knowledge," to consider cultures as something always in flux, which vary according to time, place, and social milieu. The contributions gathered in this volume will provide much evidence of this.

disease, and health. From these ways of observing, imagining, and elaborating theories on the links, the influences, or the porosity between the environment and the body, this book tries to obtain a better understanding of why these links took the form they did in these precise contexts.

1.1 Prerequisites

The understanding of the relationships between the natural environment, health, and disease that prevailed in different times and places is the major issue addressed by this book, from which several questions derive. The first is our use of the very notions of "health," "disease," and "environment." The use of these words is notoriously diverse throughout time and space; their meanings are often vague and loose; and according to the context, we may encounter other words with similar meanings. To cope with these difficulties, the authors in this book first depart from an a priori definition of "environment," "disease," and "health" and do not take the contemporary meanings for granted. Besides, they do not consider that their translation from one language to another is obvious or, on the contrary, impossible.[14] Still, we have recourse to them as a specific type of notion that allows us to designate a set of questions and problems to be investigated: how "environment," "disease," and "health," if these terms or categories even existed in the different places and times examined in the book, were defined and what they meant in these different contexts is thus an important issue addressed by all the contributions.

Some years ago, in a book devoted to a European-Chinese cross-cultural investigation on the concept of "nature," the sinologist Christoph Harbsmeier underlined that it had been customary to ask such questions as "Does Chinese have the concept of Nature?" He emphasized the extent to which this question and the more general "Do they or don't they have the concept X?" was problematic. Firstly, it implies that there has been one clear notion on the Western side of the comparison—in the particular case of "nature" addressed in that book, a clear notion of Greek *Phusis*, of Latin *natura*, or, indeed, of English Nature—which, in fact, never existed. Secondly, it implies that this particular cross-cultural investigation would start from individual Western words or concepts and would consist in finding "the presence, the absence or the equivalents elsewhere."[15] To put it in another way, Fu Daiwie, who has reflected at length on the question of comparativity in history of science, questions the relevance of a methodology that compares concepts as isolated items, without taking into account the overall knowledge system that the concept is a part of. "Suppose," he wrote, "we have two cultures C1 and C2 in different space-time coordinates, with corresponding structures S1 and S2, respectively for understanding the world. Suppose C1 wants to understand things in C2, but only through the structure of S1. We might call such a type of understanding a one-dimensional, or

[14] Blanc et al. (2017).

[15] Harbsmeier (2010, p. 224).

even 'hegemonic' (if in colonial contexts), understanding." These two scholars suggested other ways to proceed in cross-cultural investigation. Harbsmeier in fact suggested to correct the former question into a more appropriate one which, again, is associated to the concept of nature, with the following question: "What kinds of distinctions did the Chinese language make in the course of its recorded history within the general *semantic field* of *regular patterns to which things are said to conform everywhere*?" For his part, Fu Daiwie invites us to learn to be foreign. "Current historical studies of science, in the case of C1 to understand things in C2, would more or less stress the importance of 'first learning S2 as a foreign language' and try to understand things in C2 through S2 as a crucial and constitutive step in the whole process of the study. We might call this process 'understanding by learning to be a foreigner' or understanding by 'learning to be different'."[16]

We suggest that this approach—to understand by learning to be a foreigner—is appropriate for our cross-cultural investigation of such notions or concepts as "environment," "disease," or "health." In this book, the semantic field we want to address could loosely be called "milieu" in the sense once given to it by Georges Canguilhem, namely, "the whole set of actions that are exerted from the outside on the living."[17] As Valérie Chansigaud's contribution states, the term "environment" itself was used at the beginning of the nineteenth century in English to refer to the conditions under which individuals or beings live or develop, as well as all the influences on the development of the life of a character. However, as she explains, this word existed well before the nineteenth century but had other meanings, linked notably to its uses in various social circles. And as Chansigaud's contribution also stresses, other terms were used in European languages to refer to the semantic field of "environment." This point was more thoroughly developed by Ferhat Taylan in his investigation on the management and knowledge of the human environments. In his doctoral dissertation, he notably highlighted the use of several expressions converging toward the same meaning: "circonstances environnantes" (surrounding circumstances), "conditions d'existence" (conditions of existence), and "milieus."[18] It is precisely one aim of this book to offer a better understanding of how people in different times and places and in different cultural and historical contexts have understood the natural world or milieu in which they were living, as well as what they considered healthy or unhealthy. It is only through such historicized and contextual approaches that the historian or the philosopher can highlight the continuities or discontinuities of the semantics of a concept and can prevent herself/himself from imposing or projecting alien categories upon foreign cultures. While modern Chinese has an equivalent for our word and understanding of "environment" in *huanjing* (環境), other terms existed in Chinese history to designate the natural milieu, such as *shanshui* (山水) (lit. mountain-water) meaning landscape, *shuitu* (水土) (lit. water-earth), and *fengtu* 風土 (lit. wind-earth), referring to the

[16] Fu (2010, p. 286).

[17] Canguilhem (1992, p. 131).

[18] Taylan (2014).

natural and cultural features of a place (topography, climate, plants, as well as local customs).[19] What all these concepts in Chinese, what the concepts of *Iqlīm* and *Kešvar* (literally, territories with climatic conditions attached to them) in Arabic, and those of *milieu*, *nature*, *statio*, *habitatio*, *environment* in European languages meant for their users are at the fore in this book.

However, this book focuses more on the interrelation between human being and the environment. Therefore, other sets of questions have been worked out in these contributions. How was the continuity or the discontinuity between the environment and the body perceived? Or, to put it differently, how was the porosity or lack of porosity between bodies and environments imagined, described, and explained? In what terms was the environment's action on human health (impact, influence, attack, intrusion, resonance) perceived? What made a healthy or an unhealthy environment is another issue addressed in this book, as are the practices which were promoted to benefit from real or supposed good environmental influences or to avoid harm really or imagined to be caused by the natural environment. Finally, a further set of questions addresses the means used for investigation in these different places and times. Before new technology, notably the microscope, allowed the direct observation of links between the environment, health, and diseases, how did people carry out their investigations and make their claims on the relationships between the environment and the body? What value was placed on direct observations, on experimentation, and on the discourse of authorities? Which senses, if any, were used (vision, smell, hearing) to make sense of these relationships and to account for the division of healthy or unhealthy environments?

Before entering into the details and the structure of this volume, two important remarks that Andrew Wear formulated in his contribution on health and the environment in early modern England also hold true for our own investigation: "What the illiterate, who comprised the majority of the population thought about health, remains largely a blank: the history of how people perceived health and the environment in this period is based on evidence from the literate." With the exception of Ju Xi's article in this book, which includes oral testimonies from old people collected in Beijing, and Nathalie Blanc's and Laura Centemeri's contributions, which partly rely on contemporary field surveys, all the contributions in this volume rely on written sources from literate milieus. However, this does not mean a homogenous source of information, since a second remark from Andrew Wear also holds true for our research project: "It was not only doctors who might have had something to say on this question [...] Anyone could be an authority about health and the environment, because anyone could draw upon personal experience and upon traditional knowledge." [20] In addition to the learned medical theoreticians, philosophers, geographers, historiographers, urbanists, foresters, zoologists, and

[19] This association of characteristics also existed in Japanese from which the philosopher Tetsuro Watsuji (1889 – 1960) coined the Japanese concept of "*fudosei* 風土性" (that Berque translated into French as "la médiance en tant que moment structurel de l'existence"). See Tetsuro, *Fudo* 1935 translated by A. Berque (2011).

[20] Wear (1992, p. 131, 133).

agronomists, among others, had something to say. This book therefore opens a window on reflections which were elaborated within these different milieus and is based on the analysis of various types of texts and testimonies identified by their authors and/or readers as "scientific," "medical," "literary," "philosophical," "poetic," "psychological," or "institutional." It was important for us to keep this variety in mind, thanks to which the relationship between the environment, health, and disease is perceived and elaborated with a plurality of objectives: aesthetic, pragmatic, scientific, medical, and political.

1.2 Structure of the Book

Since the issue of examining, measuring, and controlling the influences of the environment on human health has, in recent decades, been put to the forefront in scientific and political terms and in different social arenas, it was of paramount concern for us to tackle this project by following an approach firmly anchored in a rigorous analysis of sources. Such an approach, we believe, is important so as not to fall into the many pitfalls open to a project of this type, namely, one that covers a wide geographical and chronological framework and where the very notions of "environment," "disease," and "health" cannot be taken for granted but have to be defined on each occasion. We thus begin with two parts dedicated to shed light on the various understandings of linkages between the environment and the body across both historical and geographical divides. The first part of this book, *Observations, Definitions, and Theories About the Environment and the Body: How to Embrace the Whole World*, brings together contributions in which particular attention is paid to the words and categories used by the actors we have chosen to present and to what these words and categories meant to them in their context. It also brings to the fore how people in these various parts of the world have imagined their links with the natural milieu that surrounded them. These contributions highlight reflections that share a common ambition of embracing the whole world, an investigation which appears to have been meaningful in these different places, societies, and times.

In their contribution on the Western Middle Ages, Isabelle Draelants and Eduard Frunzeanu focus on the wide range of meanings conveyed by the term "nature" as, at this time, the reflection on the world surrounding humankind cannot be separated from a wider reflection on "natura." To illustrate the various theoretical conceptions of nature in Middle Ages Europe, they examine different texts but pay particular attention to an encyclopedia (*Speculum doctrinale*, XV, Chap. 4) compiled by the Dominican Vincent de Beauvais in the thirteenth century. This encyclopedia synthetizes the theoretical conceptions of nature available at its time, as it includes all literary written evidence of its time as well as inputs from theology and natural philosophy brought by the translations into Latin of Arabic and Greek sources. Draelants and Frunzeanu show how, after the twelfth century, intellectuals or learned scholars paid increasing attention to the natural world, with eyes more turned to the here below than to the world beyond and its Salvatore. It is the beginning in the Western Middle Ages of a "science of nature" (*scientia* or *philosophia naturalis*) that pays increasing attention to humankind and its surrounding environment, be it climatic, cosmologic, or made of matter and of animate or inanimate elements, and to its movement. In this historical context of cultural transition, of the widening of knowledge, and of the development of a scientific discourse on nature, medieval scholars thought of space as universal, and consequently, there was no clear distinction between a terrestrial or celestial nature or between humans and gods. However, at the turn of the thirteenth century, the investigation of the various natural kingdoms took an increasingly important place in thinking on the universe, and the inquiry on nature promoted the idea that humans, who are thus endowed with a cultural and agricultural role toward nature, possess an autonomous ontological status superior in creation.

Ahmed Aarab, Kaouthar Lamouchi-Chebbi, and Mehrnaz Katouzian-Safadi invite us to look at how the relationships between humans, animals, diseases, and the environment were observed and thought of in the medieval Arabic-Islamic world. Aarab, Lamouchi-Chebbi, and Katouzian-Safadi present some of the observations and reflections on these relationships that Ǧāḥiẓ, who was a ninth-century Arab scholar, well-versed in the *Physics* (or natural sciences), the medical and zoological science of his time, expressed in his multivolume *Book of Animals*. For Ǧāḥiẓ, the natural environment was an important factor that influenced the health and behavior of animals and humans. Environmental features present in Iraq were, for instance, seen as unsuitable to the "nature" of elephants. But Ǧāḥiẓ also observed and wrote on the consequences for animals of environments changed by humans. The human habit of keeping animals in captivity, Ǧāḥiẓ noted, prevented elephants from reproducing. While, in his work, Ǧāḥiẓ examined the types of influence natural and human environments can have on animals, he also analyzed the ways in which animals, in turn, can influence human health. He notably paid particular attention to snake bites or scorpion stings, where environmental factors such as the time of the day, the season, and the temperatures were also suspected of playing an important role. Through Ǧāḥiẓ' investigation of the animal world, Aarab, Lamouchi-Chebbi, and Katouzian-Safadi illustrate how, in the eyes of this scholar, animals, humans, and the environment did not belong to isolated and independent worlds. They were interdependent, to the point that even the smallest animals, such as pests, while harmful to humans, were seen by Ǧāḥiẓ as necessary and described in terms close to the modern "food chain." The authors also offer important clues as to what the intellectual and scientific endeavors of this time meant: while drawing on the body of knowledge from the past and on the discourse of authorities, the investigation work of the medieval Arab scholar also challenged the authorities and the accepted wisdom and included personal experience, observation, and orally transmitted knowledge.

Catherine Despeux's contribution examines the terms in which the human environment, the human body, and the human health were defined in ancient China. It also explores how the relationships between humans and their environments were understood. Drawing on medical and philosophical texts written or compiled before and during the Han dynasty (208BC–220), Despeux highlights how the cosmology, which emerged at that time, imposed a correlative thinking, according to which the cosmos and the body were not separated and independent items but were in perpetual interaction. As Despeux stresses, the medical texts written at that time defined the human body through the correspondence of each of its parts with one element of the macrocosm; everything was moved by the same dynamics, that of Yin/Yang and the Five Phases. As a matter of fact, celestial elements such as the Sun, the Moon, and the Stars were known to influence human health, and some health practices existed to benefit from the light of the Stars or the Moon. Wind, cold, and other climatic features were also known to impact on human health, just like sounds, odors, and light. As Despeux shows, certain environments or landscapes were better than others for human health: climates that benefit annually from four seasons were the most beneficial to human beings, since the dynamics of Yin and Yang achieves its normal course; an environment where water and Qi circulate was better than one with stagnant waters or Qi. However, the paramount concern was for humans to be able to appropriate the surrounding world with all its visible and invisible components and to attune to its rhythms. This implied a scrupulous observation of the calendar which dictates the appropriate behaviors, often ritualized, and the choice of rhythms of life.

This first part highlights the universality of the conviction of an interdependency between natural milieus, humans, other living beings, and health. What varies lies in the explanations accounting for this conviction. The second part of the book, *Observations, Definitions, and Theories About the Environment and the Body: Questioning the Meaning of the Environment*, addresses these issues in European and Latin American contexts. Approaches based on the history of ideas, of science, and of philosophy shed light on the difficulty of proposing a single, simple definition of the environment, once the word was available. The contributions gathered in this part shed light on the structuring of new fields of research in relation to the question of the interdependency of natural environments and living beings. They also highlight how this interdependency was thought of and discussed among philosophers, notably among those who had one foot in the world of medicine and the other in the world of philosophy and among those who, through their work, were observers and managers of the environment.

Valérie Chansigaud's contribution focuses on the emergence of new words to speak of the interrelation between the natural physical environment and living organisms, at the end of the eighteenth century in Europe. She examines what is at stake in this new vocabulary, which undergoes semantic shifts through the nineteenth and twentieth centuries. As Chansigaud brings to mind, it is not that there had been no reflection on the influences of the environment on humans before the end of the eighteenth century, or on the interdependency of humans and their natural environment. Physicians, naturalists, poets, and theologians had discussed this question at length, notably after Europeans had started to travel all around the world and faced the scientific and economic issue of adaptation. What was new was the coining, in the early decades of the nineteenth century, of new terms (*habitatio*, *statio*) and the reuse of older terms, such as *milieu*, invested with new meanings. For

Chansigaud, the birth of this vocabulary and its rather rapid spread in the century is not the sign of a new field of investigation. It is rather the consequence of the scholars' attempts, naturalists notably, to use the resources of science to analyze the milieus and their influences on living organisms and to legitimate their endeavor in the new frame of modern science. These new words allowed them to distance themselves from the theological reading of the universe that had prevailed. In the mid-nineteenth century, this vocabulary included an evolutive dimension, and "milieus" and "habitation" were no longer considered as places characterized by geographic and climatic features particularly convenient for a set of living organisms but as the very reasons that the living organisms were found there. In the mid-nineteenth century, new words appeared, attempting to make sense of this interrelation—zooethics, ethology, biology, mesology, and ecology—translating the scholars' will to define ever more precisely the nascent fields of biology and ecology.

The contribution by Marie Gaille addresses the particular issue tackled by various European philosophers in the twentieth century of defining what health and pathological states meant and to what extent these different states were dependent on the relationships between human beings and their, loosely say, surrounding world. Focusing on the writings of the famous German neurologist, Kurt Goldstein, and on those of two of Goldstein's early readers, the French Georges Canguilhem and Maurice Merleau-Ponty, the contribution highlights the specific content these thinkers gave to the general and broad idea of a relationship between human health, well-being and human flourishing, and environment. Goldstein elaborated his own conception from his particular clinical experience at the bedside of patients suffering brain injuries at the end of the First World War. For him, the alteration of the relationship with the surrounding world was the very cause of his patients' distress. According to his view, medicine needed to embrace a "holistic approach" to human nature as disease endangered the whole relation of the person to the world and to others. Goldstein's ideas were actively integrated into the philosophical reflection by French philosophers Canguilhem and Merleau-Ponty as early as the 1940s. Based on his philosophy of life, Canguilhem proposed a conception of healthy and pathological states that also broke with the dominant conception in his era, promoting the idea that no absolute normalcy exists: in the fields of life and of medicine, "normalcy" was actually determined within the relationship between a living being and its environment that he called the "milieu." Making use of such verbs as "to adapt" or "to accommodate" to one's milieu and "to modify" it in order to make it favorable to one's life, Canghuilhem pictured human beings as "creators" of their own environment, in which they arranged a liveable space for themselves, according to their own normalcy. In fact, one key aspect of the convergence between both thinkers, highlighted by Gaille, is the recognition of the capacity of human beings to "tailor" (within certain limits) their own specific and individual liveable space, since universally healthy or unhealthy environments do not exist. Goldstein's reflections also influenced Merleau-Ponty who conceived of the world as a network of "atmospheres" and "milieus." He finally proposed a vision of the human relationship to the world based on the vision of each human being as a sensorial and embodied "center of perspective." Merleau-Ponty thus formulated a conception of human life as a form of presence in and knowledge of the world based on the lived (bodily) experience of this world, a conception that is used today in a variety of works in order to give an account of the role played by some specific environments for well-being and human flourishing.

Gabriel Gachelin's contribution sheds light on the particular frame of modern parasitology which, since the last decades of the nineteenth century, contributed to providing a new understanding of the relationships between the natural environment, humans, other living beings, health, and disease. Through his analysis of the complex history of the vector-borne Chagas disease, Gachelin firstly goes against the triumphalist narratives that have often described the medical scientific research of the twentieth century. While identified as a specific disease in the first decade of the twentieth century by Carlos Chagas, the first statements on the disease were invalidated very soon after their initial announcement, as was the affirmation of a single environment favoring human infection. More than one century after the disease's identification and the launching of public health programs, the conditions necessary for the propagation or reemergence of the disease are still unknown, the question of its treatment is still unresolved, and this disease is still a major public health concern in several parts of Latin America. But Gachelin's contribution goes beyond this observation. Following the history of the disease and the history of the scientific research and debate on it until today, Gachelin invites us to see the elusive feature of what constitutes the environmental conditions favoring the disease. As he recalls, several environmental causes were successively pointed out: the housing environment, notably the poor houses made of dry earth, the *cafuas*, the transformation in the housing environment following colonization, the new ecosystem resulting from an increasing demand for acai juice, and the growing exploitation of acai palm trees in certain parts of Brazil. Gachelin's contribution stresses that various environments are at work and that the very term of environment, which today is often used and associated with health issues but not clearly defined, needs to be reflected on. For Gachelin, a single, simple environment does not exist to account for Chagas disease today nor for its history. The environmental causes of the disease are rather multilayered, associating biological ecosystems, lifestyles, and economic pressures, without forgetting the scientific and political environments of medical research.

The contributions brought together in the third and fourth parts of this volume focus on a different issue captured in the question "Healthy or Unhealthy Environments: for whom and for what?" The contributions in the third part, *Healthy or Unhealthy Environments*: *Sensorial Experiences of Lands, Spaces, and Milieus*, address the issue of the perception of good or bad environments for health. They address the different sensory means various actors used to perceive and determine the healthiness or unhealthiness of an environment—smell, vision, hearing, etc. They explore the terms used to qualify good or bad milieus and examine the reasons behind such characterizations. They also shed light on the practices promoted to benefit from a good environment or to minimize the harm from a bad one.

Florence Bretelle-Establet's contribution explores what constituted for many Chinese observers an unhealthy environment in which to live. She brings to mind that one of the earliest Chinese medical treatises addresses the idea that the environment, in its broader meaning, encompassing natural and cultural specificities, has an impact on human health. This same idea is then addressed extensively in the long run of medical history. However, for many scholars, physicians or not, the southern borders of the Chinese empire, including Lingnan (lit. South of Ling mountains, covering the present provinces of Guangxi and Guangdong) and the Yunnan hinterland, were particular: they were the most harmful and terrifying places in which to live. The area was notably held to be the hotbed of "zhang," a polysemic term best rendered by the similarly vague concept of "miasma." Bretelle-Establet's analysis of narratives about the "zhang" prevalent in Lingnan and Yunnan, often written by outsiders, highlights the wide range of meanings encapsulated in this term and the particular environments, natural and cultural, that were suspected of producing them. The combination of searing heat, humidity, and mists, the absence of four regular seasons, the proximity to rivers and mountainous areas with steep slopes, the presence of thick forests with the confined fragrance of particular flowers, and, finally, the proximity to non-Han people were often blamed for the prevalence of *zhang*. For most Chinese scholars, coming notably from the central plain, such environments constituted the worst places to live. However, this viewpoint, widely shared by outsiders, was contested by a fifteenth-century scholar, Jin Gui, who considered the unhealthy reputation of Lingnan as a historical construction developed by central scholars terrified by the natural and cultural differences this environment presented in comparison with that of the central plain. For him, there was nothing special about Lingnan; its reputation was just the result of central scholars' encounters with milieus which appeared strange to them. Bretelle-Establet's analysis of medical writings by authors who lived in Lingnan, and notably their descriptions of the place, allows her to state that the unhealthiness of the southern borders was a matter of viewpoint. Through the eyes of these insiders, there was nothing particularly untoward about the southern borders, and the area was no different from the rest of the empire.

Mehrnaz Katouzian-Safadi and Meyssa Ben Saad's contribution focuses on the different ideas that circulated in the medieval Arabic-Islamic world on the natural environment's impact on living beings, plants, animals, and humans and on their interactions. As Katouzian-Safadi and Ben Saad recall, many physicians, in the continuation of ancient Egyptian and Greek medical doctrines, were interested in this issue. In the humoral medical pathology systematized by Galen and an important component of medieval Arabic-Islamic medicine, it was known that different surrounding conditions influenced health and disease and it was the physician's task to suggest the most suitable treatment to a patient, that is, by taking into account his/her external conditions and his/her temperament. However, as Katouzian-Safadi and Ben Saad underline, in the medieval Arabic-Islamic world, the knowledge of the impact of the environment on human health and disease was constantly being reconsidered with new scientific findings coming notably from the fields of geography, astronomy, botany, and zoology. Without neglecting the ideas and practices of ancient authorities, medieval physicians incorporated new ideas and constantly reshaped the knowledge of their time, based on their own travels, observations, and medical practices. What the environment was for these different scholars is broad, including water, land, air, and celestial structures. Geographical texts, which were usually known by medical scholars, offered a more precise view of the environment. We find in the geographical corpus some texts suggesting a division of the land into seven latitudes, seven *Iqlīm* ("seven climates"), while others, following the pre-Islamic-Iranian model, divide the earth into seven concentric spaces, the *Kešvar* (territories). These authors show how the specificities of these environments were known to account for variations in the body or variations in the shapes and the properties of plants. For many scholars, as they underline, some environments were better than others. The lands, occupied by the Arabic-Islamic civilization usually situated at the center of the seven spaces, were considered as a good environment for human life, compared to other zones situated to the south and to the north, and it was known that traveling from place to place, changing one's environment for another, affected human health, including a person's mood.

The first contributions of part three bear testimony to a constant desire among Chinese and Arabic-Islamic scholars to determine the specificities of natural environments and their actions on humans and other living beings. They also stress how the movement, the circulation of scholars from place to place, and the resulting change of perspective were driving forces in their efforts to make sense of health and the environment, which therefore relied on their own experience of the places encountered as much as on the knowledge of the authorities. The issue of movement and of what it offers in terms of sensory experiences of the places encountered, rural or urban, is at the fore in the last contributions of this part.

Catherine Larrère's contribution invites us to walk a while with the American forester Aldo Leopold who, in the mid-twentieth century, initiated a new ecological perspective, displayed in *A Sand County Almanac: And Sketches Here and There* (1949). This contribution highlights what made Leopold's approach different not only from the objective scientific ecology of this time and which had nourished him but also from the approach of earlier philosophers to the relationships between the environment, the body, and health. For Leopold, it was the relationship between the environment, humans, and other living beings which was his object of inquiry, and this relationship was seen from a personal perspective, namely, from one's ability to act, to be involved, and to preserve the environment. The type of narrative he chose, personal, that he thought would reveal, more than the more distant scientific ecology of this time, the ethical, aesthetic, and political dimensions of such an issue, led Larrère to qualify his approach, "in first person ecology." As Larrère stresses, Leopold defended the idea of the inextricable interdependency of the environment and living beings and tirelessly claimed the interdependency of the parts and the whole, their interconnection, and their unity. However, for him, human beings had lost their awareness of their link to their environment, and he hoped his narrative would contribute to reactivating it, not only for the sake of the good health of the Earth but also for the health of those who inhabit it. This latter claim, which would be at the basis of the later ecopsychology, had something in common, Larrère highlights, with an idea already expressed by Rousseau through his notion of "moral sensitivity" and his project to use the influence of the physical sensibility on moral sensitivity in order to improve human health. For Rousseau, Larrère recalls, environments could have an impact on our mind, since the body's limits could, on particular occasions, fade, leading to moments of fusion with the environment, ecstasies, as described in the *Rêveries*. However, as Larrère stresses, differences between Rousseau's and Leopold's views lie in the finality of the usefulness of the awareness of this link. Unlike St. Preux in Julie's Elysium, for Leopold, the objective was not to be content for oneself; it was broader and had a more collective and therefore political target: to establish a *land health* and make us aware of our condition of members of a community, which, once destroyed, notably for economic reasons, leads to individual and collective suffering.

Ju Xi's contribution focuses on urban design and arrangements in Beijing, at the beginning of the twentieth century. Through the issue of death and notably the management of corpses, from death to burial, Ju examines how Beijing citizens used to consider environments inhabited by corpses and death. Through the mapping of places devoted to death (public cemeteries, private graveyards, temples, ritual sites for funeral ceremonies) in Beijing, she shows how, until the 1950s, the living and the dead cohabited in the same urban spaces. The proximity to death and corpses was not perceived as unhealthy for the living. However, this cohabitation needed proper handling which consisted in the proper respect of the different stages of the funeral rituals, implying, in turn, a correct use of various urban places. As Ju stresses, the complexity of funeral rituals and the absolute need to find a good burial site, as well as the possibility for an outsider dying in Beijing, far away from home, necessitated that corpses and the living often shared the same spaces for long periods of time. Corpses were often kept months or years within the family house or in temples which also acted as hostels, providing accommodation for travelers or for local citizens. Furthermore, before the corpse could finally be interred, a particular care was taken to ensure the grave itself was sited in what was considered the most excellent environment for the deceased and the living. The site was often selected many years previously, for its "feng shui" (lit. wind/water), namely, for the harmony between the geographical features, the animals, and the invisible powers of the site, which together provided a quiet environment for the dead and for the living alike who were therefore eager to come, either to rest or to farm. It was also invested with a strong symbolism, that of the continuity between dead ancestors and the living. Ju's contribution underlines the strong attachment of Beijing residents to this sharing of space by the dead and the living. At the beginning of the twentieth century, in a wave of modernization, many temples were converted into schools, but the temples, which were used to keep coffins for an indefinite time, were saved from this conversion. It was only from 1946 onward, despite the Japanese attempts of changing the funeral customs as early as 1937, that these temples were prohibited and that a clear separation between the dead and the living was enforced.

The fourth part, *Healthy or Unhealthy Environments: Techniques, Practices, and Concepts to Reconsider the Relationship between the Environment, Disease and Health*, examines the promotion of specific patterns and organizations set up in order to benefit from the environment and/or to eradicate the supposed harm from the environment. The contributions brought together in this part, mostly dealing with our contemporary world, also reveal new approaches to the relationships between humans and their environments. These new approaches or concepts invite us to take some distance from the so-called disjunction of human beings and their environment that could, according to a widespread view, characterize the modern West. Furthermore, they invite us to reconsider the historical complexity of the interrelation between humans, other living beings, and the environment/milieu, as well as of its representations.

Nathalie Blanc's contribution focuses on the relationships between health and the urban environment as it has been thought of and qualified as a major issue of public health in Europe and North America, during the nineteenth and twentieth centuries. This contribution brings to mind the background against which the first urban hygienist utopias were formulated in France and in England, from the eighteenth century, and notably the widespread consensus that the countryside was far healthier than towns. Blanc's contribution analyzes the different models which were suggested and, at times, adopted to design and build new urban landscapes, from the first urban utopias of the eighteenth and nineteenth century, the urban designs by urbanists in the first part of the twentieth century, and the urban designs promoted by ecological urbanists later in the century. The contribution thus sheds light on the various types of configurations which were perceived as improving the urban environment for the needs of humans and their health, a notion whose content changed endlessly in the nineteenth- and twentieth-century Europe. Delimitation of spaces was an important aspect in all these models, starting with the removal of cemeteries and slaughterhouses from towns and continuing with a clear division between spaces for habitation and for industry. Another concern was the organization of the various elements of the city (construction of big avenues, paving of the streets), in order to improve the circulation of air, sunlight, and water. How to integrate nature in the ideal urban framework was also an issue widely debated in this period of time, even if what nature meant, as Blanc recalls, could not be summarized in one single and definite idea. By contrasting what was retained and what was rejected from nature in these various urban designs, the contribution precisely highlights what nature meant for these different actors. While the ideal urban environment in the eighteenth century underlined the idea of integrating nature, but with certain exceptions—animals, for instance—in the nineteenth century, the ideal town was understood more as a place compatible with human needs and social progress. In this sense, humans were to be shielded from the bad effects of nature like floods, miasmas, and foul odors. In the second part of the twentieth century, and notably among ecological urban planners, it was said that so far the town had been designed according to its functions and its usages but had not taken human sensibility into account. Reintroducing nature became the new challenge, while, at the same time, the issue of the relationships between human health and the natural environment was addressed more in psychiatric terms, inaugurating new fields of research.

Laura Centemeri's contribution focuses on the permaculture movement which emerged in the 1970s in a sociohistorical context marked by the rise of ecological militancy and countercultural movements, with Bill Mollison, a bio-agronomist, and

David Holmgren, one of his students, as its founding fathers. Centemeri analyzes the history of this movement, the ideas and practices which later developed all over the world, and pays particular attention to the specific understanding of health it promotes. As Centemeri stresses, Mollison and Homlgren's interest was motivated by the search for systems of a low-energy, high-yielding perennial agricultural production to sustain perennial culture, based on self-reliant villages and personal responsibility. However, their goal was not only the founding of a new type of agriculture; it was, and still is, based on the active creation of conditions for perennial human-environment coevolution, expressed in terms of ethical principles of care. This ecological movement also promotes a specific approach to human health—a distant echo of the holistic doctrines that prevailed in most parts of the world until the nineteenth century. For permaculturists, human health is inseparable from the health of the environment. The link between the environment and humankind is primarily, but not exclusively, established through food, which, needless to say, should be produced respecting as much as possible the conditions of permaculture. The crucial notion of self-reliance in permaculture is also extended to health and implies rediscovering, learning, and sharing competences in the use of herbs for healing remedies and, at the same time, de-hospitalizing birth and death. It promotes a holistic approach to health in which prevention (a healthy lifestyle) and capacities of self-care are of crucial importance with the goal of empowering people, namely, giving people the opportunity to recover a sense of control over their own life. Without rejecting the scientific evaluative framework, permaculturists promote the elaboration of a new way to measure what is of worth, what is valuable, which cannot be defined in terms of an abstract representation of the environment but considered at the aesthetic level, relating to the body-mind-environment interrelationship at a sensory level.

The focus of the contributions that end the book is the body-mind-environment interrelationship. The authors of the last two contributions shed the light on two different theoretical models that were elaborated in the twentieth century to reflect on the relationship between the body and what surrounds it. In both chapters, the authors highlight the impossibility of thinking of these two entities separately, as independent and abstract objects, not only for the sake of human health but also for the sake of the environment.

In her contribution, Anne-Lyse Chabert highlights the idea according to which the individual and his environment form a construct, a whole, and stresses the therapeutic implications of this. She does so using the conceptual framework of *affordances* forged by James J. Gibson in 1977 to reveal the adjustment that has to be made between the body and the environment in which it exists, by means of "an invitation to act."[21] The neologism was coined from the word to afford, i.e., to offer. Gibson, with his concept of affordances, describes the way individuals have the possibility of attaching themselves technically to their external environment by selecting and processing information they receive from it; this comes down to perception. As Gibson writes, "the affordances of the environment are what it offers the animal, what it provides or furnishes, either for good or ill." An object is not used in the

[21] Gibson (1997).

same way nor therefore perceived the same way by everybody. This use or perception depends on the individual's resources. How can a set of possibilities, guaranteeing a sufficient range of opportunities for action, be re-established so that the environment ceases to be neutral or even hostile and becomes hospitable and inviting? The question is crucial for those who suffer from a disability, whatever it may be. Chabert illustrates this perspective through the analysis of a specific case study, a blind football game, in which blind adolescents regained the possibility of playing soccer. On the basis of this case study, Chabert shows that the entire "space" of affordances is short-circuited, to begin with, when a disability is present. The individuals concerned can no longer communicate with the interface of their environment. Thus, taking the environment into account is indispensable when restoring the individual's well-being. To Chabert, unless this relationship is clearly understood, any attempt to think about health or treatment will inevitably lead to errors whose practical consequences, especially in the fields of health and treatment, are harmful to the individual.

In her contribution, Perrine Michon analyzes the conception of one's relationship to the living environment associated with Gestalt therapy, one of the humanistic therapies with a holistic approach that were constructed progressively in the course of the twentieth century in Europe. Part of the lineage and in a critical position in relation to psychoanalysis, the Gestalt premise and position can be taken and understood as an invitation to rethink the relationship of man and the environment and to invite another way of being in the world. What appears specific to this conception is the focus on the somehow unintentional but permanent connection between the organism and the environment, as illustrated by breathing. This position entails several consequences: first of all, it implies a constant dynamic of exchanges between the individual and the organism; then, it also implies that the individual's state of health cannot be considered as a state distinct from her/his living environment; and finally, it implies that any therapy applies to the person as embedded in his/her environment. With all these elements taken into consideration, the main goal of Gestalt therapy is to generate new dynamics of exchange between the individual and his/her environment. As a geographer and specialist in urban planning, Michon emphasizes that, for her, the interest of Gestalt therapy lies in the possibility of rejecting the vision of space, or land, as a mere abstraction, in which human beings and objects may be located indifferently and without consequences. This interest is also related to Gestalt's focus on the interval, the interrelation, that geographer Augustin Berque has developed through the notion of mesology: no entity is considered independently from its relationships with other entities. Finally, she highlights how such a frame of thought transforms, at the same time, our vision of the self, the other, and the environment.

This voyage between different periods, geographical areas, and fields of knowledge allowed us to identify different types of actors who, from various perspectives, developed reflections on the relationships between the environment, health, and disease. Let us now bring these threads together and consider some of the most important points highlighted by these contributions.

Before leaving the reader to explore this cross-cultural investigation, we would like to emphasize some of the results emerging from this work and also mention what remains to be done. First, our investigation is consistent with the view of historiographical work highlighted by Simon Schaffer, Lisa Roberts, Kapil Raj, and James Delbourgo, that it is "work that seeks to remap our understanding of theories, representations and conceptions as a global phenomenon not simply conceived of in western Europe, and subsequently diffused to the rest of the world."[22] It is also consistent with the subsequent rejection of the "view from nowhere" some philosophers may associate with the position of the objective knowledge-bearer.[23] In fact, it is in line with the historiographical perspective that seeks to understand concepts as they were thought of and elaborated in their own way, making their own path.[24] In this perspective, we hope that this volume may be a useful tool to help the reader look beyond her/his own world and get a broader awareness of globality or world history, to put it in the terms used by Roger Chartier, Patrick Boucheron, or Kenneth Pomeranz.[25] A continuation of this work would certainly be, on the basis of this remapping of our understanding and of this awareness, to study the dynamics of translation and transfer that have allowed the circulation and borrowing of conceptions, knowledge, and practices throughout the world and connect its different parts.[26]

In a more specific way, this work shows the necessity of taking a critical distance from certain typical representations or stereotypes, such as the idea that in the West there is a separation of human beings from their environment or that in China man identifies himself with his natural environment, an environment in which he will not intervene. As we show in this volume, the idea that human beings are fundamentally dependent on their environment, whatever its name, may be encountered in various conceptions of the relationship between the environment, health, and disease. It is neither specifically Chinese, Arabic, nor European, nor is it "ancient" nor "modern." Conversely, the confrontation, when it is possible, of the discourses, ideas, and practices of various contemporary actors living and writing in the same area of this world clearly shows that no archetypal ways of thinking existed in one particular area. While Canguilhem claims that anything approaching a single ideal, objective, once-and-for-all environment cannot exist, the urban planner Le Corbusier in the 1930s, on the basis of his belief in the universality of human needs, proposed his vision of the ideal dwelling house and a healthy urban environment. While many Song scholars attributed the terrifying miasmas of the southern part of the empire to the mists and fogs that occurred there, many Song painters also saw these meteorological phenomena as the embodiment of elegance, *par excellence*.

[22] Schaffer et al. (2009, p. xi).

[23] Schaffer et al. (2009, p. xv).

[24] Be it comparative or not, this historiographical endeavor is today best subsumed by the English formulation "on their own terms," see notably Elman (2005) and Hanson (2011, p. 9). Lloyd and Sivin (2002), who share the same idea, preferred to use "in its own way," p. 5.

[25] Chartier (2001); Boucheron (2012); Pomeranz (2001).

[26] Subrahmanyan (1997).

While the objective of our project was not comparatist, this volume highlights some points of convergence and divergence in ways of thinking about health and the environment between various areas. And it is on these last points that we want to retain our readers' attention. The first point of convergence is the very issue of making sense of health and the environment, which seems to have been common to the various areas studied. This research, carried out in a framework open to ancient, late imperial, and modern China, to the medieval Arabic-Islamic world, and to modern Europe, also often shows the similarity of the questions raised, the frequently recurrent ideas of what makes a natural environment healthy or unhealthy for humans, and, finally, the parallels that can be drawn between the ideas and beliefs that are thought to ensure a kind of continuity between natural environments and human beings. Furthermore, the methods used to make sense of the environment and health in different parts of the world and at different times have also much in common, making room, in all these contexts, for personal observations, the discourse of authorities, the use of the senses, and rationality.

Let us illustrate some more particular areas of convergence. We notice a shared focus on the issue of the organization of space: How should space be divided to ensure better human health? How should borders be established? We also observed the importance given to the perspective from which the relationship between environment, health, and disease may be considered and the implications of moving from one perspective to another and from one place to another. Some concerns recur throughout time and space—such as that of the quality of the air one breathes—as well as a suspicion toward some particular milieus and elements (humid places, still waters, foul odors, confined and unvented areas). On a more theoretical level, when knowledge is at stake, discourses converge to highlight the complexity and the numerous specificities of the observed environment. They also often meet on epistemological issues, such as how best to define the terms used to characterize the relationship between the environment, health, and disease: Are we to conceive of this relationship in terms of effect, influence, resonance, interaction, or impression? Finally, are we really able to account for it and in ways that may grasp both the political and epistemological dimension of the problem?[27]

It is not an easy task to assess the various aspects of this convergence, its implications, and the remaining points of divergence. We believe that it would be a mistake to consider that our work must aim for conceptual unity, called for by this convergence. Like Nietzsche, we are wary of "systematizers," as "the will to a system is a lack of integrity."[28] However, as suggested by Claude Lévi-Strauss in the *Mythologiques* and by Günter Dux in his *Historico-genetic Theory of Culture*, these recurrences may indicate some universal schemes of thought on the conception of the relationship between the environment, health, and disease.[29] The question remains open.

[27] Latour (1991, 1999).

[28] Nietzsche (2005, p. 159).

[29] Lévi-Strauss (1969, 1973, 1978, and 1981); Dux (2000).

References

Berque, Augustin. 2011. *Fudô, le milieu humain.* (translation of Watsuji Tetsuro). Paris: CNRS édition.

Blanc, Guillaume, Elise Demeulenaere, Wolf Feuerhahn (Eds). 2017. *Humanités environnementales – Enquêtes et contre-enquêtes.* Paris: Publications de la Sorbonne.

Boucheron, Patrick. 2012. *Inventer le monde : une histoire globale du XV*[e] *siècle.* Paris: La Documentation française.

———, ed. 2017. *Histoire mondiale de la France.* Paris: Seuil.

Canguilhem, Georges. 1992. *La connaissance de la vie.* Paris: Vrin.

Chartier, Roger. 2001. La conscience de la globalité (commentaire). *Annales, Histoire, Sciences Sociales* 56 (1): 119–123.

Chemla, Karine, and Evelyne Fox Keller (Eds). 2017. *Cultures Without Culturalism. The Making of Scientific Knowledge.* Durham/London: Duke University Press.

Corbin, Alain. (1982) 2008. *Le miasme et la jonquille.* Paris: Flammarion. (*The Foul and the Fragrant. Odor and the French Social Imagination.* Harvard University Press, 1986)

Dux, Günter, 2000. *Historisch-genetische Theorie der Kultur.* Weilerswist: Velbrück [translation in English, 2011, Transcript].

Elman, Benjamin. 2005. *On Their Own Terms: Science in China, 1550–1900.* Cambridge, MA: Harvard University Press.

Fu, Daiwie. 2010. When Shen Gua Encountered the 'Natural World': A Preliminary Discussion on the *Mengxi bitan* and the Concept of Nature. In *Concepts of Nature. A Chinese-European Cross-Cultural Perspective*, ed. Hans Ulrich Voguel and Günter Dux, 285–309. Leiden/Boston: Brill.

Hanson, Marta. 2011. *Speaking of epidemics in Chinese medicine. Disease and the geographic imagination in late imperial China.* London/New York: Routledge.

Hanson, Marta. 2012. *Visualizing the Geography of Diseases in China, 1870s–1920s: An Overview of the Earliest Disease Maps of China.* Paper given at the International Conference From Qing to China: Rethinking the Interplay of Tradition and Modernity, 1860–1949 (May 2012). http://www.tau.ac.il/~asi2012/qing_to_china/papers.htm. Last accessed 19 Dec 2017.

Harbsmeier, Christoph. 2010. Towards a Conceptual History of Some Concepts of Nature in Classical Chinese. In *Concepts of Nature. A Chinese-European Cross-Cultural Perspective*, ed. Hans Ulrich Voguel and Günter Dux, 220–254. Boston: Brill.

Gibson, James J. 1977. The theory of affordances. In *Perceiving, acting, and knowing. Toward an ecological psychology*, ed. R. Shaw and J. Bransford, 67–82. Hillsdale: Lawrence Erlbaum Associates.

Koch, Tom. 2011. *Disease Maps: Epidemics on the Ground.* Chicago: University of Chicago Press.

Latour, Bruno. 1991. *Nous n'avons jamais été modernes, Essai d'anthropologie symétrique.* Paris: La Découverte.

———. 1999. *Politiques de la nature – comment faire entrer les sciences en démocratie ?* Paris: La Découverte.

———. 2015. *Face à Gaïa. Huit conférences sur le nouveau régime climatique.* Paris: La Découverte.

———. 2017. *Où atterrir ? Comment s'orienter en politique.* Paris: La découverte.

Leroi, Armand-Marie. 2014. *The Lagoon: How Aristotle invented Science.* New-York: Viking Press.

Lévi-Strauss, Claude, *Mythologiques.* Trans. John Weightman and Doreen Weightman. Chicago: The Chicago University Press.

The Raw and the Cooked (1964), 1969

From Honey to Ashes (1966), 1973

The Origin of Table Manners (1968), 1978

The Naked Man (1971), 1981

Lovelock, James. 1979. *Gaia: A New Look at Life on Earth.* Oxford: Oxford University Press.

Lloyd, Geoffrey, and Nathan Sivin. 2002. *The Way and the Word. Science and Medicine in Early China and Greece*. New Haven: Yale University Press.

Monnais-Rousselot, Laurence. 1998. *Médecine et colonisation. L'aventure indochinoise, 1860–1939*. Paris: CNRS éditions.

Montgomery, Scott L., and Alok Kumar. 2016. *A History of Science in World Cultures – Voices of knowledge*. London/New York: Routledge.

Nietzsche, Friedrich. 2005. Twilight of the Idols or How to Philosophize with a Hammer. In *The Anti-Christ, Ecce Homo, Twilight of the Idols or How to Philosophize with a Hammer, and Other Writings*, ed. Aaron Ridley and Judith Norman. Cambridge: Cambridge University Press.

Picheral, Henri. 1982. Géographie médicale, géographie des maladies, géographie de la santé. *Espace Géographique* 11 (3): 161–175.

Pomeranz, Kenneth. 2001. *The Great Divergence China, Europe, and the Making of the Modern World Economy*. Princeton: Princeton University Press.

Rosenberg, Charles E. 1992. *Explaining epidemics and other studies in the history of medicine*. Cambridge: Cambridge University Press.

Schaffer, Simon, Lisa Roberts, Kapil Raj, and James Delbourgo. 2009. *The Brokered World – Go-Betweens and Global Intelligence, 1770–1820*. Sagamore Beach: Science History Publications, U.S.A. a Division of Watson Publishing International LLC.

Shelton, Dinah. 2015. Nature as a Legal Person. *VertigO – la revue électronique en sciences de l'environnement* [online], Hors-série 22. Last access 30/01/2018. http://journals.openedition.org/vertigo/16188; https://doi.org/10.4000/vertigo.16188.

Subrahmanyan, Sanjay. 1997. Connected histories: Notes towards a reconfiguration of Early Modern Eurasia. *Modern Asian Studies* 31 (3): 735–762.

Taylan, Ferhat. 2014. *Introduction à la "Rationalité Mésologique. Connaissance et gouvernement des milieux de vie (1750–1900)"*. PhD in Philosophy, Université Bordeaux Montaigne.

Unschuld, Paul U. 2003. *Huang Di Nei Jing Su Wen. Nature, Knowledge, Imagery in an Ancient Chinese Medical Text*. Berkeley: University of California Press.

Vigarello, Georges. (1993) 1999. *Histoire des pratiques de santé. Le sain et le malsain depuis le Moyen-Age*. Paris: Le Seuil.

Wear, Andrew. 1992. Making Sense of Health and the Environment in Early Modern England. In *Medicine in Society*, ed. Wear, 118–141. Cambridge: Cambridge University press.

Wild, Christopher Paul. 2005. Complementing the Genome with an 'Exposome': The Outstanding Challenge of Environmental Exposure Measurement in Molecular Epidemiology. *American Association for Cancer Research* 14: 1847–1850. https://doi.org/10.1158/1055-9965.EPI-05-0456.

Part I
Observations, Definitions, and Theories About Environment, Disease, and the Body: How to Embrace the Whole World

Chapter 2
Creation, Generation, Force, Motion and Habit: Medieval Theoretical Definitions of Nature

Isabelle Draelants, with the collaboration of Eduard Frunzeanu

Abstract In order to illustrate the theoretical definitions of nature commonly shared in the mid-thirteenth century in Europe, we translate and comment on the definition of nature given in the "Mirror of Sciences" (*Speculum doctrinale* XV, ch. 4) compiled by a Dominican friar, Vincent of Beauvais. While complex, this European fourfold definition reflects to a degree all the theoretical conceptions of nature used at this time, including the literary heritage available, some theological definitions and the new inputs from natural philosophy (Arabo- and Greco-Latin translations). It synthetizes the conceptions of nature, some of which go back to Augustine of Hippo (fifth century), to William of Conches (twelfth century), to Aristotle and to the medical school of Salerno until the beginning of the thirteenth century.

Keywords Nature · Creation · Generation · Corruptibility · Middle Ages · Encyclopedias · Philosophy · Scholastics · Augustine of Hippo · Hugh of Saint-Victor · William of Conches · Albertus Magnus · Vincent of Beauvais · Vermin

For medieval man, *natura* was a multifaceted, universal and all-encompassing concept that covered the heavenly, terrestrial, divine and human environment from the creation of the world to the end of time. The long medieval millennium, however, presented various conceptions of nature, according to the time, the place and the social, economic and intellectual conditions and the type of historical, archaeological or literary traces bearing witness to it. This requires those interested in medieval

Some of the material presented here is discussed in Eduard Frunzeanu's doctoral thesis, University of Montréal (Frunzeanu 2007).

I. Draelants (✉) · E. Frunzeanu
Institut de recherche et d'histoire des textes IRHT, UPR 841, CNRS, Paris, France
e-mail: isabelle.draelants@irht.cnrs.fr

F. Bretelle-Establet et al. (eds.), *Making Sense of Health, Disease, and the Environment in Cross-Cultural History: The Arabic-Islamic World, China, Europe, and North America*, Boston Studies in the Philosophy and History of Science 333, https://doi.org/10.1007/978-3-030-19082-8_2

nature to choose a point of view. If one keeps to the scholarly texts, it is possible to isolate, for example, for the thirteenth century, some philosophical definitions. In Europe, this was the era of the birth of universities, a period called "scholastic" because of the importance taken by teaching in "schools"; it is also the time when the natural law emerged. We have chosen to present a few rich definitions of nature drawn from encyclopaedias and dictionaries from the North of France and Germany. Taking this hub as the starting point has the advantage of giving us access retrospectively to the broad scholarly culture of the medieval West through compilations that amass, bring together and order a large number of ancient and medieval works and authors.

At the same time, the philosophical character, in the broad sense, of the texts proposed here does not make it possible to illustrate the material, economic or poetic medieval conception of nature or man's everyday environment in the Middle Ages, to which a first part of this chapter is devoted, presenting an outline of the cultural context. The second part consists in the translation and analysis of the content of the definitions of nature taken from an encyclopaedia from the middle of the thirteenth century intended for the intellectual training of Dominican monks. In order to do so, we analyse the complex network of textual, theological and philosophical sources of inspiration that informed these definitions. In the third part, two other definitions drawn from medieval dictionaries, present the various meanings of the word "nature" that were current around 1200 and the end of the thirteenth century. Nevertheless, all these sources, together with their different origins and ages, adopted, as it was always the case in the Middle Ages a very respectful attitude toward the authority of the written word and a broad intertextuality. Finally, as a counterpoint, come into discussion some of the spaces that are "out-of-nature", supernatural or beyond the reach of human understanding.

2.1 Cultural Context of the Eleventh and Twelfth Centuries

In the Middle Ages, when the population of our planet is estimated to have been between two and four hundred million, the influence of human activity on nature was considerably less than it is today. In the Western Christianity built on the ruins of the Roman Empire, whose monasticism was the new and foremost cultural, social and economic network, man was not perceived as a threat to his earthly environment. The biblical prescription "go forth and multiply" was, on the contrary, an encouragement for humanity on earth to continue the work of the sixth day of Creation, that is to say, to order, name and take dominion over nature:

> Then God said, "Let us make man in our image, after our likeness. And let them have dominion over the fish of the sea and over the birds of the heavens and over the livestock and over all the earth and over every creeping thing that creeps on the earth."
>
> So God created man in his own image, in the image of God he created him; male and female he created them.
>
> And God blessed them. And God said to them, "Be fruitful and multiply and fill the earth and subdue it, and have dominion over the fish of the sea and over the birds of the

> heavens and over every living thing that moves on the earth." And God said, "Behold, I have given you every plant yielding seed that is on the face of all the earth, and every tree with seed in its fruit. You shall have them for food. And to every beast of the earth and to every bird of the heavens and to everything that creeps on the earth, everything that has the breath of life, I have given every green plant for food." And it was so. (Genesis 1:26–30: English Standard Version)

The hereafter was perceived as temporally and spatially infinite, and the limited sublunary universe was seen as the pale, earthly reflection of divine ideas. As such, it fell into a linear timescale whose end would be marked by the last judgement, and was placed in a post-historical space. In this perspective, the rare meteorological and political records transcribed in the annals of the abbeys in the early Middle Ages show that, on a daily basis, dramatic terrestrial events such as natural disasters, excessive weather, earthquakes, volcanic eruptions and eclipses were linked to famines,[1] were often perceived as divine punishments in response to man's sins. Similarly, the wonders observed in the world were seen as signs, visible symbols of the divine majesty which needed to be interpreted unceasingly in order to attain the invisible reality of the hereafter.[2] The Apostle Paul's adage (Romans I, 20), relayed by Augustine (354–430), the Father of the Church, was in this respect a powerful speculative stimulus for the symbolic exegesis of nature in Biblical commentaries and sermons: "For his invisible attributes, namely, his eternal power and divine nature, have been clearly perceived, ever since the creation of the world, in the things that have been made".[3]

Historians have time and again observed that from the twelfth century onwards, a growing interest in the natural world of the "earth below" changed the view: Western man was more inclined to set his eyes on what surrounded him rather than towards the redeeming afterlife; he was interested in the *ornatus,* the stage of Creation concerned with the peopling and "ornament" of the terrestrial world. This period of favourable climate, relatively stable politics and economic prosperity due to better yields distanced itself it from the "dark" centuries of the Middle Ages, so-called because of the precariousness of the institutions and the frequent wars and food shortages. Cities were founded or grew, and urban schools thrived in the West, taking over the function of cultural stockpiling that the abbeys had exercised thus far. In education, the development of logic and the demand for rationality grew, and a new function emerged in society, namely that of *thinking* pertaining to those whom Jacques Le Goff called "intellectuals" or Jacques Verger more recently "men of learning".[4] At the same time, a new conception now presented nature[5] as the vicar of God, the agent of the physical continuation of the Creation from day to day.

[1] See Draelants 1995 and Draelants 1996.

[2] The Augustinian conception of nature was extremely influential in this respect. See Thonnard 1965.

[3] Saint Paul, *Epistle to the Romans*, I.20: *Invisibilia Dei a creatura mundi, per ea quae facta sunt, intellecta conspiciuntur*. On the exegetical current and its meaning in connection with the perception of nature, see Chenu 1957.

[4] Le Goff 1957, a landmark work, often reprinted. For the end of the Middle Ages, see Verger 1997.

[5] As shown by Tullio Gregory (Gregory 1975).

Intellectual curiosity – today we would say the scientific viewpoint – hitherto vilified by theologians in favour of contemplation was better received if not approved.[6] This perspective of nature encouraged, and this time also outside the monasteries, the development of a scholarly medicine, which applied to the body and its disorders and little by little, the development of the "natural sciences" *scientia* or *philosophia naturalis*, concerned with man and his meteorological, or more broadly with his cosmological environment, matter, and animate and animate bodies such as stones, plants, animals, and in motion.[7]

While the discourse on Creation had hitherto been dominated by exegesis on the *Hexaemeron* (Genesis), this perspective of nature would now allow the reception and commentary of "philosophical" texts – by this we mean all the non-religious texts intended for study – and, in the twelfth century, encouraged the reputation of schools, such as Salerno for medicine, Chartres for cosmology and Saint-Victor for theology. While it welcomed philosophical texts, another intellectual feature of the twelfth century was represented by various translation movements from Greek and Arabic into Latin, which made scientific works from Antiquity and the Arabic-speaking East available in Europe for the first time. In the following century, the foundation of colleges by religious orders (*studia*) and universities was one of the most evident testimonies of the exploitation of this new intellectual subject.

For a century and a half, many works were translated from Arabic and Greek into Latin: at the end of the eleventh century in Southern Italy for Arabic medicine, thanks to Constantine the African at the Abbey of Monte Cassino, in the twelfth century in the Iberian peninsula (Barcelona and Toledo, among others) and in the Latin states of the Near East (e.g. Antioch) for natural philosophy, medicine, astronomy, astrology and magic and then in the first half of the thirteenth century in Sicily, in the north of France, in Brabant and in the south of England for Greek philosophy.[8] All of these texts, issued from of a very long intellectual tradition which combined successive civilizations from the eastern Hellenistic, Roman, Byzantine and later an Arabized world where they were translated or created – particularly in the

[6] Until quite recently, following Chenu 1957 and Gregory 1975, we tended to consider that the Platonic vision, supported by Augustine's influence, dominated the early Middle Ages. Recently, the perception of nature in the early Middle Ages has been re-evaluated from the point of view of astronomical and cosmological representations, as highlighted by Obrist 2004, 11–13 (and see article note in this chapter).

[7] See Speer 1995 for the new investigation of nature by Adelard of Bath, Bernard of Chartres, William of Conches and Thierry of Chartres in the twelfth century.

[8] The literature on medieval translations is vast and growing rapidly, after the fundamental and pioneering work of M. Steinschneider at the end of the nineteenth century, as well as Grabmann 1916 and Grabmann 1928. On the subject of Arabic-Latin translations, see the works of Charles Burnett, who continued the work of Marie-Thérèse d'Alverny. E.g. Burnett 2009 and the following collected studies: d'Alverny 1994, 1998. Mention should also be made of Lindberg 1978 and the catalogue of Cranz and Kristeller 1960–1980, as well as Van Riet 1987; Jayyusi and Marin 1992; Jolivet 1995; González 1997; Kischlat 2000; Gutas 2005. Among the conferences, *Oriente e Occidente nel Medioevo: Filosofia e Scienze* 1971; Scarcia Amoretti 1987; Endress hrsg. 1989; Hamesse and Fattori 1990; Butterworth 1994; Draelants et al. 2000; Lau and Cobet 2000; Speer and Wegener 2006; Jenkins 2007; Goyens et al. 2008; Lejbowicz and Bourin 2009; Tischler and

Abbasid period in the eighth and ninth centuries – became, in the thirteenth century, the base of teaching in the nascent Western universities.

These "philosophical" texts, which circulated widely from the thirteenth century onwards, were based on various anthropological and cosmological conceptions. However, this diversity often went unnoticed by the medieval "men of learning", so that Hippocrates, Aristotle, Galen and Avicenna were evoked on the same realities, without necessarily distinguishing their respective modes of thinking about the world.

The growing desire to explain everything resulted in notions linked to Greek and Arabic physics such as matter and form, generation, force, motion and properties gradually emerged from this set of treatises and became, in the thirteenth century, the fundamental physical notions in the study and explanation of the created world. With the influence of Aristotelianism, the search for causes became a criterion of scientificity applied even in theology: one sought to understand both the principles of the generation of natural bodies and the possibility of demonic generation; one studied the motion of entities endowed with bodies as well as those which lacked them, questioned the various levels of the soul which *animate* plants, animals, human and angelic bodies; one asked questions on the nature of motion in the sublunar sphere as in the supralunar world. The identity or "nature" of bodies was related to their "specific force" and to the sensitive or occult properties attached to them; these properties are listed in collections that combine the multiple legacies of ancient and medieval, Greek, Latin and Arabic cultures.

The habituation to certain diets or climatic conditions, the influence of the stars and divine intervention were considered as forces capable of modifying the state of bodies. During the thirteenth century, the notion of *medium* (milieu) gained a growing position in the explanation of phenomena – particularly optical phenomena – but it was only after the Black Death (1317–1348) that the environment and contagion began to be seen as a vector in the transmission or aggravation of diseases.[9]

2.2 Some Definitions of Nature in Encyclopaedias on the Nature of Things

The accumulation of new knowledge caused the desire to structure, order and make it easily accessible, which is why, from the end of the twelfth century, an abundance of new intellectual tools appeared, intended to master these multiple types of knowl-

Fidora 2011; *Entre Orient et Occident: la philosophie et la science gréco-romaines dans le monde arabe* 2011; Köpf and Bauer 2011; Van Oppenraay and Fontaine 2012; Wisnovsky et al. 2012; Federici Vescovini and Hasnawi 2013; De Leemans 2014.

[9] There has recently been renewed interest in studies on the Black Death and epidemics. Before the Black Death, see García-Ballester et al. 1994; for changes in attitude following major epidemics, Green 2015 and Paravicini Bagliani and Santi 1998; Aberth 2010, 2013, esp. 49, 69–70; Herlihy 1980. For a specific medical example, Gottschall 2006.

edge: the generalization of alphabetical order, the frequent elaboration of tables of contents, indexes and glossaries and also collections of "distinctions" offering definitions and correspondences between works and authors concerning the same key word and a profusion of florilegia and other anthologies which in a way provided "portable libraries".[10] This was also the case for encyclopaedias, which aimed to embracing and organizing global knowledge of all areas. Among these summae we see, in particular between 1200 and 1260, the appearance of encyclopaedias devoted to nature, whose authors applied the recently enriched corpus of ancient and modern philosophical works recently translated and disputed in the schools. They were initially conceived to educate monks, especially in the new itinerant Dominican and Franciscan religious orders,[11] and in particular to provide material for preachers, that is to say, those whose task was to disseminate knowledge through sermons. By applying a method which proved its worth in the reading of holy texts, scholars interpreted these natural realities by means of the same exegetical process that returns the gaze towards heaven through comparisons, correspondences and metaphors.

Most encyclopaedias were made by compilation of various quotations and explanatory glosses called "authorities" (*auctoritates*)[12]; these quotations function like so many reclaimed bricks, laid together in a new edifice of erudition thanks to the encyclopaedist's words and organization; this was the author's main and indispensable intervention. Another more medieval image used for depicting his intervention is that of the florilegist who gathers "flowers" in a garden and brings them together in new bouquets. In the same vein, one can mention the *De naturis rerum* by Alexander Neckam (*ca.* 1200), *De floribus rerum naturalium* by Arnold of Saxe (*ca.* 1230–1240), the *De natura rerum* by Thomas of Cantimpré (*ca.* 1225–1250), the *De proprietatibus rerum* by Bartholomew the Englishman (*ca.* 1230–1247) and the very extensive *Speculum naturale* by Vincent of Beauvais (*ca.*1243–1259). All these works, combining tradition and novelty, had several recensions, indicating a constant interest in updating knowledge on nature, and all had also become in turn, before the end of the thirteenth century, a reservoir of subjects and the templates for new cathedrals of encyclopaedic knowledge, continously rewritten in a cumulative process until at least the sixteenth century.[13]

In this historical context of cultural transition, of widening of knowledge and of emergence of medieval scientific discourse on nature, some definitions of nature gained momentum in the twelfth and especially in the thirteenth century. These definitions refer in a way to all valid theoretical concepts of nature at the time, taking

[10] On this subject, see Rouse 1976, 1981.

[11] Paulmier-Foucart and Duchenne 2004, 11: "la plus grande encyclopédie médiévale, le Speculum maius ou 'Grand miroir' est destinée aux *fratres communes* – soit neuf frères sur dix – qui ne sont pas destinés à poursuivre un enseignement au-delà de celui dispensé au studium de leur couvent".

[12] See the papers collected by Zimmermann 2001.

[13] For an overview on the medieval encyclopaedic genre and bibliographical information: Draelants 2013, 2015.

into account the various textual and conceptual inheritances available in Christian theology and natural philosophy and transmitting the new contributions of recent Arabic-Latin and Greco-Latin translations. They synthesize different conceptions of nature, some of which go back to the philosopher Aristotle (fifth c. BC), others to the Father of the Church Augustine of Hippo (354–430), others to the Chartrain theologian William of Conches (*ca.* 1080-*ca.* 1150/4) and finally to the Salernitan medicine, which still prevailed at the beginning of the thirteenth century.

The majority of the definitions considered are derived from the encyclopaedic work by Vincent of Beauvais, a friar, preacher and Dominican scholar. He was close to Louis IX, king of France, and taught at the Cistercian royal abbey at Royaumont in the years 1244–1264. He wrote a vast Latin encyclopaedia called *The Great Mirror* (totalling about four million words).[14] The second, more extensive version of this encyclopaedia, written *ca.* 1245–1255, with the help of a team of Dominican friars sent to collect quotations from the various monastic libraries, was developed in three parts. The first contains the story of the Creation and therefore of nature under the name of *Speculum naturale* or the *Mirror of Nature*, and the second the history of man and literature (including hagiography) under the name of *Speculum historiale*, the *Mirror of History*. Before introducing the third part, which contains the definitions that will be examined here, let us emphasize that Vincent of Beauvais drew a distinction between *historia naturalis*, "natural history", and *series temporum*, "the unfolding of events in time", in order to name the two parts of his encyclopaedic enterprise, one dedicated to nature and the other to the history of human time. The word *historia* in the wording "natural history", is a Greek word that means "inquiry" and refers rather to the idea of collecting testimonies, in an intention that reflects *Natural History*, the great model written by Pliny the Elder in the first century. As for the third part of Vincent's encyclopaedia, it was devoted to "doctrine", that is to say, to what one had to learn regarding sciences (*Speculum doctrinale*). At the end of this part in 15 books devoted to all the sciences, their subjects and their classification, we find the first complex definition which we will study; the author himself says it, calling himself *actor*, in a chapter entitled "in how many ways one speaks of *nature*", *Quot modis dicitur natura.* We shall then see that the central preoccupation of defining nature also appears in the *Speculum naturale* and in the *Speculum historiale.*

In the table below, the text is organized in such a way as to highlight the structure of the definition, made up of two main parts, the second again subdivided into four sub-definitions: *primo*, we find a double definition (*dupliciter*), and then another (*aliter*), and that one is itself subdivided (*multipliciter*). The key words appear in bold, and small capitals are used for the sources, that is to say, the textual authorities to which Vincent of Beauvais makes constant reference.

It is important to emphasize that in the medieval era, compilation was an honourable activity, practiced by all scholars in reference to the Ancients; in a way this fits into a medieval way of making bibliographies, of collecting literature, and had more

[14] For an accessible synthesis on the contribution of this author and his work, cf. Paulmier-Foucart 2004.

value than the word of the author himself. However, not all author-compilers felt constrained to cite their sources systematically and at the same time, the sources do not generally appear under the same titles and authors' names as today. But above all, the authorities mentioned there have sometimes passed via one or more intermediaries; giving the name of an author did not mean that his text was read by the compiler, who may instead have read it via excerpts in florilegia or reported quotations[15]:

Doc. I: *Speculum doctrinale*, XV, Chap. 4, *Quot modis dicitur natura*[16] (Ed. Douai 1624: *Vincentii Burgundi Speculum quadruplex* […], vol. 2, col. 1372–73)

ACTOR. *In summa vero nota, quod natura primo dicitur dupliciter.*	ACTOR. In short, take note that we say "nature" with a double meaning.
I. *Uno modo* **natura naturans**, id est *ipsa summa lex nature, que deus est; de qua dicit* AUGUSTINUS: *Contra summam illam nature legem a notitia remotam, sive infirmorum, sive impiorum, tam deus nullo modo facit, quam nec contra seipsum facit.*	I. In one sense, by this is meant "nature naturing" [**nature that begets nature**] (*natura naturans*), that is to say, the supreme law of nature, which is God. In this respect, AUGUSTINE says: "God does not act in any way against the supreme law of nature, which is foreign to weak and ungodly minds, just as he does not act against himself".
II. *Aliter vero dicitur* **natura naturata,** *et hec multipliciter.*	II. In another sense, we speak of "nature natured" [**begotten** nature] (*natura naturata*) and this in many ways.
II. 1. *Uno modo natura dicitur* **vis insita** *rebus, ex similibus similia procreans, ut ex grano granum eiusdem speciei.*	II.1. In the first place, nature is called the **force** which, **intrinsic** to things, procreates similar beings from the beings that are like them, as a seed germinates a seed of the same species.
II.2. *Alio modo dicitur natura* **principium motus** *et* **quietis**, *ut superius dictum est.*	II.2. Elsewhere, we call nature the **principle** of **motion** and **rest**, as has been said above.
II.3. *Tertio modo,* **communis** *vel usitatus nature* **cursus** *mortalibus notus, secundum quod dicuntur* **miracula** *fieri* **contra naturam**: *unde et dicit* APOSTOLUS AD ROMANOS: *Tu autem* cum *oleaster esses, contra naturam insertus es in olivam bonam.*	II.3. In a third sense, it is the usual or **accustomed course** of nature known to mortals, in relation to which **miracles** are said to occur **against nature**. Whence THE APOSTLE [IN THE LETTER] TO THE ROMANS says: "if you were cut out of an olive tree that is wild by nature, and contrary to nature were grafted into a good olive tree".
II.4. *Item quarto modo dicitur natura* **possibilitas** *creature, quam indidit ei natura naturans,* id est **deus**, *ut ex ea fiat quod ipse vult. Et* sic *accipitur* ibidem IN GLOSA cum *dicitur: id nature est cuique rei, quod de illa fecerit deus.*	II.4. According to a fourth sense, nature is called the **capacity** which nature naturing, namely God, has conferred on the creature to bring about what He wants from it. And this very meaning is in the GLOSS, where it is said that "the nature of everything is what God has made of it".

[15] Cf. Zimmermann 2001.

[16] Translations of the Latin excerpts in this article, when not borrowed from existing publications (indicated), owe much to Eduard Frunzeanu and Monique Paulmier-Foucart (I am also grateful to Patricia Stirnemann for the revision of the English version). For each Latin extract, I have added punctuation, made typographic enhancements and presented the text in subdivisions to facilitate understanding.

It should be noted that the juxtaposition of the various patristic and philosophical sources assembled by Vincent of Beauvais leads to multiple meanings of the term *natura*. Most of these are not modern but inspired by the Christian tradition, and nearly half of them, in particular item II.3 with the quotation on the graft on the olive tree and the end of II.4, are inspired by the treatise *Contra Faustum* written at the beginning of the fifth century by the most influential of the Fathers of the Church, Augustine of Hippo. Apparently, Vincent of Beauvais found an extract of the *Contra Faustum* in a gloss accompanying a biblical text, without knowing that this exegetical note was taken from Augustine.[17] The whole passage affirms the theocentric character of created natures, since the nature of each thing is "what God will have made of it" with an idea of the future which is, in Vincent of Beauvais' adaptation, brought to mind by the term *possibilitas*, potentiality, capacity. The *Contra Faustum* is concerned here with the design of creation by divine will, with the natural course of nature and with miracles: "God, the Author and Creator of all natures, does nothing contrary to nature; for whatever is done by Him who appoints all natural order and measure and proportion must be natural in every case".[18]

In a way that is more modern and foreign to the patristic tradition, the beginning of the definition, as well as the beginning of item II.4, uses syntagms derived from natural philosophy quite close to the time of Vincent of Beauvais: the expressions

[17] Note that Thomas Aquinas, in the *Summa theologiae* I, qu. 105, Art. 6, (reply to objection 1), quotes more fully and comments on the passage of Augustine's *Contra Faustum* (XXVI): *Unde Augustinus dicit, XXVI contra Faustum, quod* id est *cuique rei naturale, quod ille fecerit a quo est omnis modus, numerus et ordo naturae.* ("Wherefore Augustine says (*Contra Faustum* xxvi, 3): 'That is natural to each thing which is caused by Him from Whom is all mode, number, and order in nature.'" [The *Summa Theologiae*, Online 2016]).

[18] Items I and II.3. are inspired by Augustine's *Contra Faustum* XXVI.3 in the chapter concerning what is "unnatural" or "against the truth" (On this excerpt, see recently Müller 2012, 136–8), as shown by what is emphasized in the following quotation, drawn from chap. *Numquam omnipotens Deus aliquid contrarium naturae sive veritati facit. (…)* "Whether the almighty God does something contrary to nature or to truth": (…) People in error, as you are, are unfit to decide what is **natural** (*secundum naturam*) and what contrary to **nature** (*contra naturam*). We admit that what is contrary to the ordinary course of human experience is commonly spoken of as contrary to nature. Thus the apostle uses the words, "If you are cut out of the wild olive, and engrafted contrary to nature in the good olive". Romans 11:24 Contrary to nature is here used in the sense of **contrary to human experience of the course of nature** (*contra consuetudinem naturae*) (…). But God, the Author and Creator of all natures, does nothing contrary to nature, for whatever is done by Him who appoints all natural order and measure and proportion must be natural in every case. And man himself acts contrary to nature only when he sins; and then by punishment he is brought back to nature again (*redigitur ad naturam*). The natural order of justice requires either that sin should not be committed or that it should not go unpunished. In either case, the natural order is preserved, if not by the soul, at least by God. For sin pains the conscience and brings grief on the mind of the sinner, by the loss of the light of justice, even should no physical sufferings follow, which are inflicted for correction, or are reserved for the incorrigible. There is, however, no impropriety in saying that God does a thing contrary to nature, when it is contrary to what we know of nature. For **we give the name nature to the usual common course of nature** (*appellamus naturam, cognitum nobis cursum solitumque naturae*) and whatever God does contrary to this, we call a prodigy (*magnalia*) or a miracle (*mirabilia*). But against the supreme law of nature, which is beyond the knowl-

natura naturans and *natura naturata* constitute a lexical innovation widely circulated from the beginning of the thirtheenth century. This distinction separates created nature, that is to say, creation as a whole and the principles that govern it from creative nature identified with God Himself. It was prepared by the interest in nature that developed in the twelfth century and which we mentioned earlier.

In fact, all the definitions of nature, in the twelfth and thirteenth centuries, are explained in terms of the relationship between God, nature and man. In the *Speculum naturale*, Vincent of Beauvais, a Dominican, takes into account the entire philosophical and theological legacy available to him in order to give a complete idea of these relationships. In the twelfth century, in a theology influenced by Neoplatonic philosophy as transmitted and Christianized by Augustine, the world is an image of God, a book accessible to all, both literate and illiterate. The conception of nature reinforces revealed theology and thus strengthens divine power. God, nature and man are conceived as creative forces with delineated spheres of action.

In this context of the repartition of creation, another passage quoted by Vincent of Beauvais in the *Mirror of Nature* clarifies the previous one. It is found in Book XXIX, Chap. 46, concerning the "four distinct operations", that is to say, the "four differences between the works of creation". This chapter consists of a quotation from the theologian Hugh (1096–1141), canon of the Abbey of Saint-Victor, in Paris, in his homily 16 on *Ecclesiastes*, which covers all cases regarding the distinction between:

- The work of God which operates *aliquando sine natura, aliquando in natura, aliquando supra naturam* (sometimes without nature, sometimes in nature, sometimes beyond nature)
- The work of nature
- The work of man that is accomplished "sometimes with God without nature, sometimes without God and with nature, sometimes with God and nature together, and sometimes without God and without nature" (*aliquando* cum *Deo sine natura, aliquando sine Deo et* cum *natura, aliquando* cum *Deo simul et* cum *natura, aliquando sine deo pariter et sine natura*)

These four cases have to cover the explanation of all "temporal things":

edge both of the ungodly and of weak believers, God never acts, any more than He acts against Himself. As regards spiritual and rational beings (*rationalis creatura*), to which class the human soul belongs, the more they partake of this unchangeable law and light, the more clearly they see what is possible and what is impossible; and again, the greater their distance from it, the less their perception of the future, and the more frequent their surprise (*miratur insolita*) at strange occurrences. (Augustine, *Contra Faustum*, transl. by R. Stothert, rev. and ed. for New Advent by Kevin Knight. http://www.newadvent.org/fathers/1406.htm).

Doc. II: Hugh of Saint-Victor, Homily 16 on *Ecclesiastes*, after *Speculum naturale*, XXIX, Chap. 46, *De quattuor differentiis operationum* (ed. Douai 1624, col. 2096e-97c)

Hugo super Ecclesiasten homilia XVI*a:* ***Quatuor*** *sunt opera quibus omnia temporalia explicantur.* *<u>Primum</u> est opus dei, <u>secundum</u> nature, <u>tertium</u> artificis* cum *natura, <u>quartum</u> solius artificis sine natura.*	Hugh on Ecclesiastes, Homily XVI: All things that exist in time are explained by **four works**. The first is the work of God, the second is the work of nature, the third is the work of the craftsman with the help of nature and the fourth is that of the craftsman alone without nature.
(1) ***Opus dei*** *est essentiam rerum de nihilo creare, materiam rerum informem disponere, motum autem rerum sub certo ordine temperare, propterea tria hec ad opus dei pertinent* id est *essentia rerum et forma et ordo. Hec autem tria de operibus dei stabilia sunt nec temporis capiunt mutabilitatem et rerum essentie hoc quod sunt nunquam esse desinunt et rerum forme secundum primam conditoris sui institutionem in suis generibus perpetuam identitatem custodiunt et motus rerum primum ordinis legem nunquam transcendunt. Neque enim vel essentie rerum nihil esse vel forme rerum aliter esse vel ordo rerum dispositioque* ab initio *mutari potuerunt.* ***Opus itaque, dei est, creare, formare, disponere universa.***	(1) The **work of God** is to create the essence of things from nothing, to arrange the shapeless matter of things and to give a certain order to the motion of things, which is why these three things belong to the work of God, namely, the essence of things, their form and order. These three Godly operations are stable and do not undergo the variations of time; the essence of a thing never ceases to be what it is; in their form, things preserve a perpetual similarity that conforms with what was originally instituted by the Creator for each genre, and the motion of things never surpasses the order originally decreed. For the essence of things has never not existed, nor have their forms been different, nor have their order and initial arrangement been modified. **God's work therefore consists in creating, giving form to all things and arranging them**.
(2) ***Opus nature*** *est semina rerum de occulto per incrementum producere eademque rursum marcentia cum concidunt per defectum ad occultum sinum unde prodierant revocare.*	(2) The **work of nature** is to bring out gradually, from the hidden place, those things that are sown, and, when they lose their vitality and die out, to return them to the secret breast whence they came.
(3) ***Opus artificis*** **cum** ***natura*** *est, ea que oriuntur de terra studio et industria adiuvare.*	(3) The **work of the craftsman with the help of nature** is to treat with care and skill all that is born of the earth.
(4) ***Opus solius artificis*** *in subiecta rerum materia operari vel disiuncta componendo vel coniuncta separando.*	(4) The **work of the craftsman** alone is to act upon the matter of things subject to him, either by uniting that which is separated or by separating that which is bound together.

(continued)

Tres itaque opifices *sunt in hoc mundo.* ***Deus, natura*** *et* ***artifex imitans naturam.*** *Sed hi tres valde dispari potentia id quod ad effectum perducunt, efficiunt. Nam deus in opere suo nec nature opera indiget nec opificis imitantis naturam. Operatur* ***enim deus aliquando sine natura, aliquando in natura, aliquando supra naturam.*** *Nam sine natura primum fecit ipsam naturam,* cum *natura facit ea que de natura secundum naturam producit. Supra naturam quando in natura preter cursum nature solitum et posse primum maiori addita potentia aliquid ad effectum perducit.* ***Omne ergo quod natura facit deus facit, sed non omne quod deus facit, natura etiam facit.*** *Tertio loco sequuntur opera artificis imitantis naturam. Et ipse quidem aliquando* cum *deo operatur sine natura. Aliquando sine deo et* cum *natura. Aliquando* cum *deo simul et* cum *natura. Aliquando sine deo pariter et sine natura.* Cum *deo operatur, quando opera iustitie, sine deo, quando opera iniquitatis operatur.* Cum *natura quando in seminibus rerum ac fetibus propagandis quibus natura incrementum subiicit: foris industriam et studium apponit. Sine natura, quando preterea in subiecta materia studium explicat ut aliquid quodcunque ad effectum promoveat, in quo natura patitur tantum, non operatur, quia materiam operandi prebet, non autem effectum operantis exercet.* ***Talia sunt omnia opera hominum****, que fiunt super terram, ex quibus multa mortalis vite necessitas cogit, multa suadet cupiditas, multa vanitas operatur.*	There are therefore **three kinds of creators** in this world: **God, nature and the craftsman who imitates nature.** However, whatever they bring about, they do so by a very different power. For God in his work needs neither the work of nature nor that of the craftsman who imitates nature. Indeed, **God sometimes operates without the help of nature, sometimes within the limits of nature, sometimes beyond nature**. For it was without the help of nature that he first made nature herself; with the help of nature, he makes all of nature's creations according to the laws of nature. He operates beyond nature when, within nature, he performs something that goes beyond the usual course of nature and its primary capacity, by imparting greater power to it. **Thus, whatever nature does, God does, but all that God does is not necessarily done by nature.** In the third place are the works of the craftsman who imitates nature. The craftsman sometimes operates with God without the help of nature, sometimes without God and with nature and sometimes with both God and nature, and sometimes without God and without nature. He works with God when he performs works of righteousness and without God when he performs works of iniquity; with the concurrence of nature, when he applies himself with care and skill to propagating the seeds of things and the offspring by which nature grows; without nature, when, in addition, he toils with the matter submitted to him in order to obtain something, an operation in which nature plays a passive role, she does not act, because she provides the matter for the operation and does not play the role of the worker. **Such are all the works of men**, which are made on earth: the necessities of mortal life require many of them, many are aroused by avarice and vanity provokes many.

Earlier, in Book II, Chap. 5 of the *Mirror of Nature*, Vincent of Beauvais had already collected the words of a contemporary of Hugh of Saint-Victor, the canon, scholar and theologian William of Conches, active in the Chartres canonical school.[19] In his *Dragmaticon*, William delineates the *opus creatoris* and the *opus naturae*, as well

[19] *Speculum naturale* II, Chap. 5, ed. Douai 1624, 82c-d: Guilhelmus de Conchis: *Sed fortasse dicet aliquis: creator, qui omnia solo dicto creavit et antequam fiant cognoscit. Cur* ***ex quattuor elementis*** *illud composuit quod non diu permansurum cognovit? Opera inquam creatoris videmus: sed causas ignoramus: quod tamen nobis inde videtur, dicemus. Omne opus vel est* ***creatoris*** *vel* ***nature*** *vel* ***artificis****. Opus creatoris est elementorum et animarum ex nihilo creatio, mortuorum resuscitatio: partus virginis et similia. (=* *Dragmaticon*, I, Chap. 7.2, l. 16–25, ed. Ronca, Badia,

as the *opus hominis*, that is to say, the respective creative works of God, nature and man. He specifies the status of man as *artifex imitans naturam*: a craftsman imitating nature, in the making of clothes to compensate for nature's indigence, for example. However, he added that neither nature nor the craftsman could equal God. In a desire to link physics and creation, Hugues also says that God created everything from the four elements, in ways that remain hidden from us, but, in order to distinguish this from the course of physical nature, he adds that He had also created the souls, the resurrection of the dead and the birth through the Virgin:

> But someone could argue: The Creator created everything with his word alone and knows everything before it comes into being; why then did He combine something out of four elements, which He knew would not last for long. I'll say that we see the works of the Creator, but we are ignorant of the causes. We shall say what we think about this question. Everything is the work of either the Creator, or nature, or a craftsman. The work of the Creator is the creation of the elements and the souls from nothing bringing the dead back to life, causing a virgin to give birth, and such like. Nature is a certain force implanted in things, producing similar from similar. It is, therefore, the work of nature that men are born of men, asses from asses, and so on. But whatever is constructed by man to provide for his natural needs, such as clothes against cold or a house against inclement weather, is the work of a craftsman. But when nature produces a certain thing, she produces something rough and confused at first, then gradually forms and divides it: first she produces must, then she drags what is sedimentary and heavy in it to the lowest place, whatever is light to the top, and what is in between to the middle place. (…) Therefore, because nature and the craftsman were unable to come up to the Creator's work, the Creator determined to come down to their standard. For, if this were not so, it would be thought to be a weakness in nature whenever things were created mixed by her. Or, as others say, God created a mixed thing to show how much confusion of things was possible if his own love were not ordering them. When Plato said, "every physical object floating with an aimless movement », he had in mind the natural motion of the elements, by which two move toward the center, two away from the center. And when he said, "He reduced to order from a disordered agitation", he did not mean an agitation that actually existed, but one that might have existed, had He not ordered things in this way. (William of Conches, *A Dialogue on Natural Philosophy*, transl. by I. Ronca and M. Curr, Notre Dame 1997, p. 18–19)

Pujol 1997, 29–30). ***Natura*** *vero est, quedam* ***vis rebus insita****, similia de similibus procreans.* ***Opus*** *igitur* ***nature****, quod homines ex hominibus nascantur, asini de asinis et* sic *de aliis:* ***opus*** *autem* ***artificis*** *est, quod ab homine contra naturales indigentias componitur: ut vestis contra frigus: contra aeris intemperiem domus. Sed* cum *natura aliquod operatur, primo quoddam commixtum operatur, deinde paulatim format et dividit: prius namque operatur mixtum: deinde quod in eo est feculentum et grave ad infimum locum trahit quod vero leve est, ad supremum, quod mediocre ad medium locum. (…)* ***Quia igitur natura et artifex non poterant ad operationem creatoris ascendere****: voluit creator ad illorum operationem condescendere. Si enim hoc non esset: debilitas nature putaretur quotiens ab ea aliqua mixta crearentur. Vel ut alii dicunt,* ***mixtum*** *creavit, ut significaret quanta* ***confusio rerum*** *esse posset: nisi sua dilectio res ordinaret?* (I, Chap. 7.3–4, l. 38–45, ed. Ronca, 30–31) *Hinc omne corporeum importuno motu dixit Plato fluitans, naturalem motum elementorum considerans: quo duo a centro duo ad centrum moventur et hec inquit ex inordinata iactatione, deus in ordinem redegit scilicet non ex inordinata que fuit: sed que esset, nisi res* sic *ordinavisset.* (*Dragmaticon* I, Chap. 7.6, l. 55–59, ed. Ronca, 32).

William had already quoted the phrase subsequently taken up by Vincent of Beauvais in item II.1. of our document I: *vero est, quedam* ***vis rebus insita****, similia de similibus procreans*: “nature is called the force which, intrinsic to things, procreates similar beings from the beings that are like them, as a seed germinates a seed of the same species”. In this extract II.1 borrowed from Hugh of Saint-Victor, the term “nature” is understood in its biological sense as the natural force intrinsic to every creature, to every individual, which allows the multiplication of the species through similar individuals. This principle ensures the ontological condition for each species by making it impossible for new forms of life to appear. Consequently, any crossing between different species constitutes a failure in this continuum of generation, since the resulting progeny do not have the ability to procreate (such as mules). The encyclopaedias, however, note a few exceptions to this rule, drawn from Pliny’s *Natural History* or Aristotle’s zoological works recently available in translation[20] but rarely explain them and they leave no room for chance in nature.

These founding principles of Aristotelian zoology are able to rationally explain every animal generation, even those which are exceptions to this rule, namely “imperfect” animals such as certain insects. In Vincent of Beauvais’ time, the Dominican master Albert the Great (*ca.* 1200–1280) wrote commentaries on the *History of Animals*. He was interested in the generation of all animals; out of 26 books, the last is entirely devoted to *vermes*, an appellation that coined the word “vermin” which covered all small crawling animals. In their regard (in his commentary on plants this time, *De vegetabilibus*), Albert says that “it is rationally explained that there cannot be generation after generation of young of the same species, in imperfect animals born out of putrefaction or of putrid fluids, while nature tends to perfection”.[21] The difficulty of examining the mode of generation of insects without the availability of magnifying instruments led to a degree of uncertainty as to their mode of apparition. Taking up a passage from the famous encyclopaedia *Etymologiae* (written at the beginning of the seventh century by the Visigoth Bishop Isidore of Seville) on the birth of worms in putrefaction, Vincent of Beauvais says in the *Mirror of Nature*, XX, Chap. 68: “Vermin (*vermis*) are animals that are generated for the most part from flesh or wood or some earthy substance, without any sexual congress – but sometimes they are brought forth from eggs, like the scorpion”.[22] This is why, as a result of an exegetical reading of this passage taken from Isidore

[20] In the first half of the thirtheenth century, Aristotle’s zoology reached the West through the Arabic-Latin translation known as *De animalibus*, made by Michael Scot around 1210–20 (from a ninth-century Greco-Arabic translation). It includes three works: *History of Animals* (Books I to X), *Parts of Animals* (Books XI–XIV) and *Generation of Animals* (Books XV–XIX).

[21] *De vegetabilibus* I, tr. 1, *An vivat planta vel non*, Chap. 4, *De positionibus eorum, qui negant vitam inesse plantis*, ed. Meyer, Jessen 1867, 15. On these passages concerning the generation of worms, see Draelants 2016, sp. 206–9, 2019b.

[22] Cf. Isidore of Seville, *Etymologies*, XII, Chap. 5. (Transl. Barney et al. 2006, 258).

of Seville, we see a positive virtue of purity in the worm born without prior coupling, which presupposes a birth without concupiscence comparable to that of Christ. Thomas of Cantimpré says in his *Liber de natura rerum*: "Although the name worm is suitable for all worms, strictly speaking what is specifically called a worm, is that which is generated from the ground pure and intact, without any mixing of seed".[23] This rapprochement between the generation of the worm and that of Christ would be a boon for the exegesis of the passage in Psalm 21:7: *ego autem sum vermis et non homo*, "But I am a worm and not a man, scorned by mankind and despised by the people".

All this remains for the course of natural procreation. Conversely, deviation from the natural course of things can only be the work of a miracle, as noted in Part II.3 [Doc. I] of the definition echoing the *Contra Faustum*.

Let us now examine a description of "physics" as a science and no longer as a created universe. This description is given by Vincent of Beauvais himself who specifies that he speaks this time as the author (*actor*), not the compiler. It is found in both the *Mirror of History* and the *Doctrinal Mirror*.[24] In the first, the passage appears under the chapter title "concerning the division of the sciences which are given as a remedy (to sin)". This title was inspired by the way in which the theologian Hugh of Saint-Victor considered knowledge in his *Didascalicon*, a work on learning in which he affirms that knowledge is given to man by God in order to enable him to regain his state before the Fall (original sin), that is to say, the plenitude of his likeness to God. The rest of the chapter, however, distances itself from the theologians of the twelfth century, since it lists, rather than defines, the characteristics of the fundamental concept of "body" according to Aristotelian physics, from the books on nature by Aristotle, translated a century earlier and who is referred to systematically in this passage.[25] The different conceptual elements are subdivided in the scholastic way by successive binary divisions – body, form and matter, corruptibility and incorruptibility, simplicity and composition and inanimation and animation – which interact in a terrestrial world now considered to be dominated by motion. This last concept includes changes of state. These notions return in force in medieval philosophy following the translation of Aristotle's treatises on nature.

---

[23] Thomas of Cantimpré, *Liber de natura rerum* IX, Chap. 52, *De verme, qui proprie vermis dicitur*. Ed. Boese 1973.

[24] *Speculum historiale* I, Chap. 53, *De divisione scientiarum que et ipse date sunt homini in remedium*, ed. Douai 1624, col. 21a, and *Speculum doctrinale* XV, Chap. 2, *De partibus naturalis philosophie*, ed. Douai 1624, col. 1371d-e, after a first passage taken from Al-Fârâbi's classification of the sciences.

[25] See Frunzeanu 2007, sp. 69–74, for the passages examined here. E. Frunzeanu discussed these elements in a lecture given in 2008 at the University of Nancy.

Doc. III: Vincent of Beauvais, *Speculum historiale*, I, Chap. 53, *De divisione scientiarum que et ipse date sunt homini in remedium* (Ed. Douai, 1624, vol. 2, col. 21a)

ACTOR: (…) **Physica** idest naturalis scientia tractat de **invisibilibus visibilium causis**, nam corpus et ea que sunt corporis principaliter considerat.	AUTHOR: **Physics**, that is to say, natural science, treats the **invisible causes of visible realities**, for it examines mainly corporal things and all that pertains to them.
Corpus autem consideratur in *generali* cum partibus suis aut in *speciali*. Si primo modo sic determinat illud ARISTOTILES IN LIBRO PHYSICORUM.	The **body** is considered either in *general* with its parts or *specifically*. ARISTOTLE adopts the first method in his book *ON PHYSICS*.
Si in *speciali* aut igitur est corruptibile aut incorruptibile.	Considered *specifically*, the body is thus corruptible or incorruptible.
Si *incorruptibile* sic est IN LIBRO DE CELO ET MUNDO. Dicit enim ARISTOTILES IBIDEM quod celum est corpus non generatum nec fabricatum nec recipiens impressiones aliquas.	The *incorruptible* body is the subject of the TREATISE ON THE HEAVENS AND EARTH. In this book, Aristotle says that heaven is a body which is neither generated nor made and which receives no imprints.
Si sit *corruptibile* aut ergo est simplex aut compositum.	As for the *corruptible* body, it is either simple or compound.
Si *simplex* sic est IN LIBRO DE GENERATIONE ET CORRUPTIONE, et loquor de *simplicitate* illa que opponitur *compositioni* naturali ex quatuor elementis.	The *simple* body is dealt with in the TREATISE ON GENERATION AND CORRUPTION, and I speak here of the notion of *simplicity* which is the opposite of natural *composition* of the four elements.
Si *compositum* aut igitur animatum aut non.	If the body is *compound*, it is either animate or *inanimate*.
Si non sic est IN LIBRO METHEORORUM ubi ARISTOTILES determinat de impressionibus aeris et de generatione grandinis et nivis et corporum mineralium et consimilium.	ARISTOTLE treats the latter in the TREATISE ON METEORS, where he discusses the disturbances of the air; the generation of hail, snow and mineral bodies; and the like.
Si vero sit *compositum animatum*, aut igitur anima vegetativa est animatum – et de tali determinatur IN LIBRO DE VEGETABILIBUS –,	If it is an *animated compound*, in this case it is animated either by the vegetative soul, which Aristotle deals with in the TREATISE ON PLANTS,
aut anima sensitiva et de tali IN LIBRO DE ANIMALIBUS,	or by the sensitive soul, which he deals with in the TREATISE ON ANIMALS,[26]
aut anima intellectiva et de tali IN LIBRO DE ANIMA. Ceteri vero LIBRI DE SOMPNO ET VIGILIA, DE MORTE ET VITA, DE SENSU ET SENSATO, DE DIFFERENTIA SPIRITUS ANIME et cetera, supponuntur LIBRO DE ANIMA et LIBRO DE ANIMALIBUS.	or by the intellective soul, which he deals with in the TREATISE ON THE SOUL. The other treatises ON SLEEP AND SLEEPLESSNESS, ON DEATH AND LIFE, ON SENSE AND THE SENSIBLE, ON THE DIFFERENCE BETWEEN MIND AND SOUL etc. must be subordinated to the TREATISE ON THE SOUL and the TREATISE ON ANIMALS.

The definition of natural science which emerges from this passage considers it both as the science of bodies and as the science of causes. We know that the

[26] In the *Speculum doctrinale*, the quotation stops here to begin Chap. 3.

Aristotelian definition of philosophical ("scientific") inquiry is specifically the search for causes; this new epistemology was gradually being replaced in the West with the search for "truth" through logic. More exactly, it coexisted with various conceptions of nature and of science, the object of which was nature.

It will be noted that the Aristotelian philosophical discourse is immediately "framed" by a recollection of the quotation from St. Paul to the Romans (I, 20) referring to the visible physical world as a reflection of the heavenly world and invisible divine purpose that also conforms with the Platonic vision of the world mentioned above.[27] However, this allusion also assumes that the causes of powerful physical phenomena can be invisible. This evokes a hermetic doctrine conveyed by Arabic scholars, the so-called doctrine of "occult" (hidden) causes in order to explain phenomena perceived through sensory experience, but whose origin is not obvious.[28]

The end of this excerpt, written as a bibliographical guide to the Aristotelian books of reference for each notion, shows the importance, for medieval authors to allocate all the knowable visible realities within classifications of the sciences and taxonomic subdivisions. All the treatises on natural philosophy mentioned here are by Aristotle or were attributed to him at the time.[29] As we see here, in the thirteenth century, their dissemination led to a conception of nature as an *intrinsic principle* responsible for the *material* and *formal* fulfilment of every body, whether this body is animated by a type of soul specific to it (vegetative, animal, rational) or whether it is inanimate – the soul being considered as the *form* of the material body.

With regard to the word "physics", it should also be pointed out that around 1230–1240, the word *physica* was still used both for medicine (hence the word "physician" in English) and for natural science. This is why Vincent of Beauvais states:

Doc. IV: Vincent of Beauvais, *Speculum doctrinale*, XV, Chap. 1, *De naturali philosophia.*[30]

Intentio igitur **Philosophie naturalis** est dare **principia** quatuor et **accidentia** et **concomitantia** in omni specie corporis. **Principia**, intellige quatuor **causas**, scilicet materiales, formales et cetera. **Accidentia** vero colorationes, pervietates, asperitates, proprietates, differentias et cetera his similia. Dicitur autem hec scientia proprie **Physica, idest naturalis**, quia *physis* grece, *natura* latine licet medici ad sui palliationem hoc sibi nomen attribuant, sed improprie.	The intention of **natural philosophy** is to explain the four **principles** and **accidents** that **accompany** them in every kind of body. By "principles", one must understand the four causes: material, formal etc.; by "accidents", one must understand the colorations, channels, asperities, properties, differences and other characteristics that are similar. In the proper sense, this science is called *physics*, that is to say, "natural", because *physis* in Greek means *natura* in Latin, although physicians use this name to cover their activity but improperly.

[27] See Note 3 for the quotation of Paul to the Romans.

[28] Nicolas Weill-Parot has definitively clarified the notion of "occult", taking into account the earlier literature on this subject. See inter alia Weill-Parot 2010.

[29] Among the pseudo-Aristotelians works are the *De vegetabilibus*, by Nicholas of Damascus (first c. BC), and the *De differentia spiritu et anima*, by Qustâ ibn Luqâ (*ca.* 820–*ca.* 912).

[30] *Speculum doctrinale* XV, Chap. 1, *De naturali philosophia*, ed. Douai 1624, 1371a.

Aristotle understood *physis* as a process *immanent* in every being, a principle that governs the development of each being, but not as the set of beings or the set of principles through which things develop and interact. Aristotle's fundamental treatise in reference to this idea is *On the Heavens and Earth.* Vincent of Beauvais used extracts from it at the beginning of the *Mirror of Nature*, in a chapter on the sensitive world and the mobility of the elements, following the quotations of Hugh of Saint-Victor [Doc. II] and William of Conches which we have already examined:

Doc. V: Vincent of Beauvais, *Speculum naturale*, II, Chap. 12, *De mobilitate elementorum*[31]

Aristoteles in libro de celo et mundo. Corporibus autem simplicibus est naturalis quies. Res enim aut quiescit **violenter**, aut **quiete** naturali necessario. Et ubi quiescit naturaliter, illuc intendit naturaliter. *Natura* enim est in unoquoque principium sui motus, *virtus* autem est principium motus eius qui est a re alia.	Aristotle in the book On the heavens and earth. Natural rest is specific to **simple** bodies. In fact, something is in a state of rest either by natural necessary **inertia** or **forcibly**. It moves naturally towards the place where it is naturally at rest. In fact, ***nature*** is the **principle of motion** for each thing, and **force** is the principle of motion that is imprinted by something else.

Aristotle defines the relationship between place and motion in relation to the notions of natural motion and violent motion. He explains that *nature* constitutes the intrinsic principle of the motion of every body, a unique natural motion, whereas *virtus*, which could be translated as "force", is the external principle of multiple motions, which belong to another body; *virtue* therefore produces accidental motion without the intervention of nature.[32]

The sense of *physis* in Aristotle ranges from an etiological (related to the causes) and an essentialist (related to the essence) definition on the one hand and a dynamic characterization as a process on the other. For Aristotle, it was indeed the laws of necessity and teleology that took the place of *natura*. Consequently, *physis* represents the "nature of the thing" only as the essential structure of a living thing connected with its purpose, while the properties represent the internal and external characteristics of a living thing which describe it as a specimen of a species. Aristotle considered species as being eternal. By respecting the "final cause", nature seeks the perfection of each species before seeking a balance between them, since the object of Aristotelian teleology is the species and not the animal realm.[33]

At the same time as the Aristotelian vision of the natural world was established, the vision of the naturalist Pliny the Elder (first century) survived in the Middle Ages. Pliny perceived nature as both the totality of all the generated forms (natural and intellectual) and the intrinsic force (the *vis insita* encountered earlier in William

[31] Ed. Douai 1624, col. 86d.

[32] On this passage, cf. Frunzeanu 2007, 183.

[33] Frunzeanu 2007, 70–1.

of Conches[34]) that supports and holds them together. This conception of nature was compatible with the Christian view. Nature thus has an innate force, specific to each thing (*res*), which determines its form and its behaviour; it is linked to a providential force that governs all beings as well as the relations between them.[35] Thus nature remains a set of principles, ordered and subject to a providential plan.

In accordance with the passage commented upon above [Doc. V], just before the *actor* passage devoted to the definition of nature (Doc. I, "in how many ways one speaks of nature"), in Book XV, Chap. 4 of the *Doctrinal Mirror*, Vincent of Beauvais quotes an extract of a summa on the soul (*Summa de anima*) by one of his contemporaries:[36]

Doc. VI: Vincent of Beauvais, *Speculum doctrinale*, XV, Chap. 4 (Ed. Douai 1624, vol. 2, col. 1372)

> *Ex summa de anima. In naturalibus idem est* ***forma*** *quod* ***natura****, sed forma dicitur respectu materie quam perficit natura vero respectu motuum quos elicit, ut forma ignis. Dicitur autem* ***natura*** *secundum Avicennam quadrupliciter. Primo modo principium movendi uno modo et non sponte, sicut in elementis et elementatis. Secundo modo principium movendi diversis modis et non sponte, sicut est anima vegetabilis in plantis. Tertio modo principium movendi diversis modis et sponte, sicut est anima rationalis in animalibus. Quarto modo principium movendi uno modo et sponte, sicut motus celi, quod est intelligentia. Primo modo dicitur natura proprie. Diffinitur autem in physicis generaliter sic. Natura est principium motus et quietis, eius in quo est, per se et non per viam accidentis.*

In fact, the study of spiritual and intellectual faculties, the object of ancient and medieval psychology, was then highly prized. This quote, relaying Aristotelian physics, identified *form* and *nature* with the following difference: one speaks of form with respect to the *matter* that it completes and of nature with respect to the *motion* that it generates in it. These notions were already well known to the philosopher Boethius (470–525) at the end of Antiquity, who also had an excellent knowledge of Aristotle, whose treatises on logic he translated and who had a strong scientific influence on medieval thought. In his celebrated treatise *On the Trinity*, he considered *physica* as the study of "the forms of bodies as well as of the matter which constitutes them".[37]

In the second sentence, the passage from the *Summa on the soul* quoted by Vincent of Beauvais, refers to the theories of Avicenna. In an awkward formulation, he translates the Aristotelian principle of nature as a principle of motion. By taking as criteria the simplicity or the diversity of motion as well as its voluntary or not voluntary (*sponte*) character, he divides all natural realities according to four pos-

---

[34] See Note 19 and the following translation in the body of the chapter.

[35] French 1994, 199.

[36] For the source used by Vincent of Beauvais, see Gauthier 1982, 34, l. 154–169.

[37] Boethius, *De trinitate*, II, ed. Moreschini 2000, 168–9: *Nam* cum *tres sint speculativae partes, naturalis, in motu inabstracta anupexairetos (considerat enim corporum formas* cum *materia, quae a corporibus actu separari non possunt: quae corpora in motu sunt ut* cum *terra deorsum ignis sursum fertur, habetque motum forma materiae coniuncta)…*

sible combinations. Each of these types of motion correspond to a type of soul – of animation – which characterizes each of the natural realm:[38]

- The simple motion (regular) and non-voluntary that is specific to the elements and the "elemented"
- The diverse (multidirectional) and non-voluntary motion that one observes in the vegetable/vegetative soul of plants
- The diverse and voluntary motions perceptible in the rational soul of animate beings
- The simple and voluntary motion which corresponds to the motion (or motor)[39] of the heavens, in other words to intelligence

The text emphasizes that the term *nature* can only be applied appropriately to the first type and therefore to the motion specific to the elements (*elementa*) and to the elemented (*elementata*). This notion of *elemented* was used to designate compound bodies is peculiar to Salernitan medicine; it is also found at the beginning of the thirteenth century among English authors who were familiar with it. *Elementum* refers to the element in its pure state, and *elementatum* indicates the same element but mixed with other elements in sensitive things: the element as it is perceived in reality.[40] These terms correspond to the notions of "nature natured" and "nature naturing" we encountered in Document I. The *nature natured* (engendered nature) can designate compounds, while the *nature naturing* is that which governs their composition.[41]

Then, two other quotations in Vincent of Beauvais' *Speculum Doctrinale* introduce definitions of nature in relation to the essential and recurring philosophical concepts of matter, substance, form and species. These quotations are found again in Book XV, this time in Chap. 3, entitled "Nature and its operation". Vincent of Beauvais took these definitions from Aristotle's *Physics* and from the *Fons Vitae* by Ibn Gabirol (1020–1054/8), a Neoplatonic philosopher from Zaragoza, with a good knowledge of Aristotle, who wrote the "Source of Life": *Fons vitae*, based on the doctrine of hylomorphism. That is to say, all creatures, both spiritually and bodily, are composed of matter and form and allow an advancement towards the knowledge of Divine Will.

As we have seen, Vincent of Beauvais is one of the authors who, in the years 1240–1250, used the syntagms *natura naturans* and *natura naturata* in order to

[38] Cf. Frunzeanu 2007, 96.

[39] Actually, the text on the soul, edited by Gauthier 1982, uses *motor* (motor) and not *motus* (motion) as in Vincent of Beauvais' text, which I have verified in the manuscripts.

[40] For example, the Earth would be an *elementatum* according to Ralph of Longchamp *ca.* 1212. On this concept, see Silverstein 1954.

[41] In his translation of the *Introductorium maius ad scientiam iudiciorum astrorum* of Abu Maʿshar, made in 1140, Herman of Carinthia used the words *naturatum et natura* where John of Seville, (in his own translation of the same text in 1133), used the words *elementata* and *elementans* to distinguish the compound bodies and the cause governing this composition. See Weijers 1978, 70.

nuance the similarities between God and nature. Yet, from 1250–1260, other Dominican voices emphasized, in the light of Christian theology, the incongruities of these doctrines influenced by Arabic philosophy.

The German philosopher and theologian Albert the Great, in his commentary on the treatise on divine names (*De divinis nominibus*) attributed to Dionysius the Areopagite, emphasized that the similarity proposed by the Cordoban Averroes (1125–98) between universal nature (*natura universalis*) and the divine art (*ars divina*) was based on "the opinion of those who distinguish a dual status of nature, namely the *natura naturans* (nature that begets nature) and *natura naturata* (begotten nature), and say that God is the *natura naturans*". However, Albert says, "we find no philosopher nor holy father who calls God 'nature' except in the case where we speak of the Father as the principle of the generation of the Son; furthermore, we cannot properly call 'nature' what is foreign to all natural things. If, however, we wanted to save the remarks of the Commentator [Averroes] and call 'nature' the divine art, it would be equivocal, just as if we called the sculptor the statue".[42]

Also in contexts seeking to explain the meaning of "universal nature" we find in the theology of Thomas Aquinas, who attended classes by Albert the Great, two critical details against the equivalence between God and *natura naturans*,[43] the partisans of which remain anonymous (*quidam*). Along with Albert, with whose text passages by Thomas bear great affinity, the Aquinate suggests bringing *natura universalis* closer to a celestial virtue/force proceeding from the principles of all natural things and their motions. This notion goes back to a hermetic doctrine.[44]

However, Vincent of Beauvais did not seem to be wary of the concept, to which he gave only limited space, presenting it as a novelty illustrating the debates between natural philosophers of his time. The reticence shown by the other two Dominicans in the use of *natura naturans* is partly explained by their attempt to differentiate and

[42] Albertus Magnus, *Super Dionysium de divinis nominibus,* ed. Simon 1972, 281: *Dicendum, quod hoc quod Commentator nominat divinam artem* ***universalem naturam****, videtur esse secundum opinionem illorum, qui distinguunt duplicem naturam, scilicet naturam naturantem et naturam naturatam, dicentes naturam naturantem deum. Sed hoc non invenitur nec ab aliquo philosopho nec ab aliquo sancto, quod deus dicatur natura, nisi inquantum in patre est principium generationis filii; nec proprie dici potest, ut quod est extrinsecum omni rei naturali, natura dicatur. Si tamen, ut salvetur dictum Commentatoris, divina ars dicatur natura, hoc erit aequivoce dictum, sicut diceretur statuarius statua.*

[43] Thomas Aquinas, *Summa Theologiae*, I-II, qu. 85, a. 6: *Natura vero universalis est virtus activa in aliquo universali principio naturae, puta in aliquo caelestium corporum; vel alicuius superioris substantiae, secundum quod etiam Deus a quibusdam dicitur natura naturans. Idem, In de divinis nominibus,* Chap. 4, l. 21: *Est autem Deus universalis causa omnium quae naturaliter fiunt; unde et quidam ipsum nominant naturam naturantem.* "On the other hand, the universal nature is an active force in some universal principle of nature, for instance, in some heavenly body, or again belonging to some superior substance, in which sense God is said by some to be 'the Nature Who makes nature'" (The *Summa Theologiae*, online 2016).

[44] Albertus Magnus found the concept of universal virtue in the works of a German collegue, Arnold of Saxony (Arnoldus Saxo), who wrote two decades earlier. (Draelants 2003). The concept of "universal nature" was recently discussed in Weill-Parot 2013, Chap. 1.

specialize as much as possible the role of God and the role of nature in the processes of creation. Furthermore, the question of preserving for God a capacity for action that went beyond that which is created naturally. Instead of *natura* id est *Deus* – nature, that is to say, God – they preferred to think in terms of several distinctions between God, universal nature, particular nature and natural things.

As shown by the various passages examined, all taken from a mid-thirteenth century *summa* intended to train preachers: without being syncretic, the Middle Ages inherited a large number of doctrines on nature, from all periods, with which scholars tried to compose a picture that accurately accounted for the multiple facets of reality, divine and human, celestial and terrestrial.

2.3 Definitions of Nature in Two Medieval Dictionaries

Another means to have access to the medieval meanings of the notion of nature is to open medieval dictionaries, tools designed in the form of alphabetic or thematic lexicons, which succeeded the glossaries of the early era. Two other definitions, about 80 years apart from each other, chronologically frame those collected by Vincent of Beauvais in the middle of the thirteenth century and provide access to all the common and general meanings of the concept of nature.

The first is provided by the theologian and philosopher Alain of Lille (1125/30–1203); taken from his theological dictionary made up of successive distinctions, that is to say, subdivided definitions, destined to be used in teaching through preaching. This work reveals the cross-fertilization of Neoplatonism and Aristotelianism in Boethius (ca. 450–524), who inspired Alain of Lille. By the authority of Boethius, he explains that the beginning of the definition adopts an epistemological perspective, based on the theory of knowledge and the doctrine of perception.

The second definition is shorter and later, taken from the dictionary by John of Genoa († 1298), written in 1286.[45] It is based on an etymology of the word "nature" derived from "to be born" (*naître* in French, *nascere* in Latin). One might expect, in view of the period when it was written, that it would take into account the fusion of Neoplatonic and Aristotelian traditions in which the scholastic thinking of the thirteenth century engaged in. In fact, its content reveals a conservative etymological tradition going back to the encyclopaedia of the Visigoth Bishop Isidore of Seville entitled *The Etymologies* (completed in 622, it had an enormous influence on the Latin West); on the other hand, it also transmits elements of Alain of Lille's dictionary.

[45] E. Frunzeanu presented the elements taken from the dictionaries of Alain of Lille and John of Genoa during a talk given for medievalists in 2008 at the University of Nancy.

Doc. VII: Alain of Lille, *Distinctiones dictionum theologicalium* (Ed. Migne, *PL* 210, col. 871a–d)

1. ***Natura*** aliquando ita large sumitur, quod omne illud, **quid quo modo potest intelligi**, natura dicatur; unde Boetius: *Natura est quidquid quid quo modo intelligi potest.* Secundum hanc expositionem, et ***hyle*** et Deus potest dici *natura*; quia, quamvis *hyle* proprie intellectum capi non possit, sed tantum per **formae abnegationem**, tamen quo modo intelligitur. Similiter *divina forma*, quamvis tantum intelligatur per materiae remotionem, tamen quo modo intellectu capitur.	1. **Nature** is sometimes understood in such a broad sense that what we call "nature" **is anything that can be grasped by the intellect in any way**. Hence Boethius: *Nature is anything that can be understood in any way*. According to this definition, **matter** (*hyle*) as much as God can be called *nature*. For, although *matter* cannot be properly understood by the intellect but only by an **abstraction of the form**, it remains that it is understood in a certain way. Similarly, the *divine form*, although it can only be understood by its distancing from matter, is nevertheless understood by the intellect in a certain way.
2. Aliquando sumitur **in designatione substantiae** tantum, unde Boetius: *Natura est quidquid agere vel pati potest;* et secundum hanc acceptionem, Deus potest dici *natura*, quia ipse est *causa* universorum *efficiens*.	2. Sometimes nature is understood only to designate **the substance**. Hence Boethius, *Nature is something that can act or suffer*. In this sense, **God** can be called ***nature*** because He is the *efficient cause* of all things.
2.1. Restringitur tamen hoc nomen *natura* **circa substantiam corpoream**, unde Boetius: *Natura est principium motus* per se *et non* per accidens; hoc enim tantummodo pertinet ad substantiam corpoream ut sit **principium motus per se**, id est ut principaliter et per se moveatur; ipsa enim sola proprie movetur aut a centro ad circumferentiam, ut levia, scilicet ignis et aer; aut a circumferentia ad centrum, ut gravia, scilicet terra et aqua.	2.1. In a restricted sense, the term *nature* applies to **the bodily substance**. Hence Boethius: *Nature is the principle of motion in itself (intrinsic) and not by accident.* This meaning pertains only to the bodily substance in so far as it is the **principle of motion per se**, which is to say that bodily substance sets itself in motion as if by its own principle. It moves effectively by itself, either from the centre to the circumference, like fire and air and light elements, or from the circumference to the centre, like earth and water and heavy elements.
2.2. Restringitur etiam **circa substantialem differentiam** et **specificam** quae adveniens generi facit speciem, ut hoc universale rationabile, unde Boetius: *Natura est reformans specificam differentiam.*	2.2 (The term nature), as a **rational** universal notion, is also restricted to **the substantial and specific difference** which occurs when the species is delineated within the genus. Hence Boethius: *Nature repeats the formation of the specific difference.*
2.3. Dicitur esse **substantiale rei per quod res nascitur**, id est suum esse ingreditur; unde dicitur Christus duarum naturarum, quia tam humanitas quam divinitas est esse Christi.	2.3. It is said that (nature) is **the substantial being of the thing by which this thing is begotten**, namely, by which its being is engendered. Hence, it is said that Christ has two natures, because both humanity and divinity are the essence of Christ.

(continued)

3. Dicitur **origo**, unde dicitur quod angelus *de natura* habuit peccare, id est ab origine habuit libertatem arbitrii ad bene agendum vel male; unde PLATO IN *TIMAEO* introducens Deum loquentem ad angelos ait: *Dii deorum natura quidem indissolubiles.*	3. (Nature) is said to be **the origin**, whence it is said that the angel had *by nature* the ability to sin, which is to say that from the origin he had free will to do good or evil. Hence PLATO, IN THE TIMAEUS, introduces a god who says to the angels: *The gods are by the nature of gods incorruptible.*
4. Dicitur etiam **complexio**, unde: *Physica res diversas diversarum naturarum asserit,* id est complexionum.	4. It is also said to be **temperament**, hence this definition: *medicine deals with the different states of different natures*, that is to say, temperaments.
5. Dicitur **vitium inolitum pro natura**, unde in iure consuetudo dicitur altera natura; et homo dicitur mori de natura, id est ex vitio inolito pro natura.	5. It is said to be **corruption that takes the place of nature,** so that it is rightly said that habit is a "second nature"; and it is said that man dies "by nature", that is, because of corruption that takes the place of nature.
6. Dicitur **naturalis calor**, unde physicus dicit esse pugnam inter morbum et naturam, id est naturalem calorem.	6. It is called **natural heat**, whence the physician says that there is a struggle between disease and nature, that is to say, natural heat.
7. Dicitur **naturalis ratio**, unde APOSTOLUS ait quod *gentes, quae legem non habent, naturaliter quae legis sunt faciunt,* id est naturali instinctu rationis; et secundum hoc solet dici quod natura dictat homini ut non faciat aliis quod sibi non vult fieri, id est naturalis ratio.	7. It is called **natural reason**, whence the APOSTLE says that *the peoples who know no law, naturally do what belongs to the law*, that is to say, through the natural instinct of reason. Hence we are accustomed to saying that nature – that is to say, natural reason – dictates that man should not do to others what he does not wish to be done to himself.
8 [cf. 2.2.]. Dicitur **potentia rebus naturalibus indita ex similibus procreans similia**, unde ALIQUIS dicitur fieri secundum naturam; unde HILARIUS ait quod *Creator factus est creatura, non est naturae ratio, sed potestatis exceptio.*	8. It is said to be **the power which is in natural things, procreating similar beings from the beings that are like them.** Whence it is said that someone is made "according to nature". Hence HILARY [of Poitiers] says that [if] *the Creator made himself a creature, it is not by reason of nature, but by an exception within (his) power.*

In item 2.2, we find the explanation given by the medieval scholastic of "specific natures", based here on a quotation from Boethius going back to the end of the fifth century. These specific natures correspond to the ontological value specific to each "substantial form". The operations specific to each of these forms, through their specific properties, make it possible to infer, by means of reason, the substantial form of every material thing. This special/specific essence of the being, manifested in its operations meant to differentiate it from other beings, is called *nature*, distinguishing the species from the genus. There is here an identification between *substance* and *nature*. Substance is the being which exists in itself, acts and suffers and which is the base of properties and operations which do not have this constancy but which are seen as "accidents" of the form.

Item 4 shows that, following the circulation of medical theories of Greek and Arabic origin, the nature specific to each biological creature, to each body, had been translated in terms of temperament or *complexion*, that is to say, of composition made up of a combination of the qualities derived from four primary elements: cold, hot, wet and dry. In the school of Chartres, where fragments of ancient Galenic medicine circulated at the beginning of the twelfth century, the theologian William of Conches had already said that *physica* concerned "nature and the complexion of bodies" (refering to the end of Doc. IV on medicine as *physica*) and that Plato's *Timaeus* contributed to this science when speaking of the "four elements, the creation of animals, and primordial matter".[46]

From the fifth part of the definition, the notion of unnatural nature or at least of obstacles to nature is introduced in various ways, the first, already encountered, being miracles (item II.3 of the first definition in, Document I). The course of nature, dependent on intrinsic properties, is constantly threatened by several factors. Thus, the confluence of certain circumstances may alter the original nature and thus institute a "second nature", whether in the domain of moral conduct, regulated by the canon law, or in the field of biology.

In the case of moral conduct, as mentioned in item 7 (in Alain of Lille), "natural reason" makes it possible to know what should not be done to others as we would not want done unto us; we may be surprised that in Alain of Lille, as it was the case with Vincent of Beauvais later, it is not explicitly stated that this is about sin, whereas Augustine's *Contra Faustum* (which we have identified above as the source of a part of the first definition of nature in the *Doctrinal Mirror*) clearly defined sin as acts "against nature".[47]

In the case of biology, the "other" or "second nature" may be disease, due to imbalances in temperament (items 4 and 6) or it could be, from the ontological point of view, a kind of resistance of matter to the action of the form which would make an individual dissimilar to his parents, but these cases concerning exceptions to "Aristotelian" generation (such as the "imperfect" worms mentioned above), are not taken into consideration. The exception mentioned in item 8 is *divine* nature, which engendered the son of God made man.

Doc. VIII: John of Genoa, *Catholicon* Ed. Venetiis 1487, revised with the manuscripts Paris, BnF, Latin 7629, f. 234vb; BnF, Arsenal 978, f. 230rb; Montpellier, B.M., 8, f. 365va

Natura a *nascor, nasceris* ; dicitur haec *natura, naturae*, id est **nativitas**.	(2.3) The word nature derives from the verb *nascor, nasceris* and one says *natura, naturae*, in the sense of **birth**.
Et natura dicitur deus quia omnia creat et nasci facit.	(2) God is called nature, because He creates everything and gives birth to everything.
Et natura dicitur quaelibet **creatura**.	Every **creature** is also called *nature*.
Et natura dicitur **complexio**.	(4) Nature also has a sense of **temperament** (*complexion*).

(continued)

[46] William of Conches, *Glosae super Platonem*, X.18–19, ed. Jeauneau 2006: ***Phisica*** *vero est de* ***naturis*** *et* ***complexionibus*** *corporum: 'phisis' enim est natura*; XI.7–8: (…) *ubi vero de quatuor elementis et creatione animalium et de* ***primordiali materia***, *de phisica*.

[47] See the *Contra Faustum*, Note 18.

Et natura dicitur quedam **vis** naturaliter rebus insita de similibus similia procreans.	(8) And we call *nature* a **force** which, naturally intrinsic to things, procreates similar beings from the beings that are like them.
Natura etiam dicitur quedam **significatio** vocabuli, quia significatio causa est quare vocabulum nascatur, id est imponatur.	(9.1) We also call *nature* the **meaning** of a word, for the meaning is the reason why a word originates, namely, that it is attributed (to a reality).
Natura etiam dicitur **impositio** vel inventio vocabuli quia vocabulum nasci nihil aliud est quam ipsum inveniri vel imponi. Vide supra in *miraculum.*	(9.2) *Nature* is also called the **attribution** or the invention of a word, because the birth of a word is nothing other than the fact of finding or imposing it. See above the "miracle" section.

The respective definitions by Alain of Lille and John of Genoa consider (like the definition taken from the *Speculum Doctrinale* examined above, Doc. I) nature as characteristic to all created reality and as the sum of all the visible realities. It may be also noted that John of Genoa simplified the various meanings of the word "nature" given by Alain of Lille. As a good lexicographer, heir to Isidore of Seville, he added the relationship between the word and its meaning and also the reality to which it relates. He says that both the emergence of what the word means and the attribution of a reality to this word are governed by nature. Pursuing creation – as *nature naturing* – it gave rise to both reality and the term to designate it in the mind of man who, like Adam the *onomatothete*, *names* things and is therefore also the continuator of creation. In reality, the reason why the definitions by Alain of Lille and John of Genoa are so close to one another is that John took, almost literally, the text of the *Derivationes*, a kind of dictionary written by an Italian lexicographer, Hugh of Pisa (1140–1210), who wrote them at the same time as Alain of Lille's *Distinctiones*.[48]

It can also be noted that John of Genoa did not keep a single notion derived from natural law. The concept of natural law was born from theological reflection and was developed during the thirteenth century. The notion derived from it evoked the sentence in Alain of Lille: "(nature) is said to be *corruption which takes the place of nature*, so that it is rightly said that habit is a 'second nature'; and it is said that man dies 'by nature', that is to say, because of corruption that takes the place of nature". In fact, at the turn of the twelfth and thirteenth centuries, the natural law was distinguished from both the law of the Scriptures and the divine law, in order to identify it with a fundamental disposition of human cognition.

Citing Guerric of Saint-Quentin, a Dominican master in the 1240s, natural law refers to the universal concepts of the mind (*communes animi conceptiones*), which are distinct from the *cultural* principles of laws, for, as Guerric remarks, those who hear that stealing is forbidden do not necessarily understand the illegality of the act. By means of these common concepts, men are able to apprehend and think what to do or not to do. *Natural* law, which is also distinct from codified norms and legal customs, acquires a content of universality, connected to man as a *species* and not only as people of God.

[48] Hugh of Pisa, *Derivationes*, ed. Cecchini 2004, 821.

2.4 The Place of Wonder and Space Outside of Nature

The definitions of nature are all based on the analogy and comparison enabling man to understand the whole universe, macrocosm and microcosm at the same time: as Boethius said, taken up by Alain of Lille, nature is "all that can be understood in a certain way".

Alongside the positive definitions of nature discussed thus far, there are, however, several forms of *negative* definitions, in the photographic sense of the word, which make it possible to reveal an intaglio image of nature between the similar and the different. This is especially the space occupied by "wonder" that arouses astonishment and admiration or was seen as a sign of the intuition of the existence of God or that of the hereafter. In fact, the medieval universe, as it appears through the written sources, is not only natural. It comprises several regions that remain unfathomable, unattainable through human cognition, atemporal and irreducible to the categories of intrinsic property and resemblance. These are, firstly, the three eschatological spaces, hell, purgatory and heaven, although some believed them to be on Earth. Nevertheless, on Earth, too, there were unnatural regions: parts of the unknown world, inhabited by monstrous races, dealt with in various encyclopaedias, especially that by Thomas of Cantimpré, written between 1230 and 1255, who devoted a whole book to them.[49] This is the *finis Africae* (the boundary of Africa) of many Renaissance era portulans and geographical maps.

And even within the natural order of species, a fantastic world of mythological or Christian inspiration attributed properties to certain species which transgressed the course of nature, either because of the impossibility of being transplanted, cultivated or raised, as is the case with balsam and barnacle geese,[50] or because of their uniqueness, as is the case with the phoenix which does not reproduce. In view of these particular cases, one can conclude that the course of nature does not correspond to a perfect mechanism but remains vulnerable to certain upheavals due to the vagaries of matter or to the direct and rare intervention of God through miracles. As a result, nature understood as the totality of the created things can neither be understood nor controlled in its entirety.

However, learned thirteenth-century scholars had the desire to explain everything with the help of reason. In the tradition of treatises on the "nature of things" written since Antiquity, they asserted that there are visible and undeniable properties which make it possible to understand the "operations" specific to compound bodies (animals, plants and stones) and thereby their nature; these are illustrated, for example, in their therapeutic virtues. Their cause, if not obvious, was supposed to originate in the intrinsic "force" (*virtus*) that characterizes bodies and specifies them. Properties that manifest this specific nature could generally be explained by the interaction of the elementary components of the complexion of the thing (quali-

---

[49] This is Book III, *De monstruosis hominibus*, by Thomas Cantimpratensis, *Liber de natura rerum*, ed. Boese 1973.

[50] Cf. Van der Lugt 2004.

ties of coldness, heat, humidity, dryness and their proportions) thanks to the Aristotelian and the Galenic theory of humours and complexions. However, certain "wondrous" properties that were well known and experienced, but unusual, such as the attraction of magnets, the ability of aetite to prevent drunkenness or the dissolving virtue of rhubarb, did not allow the identification of an elementary cause. This is why, under the influence of the Arabic philosophy of Avicenna in particular, the doctrine of the "occult cause" was born during the thirteenth century. In a way, it would bring back *into* nature phenomena that would have been destined to be beyond it; in the same way, magical virtues could be explained through "sympathetic" properties brought back into the course of nature.[51]

2.5 Conclusion

Without reiterating each of the definitions of nature that have just been covered, some highlights are to be emphasized in the theoretical investigation of nature carried out in the medieval era. First, in comparison with today, the space considered by medieval man's thinking was endowed with a more complete universality: finite time and eternity were both part of it. Consequently, *nature* was as earthly as it was celestial, as human as divine, and even angelic. Nevertheless, during the twelfth and thirteenth centuries, thanks to the evolution of science informed by Greek and Arabic sources, the view became more terrestrial and anthropocentric: the various natural terrestrial realms of minerals, plants and animals took up more space in the explanation of the universe, while in anthropology and psychology, focused on the animation of the living, was developed the study of the rational soul and its faculties.

Once the assimilation of ancient thinking and that of the Arabic commentators was completed, the definitions inspired by theology, natural philosophy and medicine merged in the thirteenth century around the central notion of "force" as the explanatory foundation of the dynamics of nature, whether it be celestial (the *vis celestis* of Thomas Aquinas, which sets the universe in motion), intrinsic (the *vis insita* in every developing being), particular (the *virtutes* or properties that characterize compound bodies) or external (forces that cause unnatural motion).

Finally, the inquiry into nature at the turn of the thirteenth century also promoted the idea that man has an autonomous and superior ontological status in creation. As such, he has a special responsibility towards other creatures and a moral obligation to behave well and to try to find, through knowledge, the primordial resemblance with God. The study of his condition led to the development of a true anthropology, as exemplified by the treatise *De homine* (*ca.* 1242) by Albert the Great.[52]

---

[51] Weill-Parot (2013) deals with the question of the scholastic necessity for the rational explanation of nature.

[52] *Alberti magni Ordinis Fratrum Praedicatorum De homine*, ed. Anzulewicz, Söder 2008.

As an epilogue, the spirit in which man's cultural – and agricultural! – role in nature was understood is well illustrated by the following extract from Albert the Great's Summa Theologiae commenting on an excerpt from Genesis, which underlines that it is man's duty to cultivate and to understand all that surrounds him. Here Albert considers the difference between the "divine science of philosophers",[53] which is the consideration or speculation typical for philosophers, and contrasts it to the theological consideration of the world, both of which are given by divine providence. Since Adam left paradise where harmony was total, man has been entrusted, through the "liberal arts", that is to say, theoretical education, and through work ("operation") and the experience of things, with the learning of "nature and the science of things":

Doc. IX: *Summa theologiae,* Pars II, tr. 14, qu. 89, m. 1, ed. Borgnet 1894–1895, 163–4a

Super illud enim GENESIS, I, 15: Ut operaretur et custodiret illum, distinguit GLOSSA, quae accipitur EX LIBRO VIII AUGUSTINI SUPER GENESIM AD LITTERAM. Duplex est **operatio** sive **cultura**.	Concerning the verse from Genesis I, 15: "[The Lord took man, and placed him in the Garden of Eden] so that he might labour and preserve it", the GLOSS, which takes up AUGUSTINE'S BOOK VIII ON GENESIS AD LITTERAM,[54] says that **work** or **culture** is to be understood in two ways.
Una est cum afflictione, quam Deus homini non indixit tunc cum esset in paradiso: quia DICIT DAMASCENUS, quod in paradiso non debuit nisi solatiosam vitam agere, et in jucunditate esse cum Deo.	On the one hand, it is work with much suffering, which God did not make known to man while he was in paradise, because, SAYS JOHN OF DAMASCUS, in paradise he had only the duty to live a pleasant life and to be joyful with God.
Alia **disciplinalis** est et liberalis, per quam scilicet per fructum opere **discitur** quid virtutis sit in radice. Et haec disciplina indicta est Adae, UT DICIT AUGUSTINUS: et hoc modo per intervalla temporum **discitur natura rerum per experimenta**. Ergo videtur, quod Adam per intervalla temporum **scientiam** accepit **rerum**	On the other hand, it is the work of **learning** and with the help of the liberal arts, by which we **learn** with effort, from the fruit, what virtue is found in the root. And this work of learning was assigned to Adam, AS AUGUSTINE SAYS, and it is in this way that, over time, we learn the nature of things by **experiencing** them. It thus seems that Adam, over time, acquired the **science** of **things.**

[53] It is probable that Albert borrowed the concept from the Baghdadi scholar Al-Ghazzâli (Algazel, 1058–1111), who wrote on Hellenistic Islamic philosophy, logic and religious philosophy. His works were translated by Gundisalvi in the third quarter of the twelfth century. See also Draelants 2019a.

[54] Augustinus, *De genesi ad litteram*, VIII.9, ed. *Patrologia latina*, vol. 34, 376–7. Augustine (354–450) discusses the existence of philosophy as wisdom. He refers to the acquisition of agriculture but also to all knowledge on nature, that is to say, human culture.

References

Sources

Alanus de Insula, *Distinctiones dictionum theologicalium,* ed. J.-P. Migne, *Patrologia latina,* vol. 210. Paris: 1855, col. 685–1011.

Alberti Magni ex ordine praedicatorum De vegetabilibus libri VII, ed. E. Meyer, C. Jessen. Berlin: G. Reimer, 1867 (re-ed. Frankfurt/Main, 1982).

Alberti magni Ordinis Fratrum Praedicatorum De homine, ed. Heinrich Anzulewicz, J.R. Söder. Monasterii Westfalorum [Münster]: Aschendorff, 2008 (*Alberti Magni Opera omnia* XXVII.2).

Albertus Magnus, *Summa theologiae*, ed. Auguste Borgnet, *Alberti Magni Opera omnia.* Paris, 1894–1895.

Albertus Magnus, *Super Dionysium de divinis nominibus*, ed. Simon, Paul. Münster: Aschendorff, 1972 (Alberti Magni Opera omnia XXXVII.1).

Anonymous (13th century), *Summa de anima*: ed. Gauthier 1982.

Augustinus, *De Genesi ad litteram,* I–VIII, ed. J.-P. Migne, *Patrologia latina,* vol. 34. Paris: 1845, col. 247–394.

Boethius, *De consolatione Philosophiae; Opuscula theologica,* ed. C. Moreschini, Leipzig, Teubner, 2000 (Bibliotheca Teubneuraria).

Guillelmus de Conchis, *Dragmaticon,* ed. I. Ronca, L. Badia, J. Pujol. Turnhout: Brepols 1997 (Corpus Christianorum Continuatio Mediaevalis 152).

Guillelmus de Conchis, *Glosae super Platonem,* ed. E. Jeauneau. Turnhout: Brepols, 2006 (*Corpus Christianorum, Continuatio Medievalis* 203).

Iohannes Balbus de Genoa, *Catholicon*, ed. Venetiis: Hermann Liechtenstein, 1487.

Isidori Hispalensis episcopi Etymologiarum sive Originvm libri XX, ed. W. M. Lindsay, Oxford, 1st ed. 1911, 2d ed. 1957. Text online: http://www.thelatinlibrary.com/isidore.html.

The Etymologies of Isidore of Seville, Trans. S.A. Barney, W.J. Lewis, J.A. Beach, O. Berghof. Cambridge University Press. 2006.

Thomas Aquinas, *Summa theologiae*, "editio leonina": *Sancti Thomae de Aquino Opera omnia iussu Leonis XIII P. M. edita*, 1882– . Text and bibliographic references online: http://www.corpusthomisticum.org/repedleo.html.

The *Summa Theologiae*, Online 2016: The *Summa Theologiae* of St. Thomas Aquinas, literally translated by Fathers of the English Dominican Province (http://www.newadvent.org/summa/index.html Online Edition) Copyright © 2016 by Kevin Knight.

Thomas Cantimpratensis, Liber de natura rerum, Teil I: Text, hrsg. von H. Boese. 1972. De Gruyter: Berlin-New York.

Uguccione da Pisa, Derivationes, ed. E. Cecchini (et al.). 2004. 2 vol. (Edizione nazionale di testi mediolatini 11). Firenze: Sismel.

Vincentii Burgundi Speculum quadruplex, naturale, doctrinale, morale, historiale.(…). Opera et studio Theologorum Benedictorum Collegii Vedastini in alma Academia Duacensi, Duaci, I–IV, 1624 (ed. anast. Graz, 1961/65). [in the notes: *Speculum*, ed. Douai 1624].

Works

Aberth, John. 2010. *From the Brink of the Apocalypse: Confronting Famine, War, Plague, and Death in the Later Middle Ages*. 2nd ed. New York: Routledge.

———. 2013. *An Environmental History of the Middle Ages: The Crucible of Nature*. London/New York: Routledge.

Burnett, Charles S.F. 2009. *Arabic into Latin in the Middle Ages. The Translators and Their Intellectual and Social Context*. Farnham: Ashgate.

Butterworth, Charles, ed. 1994. *The Introduction of Arabic Philosophy into Europe*. Leiden: Brill.

Chenu, Marie-Dominique. 1957. *La théologie du XIIe siècle*. Paris: Vrin (Etudes de philosophie médiévale 45).

Cranz, F. Edwards and Paul O. Kristeller. 1960–1980. *Catalogus translationum et commentariorum. Mediaeval and Renaissance Latin Translations and Commentaries. Annotated Lists and Guides*, Washington, DC: Catholic University of America Press (Union Académique Internationale).

D'Alverny, Marie-Thérèse. 1994. *La connaissance de l'Islam dans l'Occident médiéval*. Aldershot: Variorum (Variorum collected studies 445).

———. 1998. *Pensée médiévale en Occident. Théologie, magie et autres textes des XIIe-XIIIe siècles*. Aldershot: Variorum.

De Leemans, Pieter, ed. 2014. *Translating at the Court. Bartholomew of Messina and Cultural Life at the Court of Manfred, King of Sicily*. Leuven: Leuven University Press (*Mediaevalia Lovaniensia*. Series 1/Studia 45).

Draelants, Isabelle. 1995. *Eclipses, comètes, autres phénomènes célestes et tremblements de terre au Moyen Âge. Enquête sur six siècles d'historiographie médiévale dans les limites de la Belgique actuelle (600–1200)*. Louvain-la-Neuve (Travaux de la Faculté de Philosophie et Lettres de l'Université catholique de Louvain, XXXVIII, Section d'Histoire IX).

———. 1996. *Phénomènes célestes et tremblements de terre au Moyen Age: enquête sur l'historiographie médiévale dans les limites de la Belgique actuelle (600–1200)*, In *Les catastrophes naturelles dans l'Europe médiévale et moderne. Actes des XVes journées internationales d'histoire de l'abbaye de Flaran, 10, 11 et 12 septembre 1993*, ed. B. Bennassar, 187–222. Toulouse: Presses Universitaires du Mirail, (Flaran XV).

———. 2003. La *virtus universalis*: un concept d'origine hermétique ? Les sources d'une notion de philosophie naturelle apparentée à la forme spécifique. In *Hermetism from Late Antiquity to Humanism. La tradizione ermetica dal mondo tardo antico all'Umanesimo*, Atti del Convegno Internazionale di Studi (Napoli, 20–24 nov. 2001), ed. P. Lucentini, I. Parri, V. Perrone Compagni, 157–188. Turnhout: Brepols (*Instrumenta patristica et mediaevalia* 40).

———. 2013. Le 'siècle de l'encyclopédisme': conditions et critères de définition d'un genre. In *Encyclopédire: Formes de l'ambition encyclopédique dans l'Antiquité et au Moyen Âge*. Ed. A. Zucker, 81–106. Turnhout: Brepols (Collection d'Études médiévales de Nice 14).

———. 2015. Modèles épistémologiques de l'enquête encyclopédique sur la nature des choses 'ob posteritatis utilitatem'. In *Actes du congrès de la SHMESP à Nancy et Metz, mai 2014*, 235–260. Paris: SHMESP.

———. 2016. Poux, puces et punaises chez les naturalistes du 13e siècle: de simples vermes ou des parasites nuisibles ? In *Poux, puces et punaises: la vermine de l'homme. Découverte, description et thérapeutique. Antiquité, Moyen âge, Temps Modernes*, ed. Franck Collard and Evelyne Samama, 195–225. Paris: L'Harmattan.

——— 2019a. The notion of 'Properties': Tensions between Scientia and Ars in medieval natural philosophy and magic. In *The Routledge History of Medieval Magic*, ed. Sophie Page and Catherine Rider, 169–186. Routledge: Abingdon, New York.

——— 2019b. *Ego sum vermis*: De l'insecte né de la pourriture, à la conception du Christ sans accouplement. Un exemple de naturalisme exégétique médiéval. In *Mélanges en l'honneur de Catherine Jacquemard*, ed. Brigitte Gauvin, and Marie-Agnès Avenel. Caen: Presses universitaires de Caen (coll. Mélanges): 151–184.

Draelants, I., Baudouin Van den Abeele and Anne Tihon, eds. 2000. *Occident et Proche-Orient: contacts scientifiques au temps des croisades. Actes du Colloque de Louvain-la-Neuve, 24–25 mars 1997*. Louvain-la-Neuve: Centre d'histoire des sciences, Turnhout: Brepols (Réminisciences 4).

Endress, Gerhard, ed. 1989. *Symposium Graeco-Arabicum II. Akten des 2. Symposium Graeco-Arabicum, RuhrUniversität Bochum 3.-5. März 1987*. Amsterdam: B.R. Grüner (*Archivum Graeco-Arabicum* 1).

Entre Orient et Occident: la philosophie et la science gréco-romaines dans le monde arabe, Genève, Fondation Hardt, 2011 (Entretiens sur l'Antiquité classique de la Fondation Hardt 57).

Federici Vescovini, Gabriella and Ahmad Hasnawi, eds. 2013. *Circolazione dei saperi nel mediterraneo: filosofia e scienze (secoli ix-xvii) circulation des savoirs autour de la Méditerranée: philosophie et sciences (IX e–XVIIe siècle)*, Fiesole: Edizioni Cadmo.

French, Roger K. 1994. *Ancient Natural History. Histories of Nature*. London/New York: Routledge.

Frunzeanu, Eduard. 2007. *Les configurations de la* natura *dans le* Speculum maius *de Vincent de Beauvais,* thesis defended in October 2007. Montreal: Faculté des études supérieures.

García-Ballester, Luis, French Roger, Jon Arrigabalaga and Anthony Cunningham, eds. 1994. *Practical Medicine from Salerno to the Black Death*. Cambridge: Cambridge University press.

Gauthier, René A. 1982. Le traité *De anima et de potenciis eius* d'un maître ès arts (vers 1225). *Revue des sciences philosophiques et théologiques* 66 (1982): 3–35.

González, S. Vegas. 1997. *La Escuela de Traductores de Toledo en la Historia del Pensamiento*. Toledo: Ayuntamiento de Toledo, Concejalía de Cultura.

Gottschall, Dagmar. 2006. Conrad of Megenberg and the causes of the plague: A Latin Treatise on the Black Death Composed ca. 1350 for the Papal Court in Avignon. In *La vie culturelle, intellectuelle et scientifique à la cour des papes d'Avignon,* ed. J. Hamesse, 319–332. Louvain-la-Neuve. Turnhout: Brepols.

Goyens, Michèle, Pieter De Leemans and An Smets, eds. 2008. *Science Translated: Latin and Vernacular Translations of Scientific Treatises in Medieval Europe*. Leuven: Leuven University Press (*Mediaevalia Lovaniensia Series* 1, *Studia* 40).

Grabmann, Martin. 1916. *Forschungen über die lateinischen Aristoteles-Übersetzungen des XIII. Jahrhunderts*. Münster: Aschendorff (Beiträge zur Geschichte der Philosophie des Mittelalters. Texte und Untersuchungen 17/5–6).

———. 1928. Mittelalterliche lateinische Aristoteles-Übersetzungen und Aristoteles-Kommentare in Handschriften Spanischer Bibliotheken, München (Sitzungsberichte der Bayer. Akademie der Wissenschaften, Philos.-philol. und hist. Kl., Jahrlang 1928, Abhandlung 5).

Green, Monica H., ed. 2015. *Pandemic Disease in the Medieval World: Rethinking the Black Death*. Kalamazoo: Western Michigan University Libraries (The Medieval Globe 1).

Gregory, Tullio. 1975. La nouvelle idée de nature et de savoir scientifique au XIIe siècle. In *The Cultural Context of Medieval Learning. Proceedings of the First International Colloquium on Philosophy, Science and Theology in the Middle Ages (September 1973),* ed. J.E. Murdoch, E.D.F. Sylla, 192–218. Dordrecht/Boston: D. Reidel Publ. Company (Boston Studies in the Philosophy of Science 26).

Gutas, Dimitri. 2005. *Pensée grecque, culture arabe, Le mouvement de traduction gréco-arabe à Bagdad et la société abbasside primitive (IIe–IVe/VIIIe–Xe siècles)*, trad. A. Cheddadi. Paris: Aubier.

Hamesse, Jacqueline and Martha Fattori, eds. 1990. *Rencontres de cultures dans la philosophie mediévale: Traductions et traducteurs de l'antiquité tardive au XIVe siècle*. Actes du colloque international de Cassino, 15–17 juin 1989. Cassino, Louvain-la-Neuve: Centre d'études médiévales, Turnhout: Brepols.

Herlihy, David, 1980. Attitudes Towards the Environment in Medieval Society. In *Historical Ecology: Essays on Environment and Social Change,* ed. Lester J. Bilsky, 100–116. Port Washington, NY: Kennikat Press.

Jayyusi, Salma K. and Manuela Marin, eds. 1992. *The Legacy of Muslim Spain*. Leiden/New York/ Köln (*Handbuch der Orientalistik. Handbook of Oriental Studies.*1. Abt., *Der nähe und mittlere Osten. The Near and Middle East* 12).

Jenkins, Jacqueline. 2007. In *The Medieval Translator. Traduire au Moyen Âge. Proceedings of the International Conference of Paris*, ed. Olivier Bertrand. Turnhout: Brepols.

Jolivet, Jean. 1995. *Philosophie médiévale arabe et latine. Recueil d'articles*. Paris: Vrin.

Kischlat, Harald. 2000. *Studien zur Verbreitung von Übersetzungen arabischer philosophischer Werke in Westeuropa 1150–1400. Das Zeugnis der Bibliotheken*. Münster: Aschendorff.

Köpf, Ulrich and Dieter R. Bauer, eds. 2011. *Kulturkontakte und Rezeptionsvorgänge in der Theologie des 12. und 13. Jahrhunderts*. Münster: Aschendorff (*Archa verbi. Subsidia* 8).

Lau, D. and J. Cobet, eds. 2000. *Europa. Die Gegenwärtigkeit der antiken Überlieferung*. Aachen: Shaker.

Le Goff, Jacques. 1957. *Les intellectuels au Moyen Âge*. 1st ed. Paris: Seuil. (Le temps qui court 4).

Lejbowicz, Max and Monique Bourin, eds. 2009. *Une conquête des savoirs: les traductions dans l'Europe latine (fin du XIe siècle – milieu du XIIIe siècle). Colloque organisé à la Fondation Singer-Polignac le jeudi 27 novembre 2008*. Turnhout: Brepols (Rencontres médiévales européennes 9).

Lindberg, David D., ed. 1978. *Science in the Middle Ages*, Chicago: University of Chicago Press.

Müller, Eugen, 2012. *Natur und Wunder, ihr Gegensatz und ihre Harmonie. Ein apologetischer Versuch*. Bremen: Unikum.

Obrist, Barbara. 2004. *La cosmologie médiévale. Textes et images*. I. *Les fondements antiques*. Firenze: SISMEL – Edizioni del Galluzzo (Micrologus' Library 11).

Oriente e Occidente nel Medioevo: Filosofia e Scienze. 1971. Atti del Convegno Internaz. dell'Accademia Naz. dei Lincei, Roma.

Paravicini Bagliani, Agostino and Francesco Santi, eds. 1998. *The regulation of Evil. Social and cultural attitudes to Epidemics in the late Middle Ages*, Firenze: Sismel – Edizioni del Galluzzo.

Paulmier-Foucart, Monique and Marie-C. Duchenne (coll.). 2004. *Vincent de Beauvais et le Grand miroir du monde*. Turnhout: Brepols (Témoins de notre histoire).

Rouse, Richard H. 1976. La diffusion en Occident au XIIIe siècle des outils de travail facilitant l'accès aux textes autoritatifs. *Islam et Occident au Moyen Age, Revue des études islamiques* 44: 115–147.

———. 1981. Le développement des instruments de travail au XIIIe siècle. In *Culture et travail intellectuel dans l'Occident médiéval*, ed. G. Hasenhor, J. Longère, 115–144. Paris: éditions du CNRS.

Scarcia Amoretti, Biancamaria, ed. 1987. *La diffusione delle scienze islamiche nel Medio Evo europeo. Convegno internazionale.... Roma, 2–4 ottobre 1984*. Roma: Accademia Nazionale dei Lincei.

Silverstein, Théodore. 1954. Elementatum: Its Appearance among the Twelfth-Century Cosmogonists. *Mediaeval Studies* 16: 156–162.

Speer, Andreas and Lydia Wegener, eds. 2006. *Wissen über Grenzen. Arabisches Wissen und lateinisches Mittelalter*, Berlin/New York: Walter De Gruyter (*Miscellanea mediaevalia*. Veröffentlichungen des Thomas-Instituts der Universität zu Köln 33).

Speer, Andreas. 1995. *Die entdeckte Natur*. Leiden-New York-Köln, Brill.

Thonnard, François-Joseph. 1965. La notion de nature chez saint Augustin. Ses progrès dans la polémique antipélagienne. *Revue d'études augustiniennes* 11: 239–265.

Tischler, Mathias and Alexander Fidora, eds. 2011. *Christlicher Norden – Muslimischer Süden. Ansprüche und Wirklichkeiten von Christen, Juden und Muslimen auf der Iberischen halbinsel in Hoch- und Spätmittelalter*. Münster: Aschendorff (*Erudiri sapientia* 7).

Van der Lugt, Maike. 2004. *Le ver, le démon et la Vierge: les théories médiévales de la génération extraordinaire; une étude sur les rapports entre théologie, philosophie naturelle et médecine*. Paris: Les Belles Lettres.

van Oppenraay, Aafke, ed., R. Fontaine, collab. 2012. *The Letter before the Spirit: The Importance of Text Editions for the Study of the Reception of Aristotle*. Leiden/Boston: Brill.

Van Riet, Simone. 1987. L'essor de la philosophie et des sciences à Bagdad, ferment de la pensée médiévale en Europe. *Mededelingen der koninglijke nederlandse Akademie van wetenschappen, Afd. letterkunde*, Niewe reeks, Deel 50, n. 7: 215–228.

Verger, Jacques. 1997. *Les gens de savoir en Europe à la fin du Moyen Âge*. Paris: PUF.

Weijers, Olga. 1978. Contribution à l'histoire des termes *natura naturans* et *natura naturata* jusqu'à Spinoza. *Vivarium* 16: 70–80.
Weill-Parot, Nicolas. 2010. Astrology, Astral Influences, and Occult Properties in the Thirteenth and Fourteenth centuries. *Traditio* 65: 201–230.
———. 2013. *Points aveugles de la nature. L'occulte, l'attraction magnétique et l'horreur du vide (XIIIe-milieu du XVe siècle)*. Paris: Les Belles Lettres.
Wisnovsky, R., J. C. Faith Wallis, and C. Fraenkel Fumo, eds. 2012. *Vehicles of Transmission, Translation, and Transformation in Medieval Textual Culture*. Turnhout: Brepols (Cursor 4).
Zimmermann, Michel, ed. 2001. Auctor *et* Auctoritas. *Invention et conformisme dans l'écriture médiévale, Actes du colloque de* Saint-*Quentin-en-Yvelines (14–16 juin 1999)*. Paris: Ecole des Chartes.

Chapter 3
The Animal Environment and Human Health: The Approach Followed by the Medieval Zoologist Ğāḥiẓ (Ninth Century)

Ahmed Aarab, Kaouthar Lamouchi-Chebbi, and Mehrnaz Katouzian-Safadi

Abstract Medieval medical studies allow us to address the issue of human health in its general environment and in particular in its geographical and climatic environment. The work of Ğāḥiẓ (an Arab scholar of the eighth and ninth centuries) allows us to focus notably on the relationship between human health and its animal environment. This author was in an exceptional scientific position for such analysis. In fact, he was very familiar with the *physics* (or natural sciences) of medieval times as well as with the body and human temperament and had a profound knowledge of the zoological science of his time. The originality of this scholar lies in the conviction that animals are worthy of specific research and must be studied for themselves. Ğāḥiẓ, with his deep knowledge of zoology, also examined the effects of animals on the physical and mental health of humans. In this chapter, we will focus in particular on his *Book of Animals*, the *Kitāb-al-Ḥayawān*, composed of 7 volumes of nearly 400 pages each. In this study, the author brings together all the zoological knowledge of his time. He reviews the data from Greek sources (notably Aristotle and Galen); he also mentions the writings of his contemporaries, particularly the Arab grammarians (such as al-Asmaʿī) who enriched zoological knowledge through personal observation and precise terminological research. Ğāḥiẓ also collected the oral testimonies of travelers. But above all, this work includes his observations, associ-

A. Aarab
Faculty of Sciences and Techniques, University Abdelmalek Essadi, Tanger, Marocco

UMR 7219 – SPHERE, CNRS/Université Paris Diderot, Paris, France

K. Lamouchi-Chebbi
Université de Paris, SPHERE, UMR 7219, CNRS-Université de Paris, Paris, France

Manouba University, Tunis, Tunisia

M. Katouzian-Safadi (✉)
Université de Paris, SPHERE, UMR 7219, CNRS-Université de Paris, Paris, France
e-mail: mehrnaz.katouzian-safadi@univ-paris-diderot.fr

F. Bretelle-Establet et al. (eds.), *Making Sense of Health, Disease, and the Environment in Cross-Cultural History: The Arabic-Islamic World, China, Europe, and North America*, Boston Studies in the Philosophy and History of Science 333, https://doi.org/10.1007/978-3-030-19082-8_3

ated at times with his own experimentation. Ǧāḥiẓ therefore brings to the zoological knowledge of his time some elements that are totally new, on occasion rejecting propositions made by the most legitimate authorities (such as Aristotle). In this chapter, we shall first focus on Ǧāḥiẓ's observations regarding the health of animals when in captivity. Then, we will examine two categories of animals: firstly, snakes, the mysterious animal featured in tales and beliefs, a symbol of strength, cleverness, and cunning, as well as an object of admiration and fear because of the power of their venom. Secondly, we focus on pests, today called "insects." Ǧāḥiẓ pays the same amount of attention to these two categories, examines their power and nuisance, and searches for scientific explanations as well as solutions in relation to human health.

Keywords Ǧāḥiẓ · Environment · Animal cohabitation · Pest control · Venom

3.1 Introduction

Medieval medical studies allow us to address the issue of human health in its general environment and more particularly in its geographical and climatic environment. To address the relationships of human health with its animal environment, it is necessary to study the texts of Arab zoology. One of the major works in this field is the *Kitāb-al-Ḥayawān (Book of Animals)* by Ǧāḥiẓ[1] that is composed of seven volumes of about 400 pages each.

The author of the book, being a Mu'tazilite,[2] had a deep knowledge of physics (or natural sciences) of the medieval period, as well as of the human body and its temperaments. He had the great advantage of having access to both oral and written zoological science of the Arab authors of his time, as well as data from Greek sources, particularly by Aristotle and Galen. Ǧāḥiẓ was also a scrupulous observer. He added in his book his personal observations and experiences; he studied animals for themselves as creatures worthy of consideration, in the same way as he studied humans.

Armed with his deep knowledge of zoology, he examined the effects of the presence of animals on the physical and mental health of humans. He gives us valuable information on the various scientific discussions that took place in his time. These discussions highlighted some of the scholars, philosophers, physicians, and Arab grammarians of that period such as Baḫtīšū' Ibn Ǧibrīl, Abū Isḥāq al-Naẓẓām, Ibn Abī 'Aǧūz, and many others (see Table 3.1 regarding the names of the authors). Ǧāḥiẓ actively participated in meetings and discussions between scholars, in

[1] Ǧāḥiẓ 1945. This book has not undergone complete translation and editing. The full name of the author is Abū Uṯmān 'Amr Ibn Baḥr al-Kinānī al-Baṣrī, born in Basra in Iraq in 776 and nicknamed Ǧāḥiẓ which in Arabic means one who has bulging eyes.

[2] Mutazilism is one of the schools of thought within Islam. This movement is one of the branches of Islam, which privileges reason over faith. See Gimaret 1981.

cenacles and discussion circles, or *mağlis*, in Arabic.[3] He frequently refers to the texts and stories of Arab grammarians who enriched zoological knowledge with their own observations and precise terminological research. This is the case, for example, of the famous grammarian al-Asma'ī. Ǧāḥiẓ also collected oral testimonies from travelers, animal breeders, and Arab nomads. His personal observations, sometimes associated with experiments, often complement the remarks and data of his peers. Ǧāḥiẓ does not hesitate to accept some of these and reject those that, in his opinion, do not seem logical.[4] He rejects certain proposals from recognized authorities. Thus, while knowing Aristotle's *Book of Animals*, Ǧāḥiẓ submits his own reflections and addresses subjects that are not treated by the Greek philosopher.[5]

In the present study, we shall discuss the multiplicity of animal-human relationships. Indeed, the cohabitation of animals and humans and what this can bring to man: leisure, care, and diversion, but also nuisance, whether physical or psychological. We shall first focus on Ǧāḥiẓ's observations regarding the health of animals when in captivity. We shall then deal with a mysterious animal, the snake as depicted in tales and beliefs, a symbol of strength and intelligence and an object of admiration and fear because of the power of its poison. Finally, we shall examine the case of small pests, now commonly designated "insects." In his observations, Ǧāḥiẓ gives the same importance to these two categories. He examines their strength and nuisance and seeks sound scientific explanations and solutions in relation to human health.

3.2 Desired Cohabitation: Animals in Captivity

Keeping animals in captivity is an ancient tradition, and it is sometimes undertaken for human enjoyment. When animals are adopted, men try to provide them with an environment[6] that they consider the best possible for the animal. These animals are generally offered adequate amounts of good quality foodstuffs; they enjoy a good level of hygiene and their masters show them consideration and affection. However, one can notice behaviors indicating various disorders in these animals. In the following section, we will examine some changes reported by our author.

In the *Kitāb-al-Ḥayawān*, Ǧāḥiẓ devotes several passages to problems of reproduction experienced by some wild animals while in captivity. He then gives a general analysis, based on various debates that took place in his time on the changes that may arise in these animals. Some, he says, mate and reproduce while maintaining their wild character. Others do not breed at all. Others stop eating and drinking.

[3] Katouzian-Safadi 2004.

[4] See particularly Aarab et al. 2003; Aarab and Lehrminier 2015.

[5] Aarab et al. 2003.

[6] In this article, we use the word milieu and environment indiscriminately to talk about what encompasses animals or humans.

Some, such as snakes, may be forced to eat by forcing food down their throats.[7] Finally, there are some, which neither mate nor become domesticated and stop taking food, drinking or fall silent until they die. The latter case, as the author points out, is observed in particular in wild birds.

The elephant is one of the species that interested Ǧāḥiẓ the most. Our author emphasizes the difficulty in raising this animal in Iraq[8] as well as the impossibility for it to reproduce in this country. This, he explains, was due to the incompatibility between this region and the very nature (*ṭabāi'*, طبائع) of the elephants that caused them to catch diseases and makes them vulnerable.[9] Also, the passage from the wild (*waḥšī* وحشي) to domestic life (*ahlī* أهلي) brings the elephants changes that might be harmful to them. Being wild, these animals feed according to their needs, following a diet that is specific to them. But once caught and domesticated, they undergo a great change in their feeding behavior. In fact, servants were assigned to them to take care of all their needs and imposed a different diet on them, thus causing them great harm.[10]

Ǧāḥiẓ then refers to a historical fact that took place before he was born and before he started his own observations. Speaking of the 950 elephants trained at the court of the Sassanid king Parvīz,[11] Ǧāḥiẓ reports that out of this large number of elephants, one exceptional mating was successful leading to the birth of a baby elephant. It was, he said, the first and last case of elephants breeding in Iraq.[12]

The author stresses the fact that the phenomenon of maladjustment extends to other animals. Thus, dromedaries could not live for long in Bilād al-Rūm (Byzantium[13]), and crocodiles perished once transported to the Euphrates or the Tigris rivers. He also states that this phenomenon existed even in plants and cites several examples of Indian essences that could not be grown in his country.[14]

To discuss the various possible causes preventing the reproduction of elephants in Iraq, Ǧāḥiẓ extends his examination to other species of animals. He notes that peacocks imported from India bred easily in Iraq. But many species of wild bird living in fields and gardens did not breed in captivity in homes even if the region and climate were not altered. However, these same birds did reproduce in houses if they were brought up there from a young age. To better understand this phenomenon, our author cites several other types of wild animal such as lions, wolves, leopards, and tigers. According to him, if these animals could benefit from more favorable

[7] Ǧāḥiẓ 1945, vol. VI, p. 25–26.

[8] It was in Iraq that the Abbasid caliphs had chosen to set up their capitals, in Baghdad and then Samara.

[9] Ǧāḥiẓ 1945, vol. VII, p. 134–135.

[10] Ibid, vol. VII, p. 134–135.

[11] Sassanide: a pre-Islamic dynasty, who ruled in Persia between 224 and 651 AD. The Emperor Parviz (known also in literature as Khosro or Khosrow II or Chosroes in Greek) reigned from 590 to 628.

[12] Ǧāḥiẓ 1945, vol. VII, p. 181.

[13] Bilād al-Rūm بلاد الروم corresponds to the definition given later to the geographical area corresponding to Byzantium.

[14] Ibid.

conditions by living in Iraqi meadows, as their needs are satisfied, they would be able to reproduce. Thus, if young elephants were released into these fields, it would probably favor their reproduction.[15] These comparisons led the author to suggest that to make animals used to captivity, they should be moved at an early age so as to reduce their attachment[16] to the wild.[17] Ǧāḥiẓ reports that in his time, and following the same procedure described above, a certain Ǧa'far Ibn Sulīymāne succeeded in breeding a few gazelles on his own farm in Basra,[18] to the astonishment of Caliph Hārūn al-Rachīd.[19]

Another interesting aspect emphasized by Ǧāḥiẓ is that of behavioral changes in some wild animals in captivity. This, for example, was the case of the nightingale (*bulbul*[20]). Our author contests certain opinions that affirm that this bird is of a very excited nature and keeps moving. According to him, such behavior does not exist in nightingales that live freely in the wild, but only in those which are in captivity, locked up in their cages.[21]

Ǧāḥiẓ was therefore aware of the fact that captivity and living conditions in artificial environments had adverse effects on the health of these animals. At present, it is established that this situation could lead to several cases of behavioral aberrations that do not exist in nature.[22] It should also be noted that while Ǧāḥiẓ was able to stress the influence of man on the health of animals in captivity, he also raised our awareness of the influence that some of these animals have had on the behavior and psychology of men. This is particularly the case of pigeons and the activity of pigeon racing. According to Ǧāḥiẓ, men belonging to all social categories (from the simple artisan to the king and especially the eunuchs[23]) were fascinated by the taming and breeding of pigeons and especially of carrier pigeons.[24] This enthusiasm led some to invest extravagant amounts of time and money to acquire pigeons of the best pedigree.[25] These people could ultimately ruin themselves. Ǧāḥiẓ emphasizes that such an uncontrolled passion toward pigeons ended up as an addiction and illness harmful to man.

---

[15] Ǧāḥiẓ 1945, vol. VII, p. 187–188.

[16] We translate by "attachment" the term "*al-Waḥšah*, الوحشة" that can equally be translated by "nostalgia."

[17] Ǧāḥiẓ 1945, vol. VII, p. 186.

[18] Ibid.

[19] Abbassid Caliph (786–809).

[20] Bulbul بلبل even today the word *bulbul* means nightingale in Arabic and Persian.

[21] Ǧāḥiẓ 1945, vol. V, p. 224.

[22] At present, and in order to avoid false ethograms, any ethological study carried out in the laboratory must be complemented by observations made in the environment in which the animal species in question lives.

[23] Ǧāḥiẓ 1945, Vol. I, p.118.

[24] Ibid., vol. III, p. 147.

[25] Ibid. Vol. III, p.212, 295–296. This passion for pigeon racing or the art of taming pigeons is part of a long tradition that is still maintained in many countries with the same enthusiasm. https://en.wikipedia.org/wiki/Pigeon_racing.

3.3 Necessary Cohabitation

Ğāḥiẓ was an admirer of the immensity of the animal world. He recalls this by quoting a Qur'anic verse.[26] He wanted to illustrate this rich diversity by an observation:

> *To fully understand this verse and to get an idea of the unimaginable number of creatures and their diversity, one has only to light a fire in a forest or in a desert and to see the impressive diversity of species attracted by the flames and that no one has seen before. Other species will appear if the experiment is carried out in another forest or in another environment such as the seas or mountains. The number of creatures observed in these experiments is far from reaching those that are unknown.*[27]
>
> من أراد أن يعرف قوله" ويخلق ما لا تعلمون" فليوقد نارا في وسط غيضة، أو في صحراء برية ثم ينظر إلى ما يغشى النار من أصناف الخلق من الحشرات والهمج، فإنه سيرى صورا، ويتعرف خلقا لم يكن يظن أن الله تعالى خلق شيئا من ذلك العالم. وعلى أن الخلق الذي يغشى ناره يختلف على قدر اختلاف مواضع الغياض والبحار والجبال. ويعلم أن مالم يبلغه أكثر وأعجب.

This experiment proposed in *Kitāb al-Ḥayawān* is easy to carry out and demonstrates the argument quite well. According to the author, this same experiment carried out in other environments reveals the existence of other animals.[28] Zakarīyā Ibn Muḥammad al-Qazwīnī[29] (XIII th) reported that a similar experiment was carried out in the presence of the Abbasid caliph Al-Mu'taḍid billah (ninth/tenth century). They were able to collect and distinguish 72 different kinds of flying insects that were attracted by the fire. The existence of this huge number of animal species caused Ğāḥiẓ to reflect on the goals and purposes (*المصلحة*) of creating harmful or despicable beings such as snakes and scorpions or devastating pests and flies and even carnivorous animals. He pointed out:

> *… the harmful species are as important as the useful species… Their harmful side is of capital importance, for all these beings are complementary and indispensable to ensure general interest and global welfare… If the elimination of one of them is allowed, it will be allowed for the others as well, because the whole consists of elements of equal importance, the first of which is attached to the second, the second to the third, and so on.*[30]
>
> فسبحان من جعل منافعها نعمة، ومضارها ترجع إلى أعظم المنافع ... وجعل في الجميع تمام المصلحة، وباجتماعهما تتم النعمة، وفي بطلان واحد منها بطلان الجميع، إنما هو واحد ضم إلى واحد وواحد ضم إليهما ... فإذا جوزت رفع واحد والآخر مثله في الوزن وله مثل علته وحظه ونصيبه، فقد جوزت رفع الجميع، لأنه ليس الأول بأحق من الثاني في الوقت الذي رجوت فيه إبطال الأول، والثاني كذلك والثالث والرابع، حتى تأتي على الكل وتستخرج الجميع.

[26] "And He has created (other) things of which you have no knowledge," Coran S. XVI, V. 8.

[27] Ğāḥiẓ 1945, vol. II, p. 110–111.

[28] In fact, the method is widely used today to catch photopositive insects (Chauvin 1967; Wiberg-Larsen 1998).

[29] Zakarīyā Ibn Muḥammad al-Qazwīnī (1203–1283) was a judge (*Cadi*), under the reign of the last Abbasid caliph. (See "Kazwini", by T. Lewicki, in *The Encyclopaedia of Islam*, 2nd ed., vol. 4, p. 865–7.) He retired from public life after the installation of the Ilkhanid dynasty and devoted himself to cosmography, astronomy, astrology, and mineralogy. At the end of his life, he wrote his important work on cosmography entitled *The Wonders of Creation* or "*Ağā'ib al-maḫlūqāt wa ġarā'ib al-mawğūdāt*." The experiment cited is related in this famous book.

[30] Ğāḥiẓ 1945, vol. I, p. 206.

In another passage and in the same vein, Ǧāḥiẓ describes the principle of what we now call the food chain:

> *... Each species constitutes food for another (species). Because each of them cannot do without food, and therefore every predator is obliged to hunt. Thus, each animal species will feed on a species weaker than itself. In the same way as each animal species is going to be the food of a species stronger than itself... Allah wanted some species to be the source of life for others, and some the cause of death for others.*[31]
>
> هذا مذهب في أكل الطيبات بعضها لبعض. وليس لجميعها بد من الطعم، ولا بد للصائد أن يصطاد، وكل ضعيف فهو يأكل أضعف منه، وكل قوي فلا بد أن يأكله من هو أقوى من منه، ... فجعل الله عز وجل بعضها حياة لبعض، وبعضها موتا لبعض.

Moreover, the author gives more details on this chain to show the amazing distribution and allocation of food among the various animal species. It is a long chain formed by ten species where the first feeds on the second, the second on the third, and so on up to the tenth species. At the top of the chain proposed in this example is the wolf, then the fox, the hedgehog, the viper and other species of snakes, the sparrow, the grasshopper, the eggs and larvae of the hornet, the bee, the fly, and finally the mosquito.[32] Thus, according to our author, no animal is excluded from this chain of life; no life is meaningless or useless.

3.4 Animal, Man, Painful Promiscuity, and Mortal Terror

3.4.1 The Case of Small Harmful Animals (الضارة)

Ǧāḥiẓ's reflection on the usefulness in nature of species harmful to man does not prevent him from observing that some species can hamper the daily life of human beings[33] to the point of them having to change their lifestyles and habits. This is the case of flies, for example, that sometimes prevent the normal course of human life and require people to modify their way of living in the preparation of meals. Thus, he tells us:

> *The inhabitants of the land of India do not cook in a pot, do not make sweets, and almost eat only at night because of the flies that rush onto their food, and this is proof of the putrid quality of the air and the soil.*[34]
>
> فأما سكان بلاد الهند فإنهم لا يطبخون قدرا، ولا يعملون حلوى، ولا يكادون يأكلون إلا ليلا، لما يتهافت من الذبان في طعامهم. وهذا يدل على عفن التربة ولخن الهواء.

[31] Ǧāḥiẓ 1945, vol. VI, p. 400.

[32] Ibid, vol. VI, p. 313.

[33] Katouzian-Safadi and Chebbi-Lamouchi 2015.

[34] Ǧāḥiẓ 1945, vol. III, p. 328. Indeed, Ǧāḥiẓ like Aristotle believed in "spontaneous generation" where certain forms of living organism arise from inanimate matter such as dust, putrefied soil, or air (Katouzian-Safadi and Chebbi-Lamouchi 2015, p. 169–171).

Our author does not hesitate to advise his reader to avoid them by using cunning tricks:

> *... Flies possess two commendable qualities, one of which is the simplicity of the trick, which enables the dismissal of their nuisance, and the repulsion of their evil. So if someone wants to get them out of the house while keeping the same brightness and quietness and avoiding the nuisance of the flies, he only has to close the door, they will then rush out and compete to reach the light and flee the darkness. […] and this is not the case for mosquitoes, as their nuisance rises, their powers are strengthened and their rage increases in the dark, like the flies in the light.*[35]
>
> *وفي الذباب خصلتان من الخصال المحمودة: أما إحداهما فقرب الحيلة لصرف أذاها ودفع مكروهها، فمن أراد إخراجها من البيت فليس بينه وبين أن يكون البيت على المقدار الأول من الضياء والكن، بعد إخراجها، مع السلامة من التأذي بالذبان، إلا أن يغلق الباب، فإنهن يتبادرن إلى الخروج، ويتسابقن في طلب الضوء والهرب من الظلمة […] وليس كذلك البعوض، لأن البعوض إنما يشتد أذاه، ويقوى سلطانه، ويشتد كلبه في الظلمة، كما يقوى سلطان الذبان في الضياء.*

Ants, animals considered as symbols of laborious activity in their natural environment, can, according to Ǧāḥiẓ, become harmful and devastating in the eyes of humans. According to him, they can oblige an entire population to leave its country.[36] The same applies to termites:

> *... termites may infest villagers' homes and eat whatever they possess. They continue this way until ants appear in these villages, God then grants these ants power over the termites and they exterminate them to the last. However the ants will cause damage later on but they will be less aggressive than the termites. And most of the time the ants disappear from these villages and thus the security of their inhabitants regarding the two species is ensured. Some report that these same termites turn into ants and that their disappearance is not caused by the fact that they are devoured by the ants but that they themselves transform into ants and that the decrease in their number and their nuisance over time corresponds to the number of those that has changed.*[37]
>
> *... وربما أفسدت الأرضة على أهل القرى منازلهم، وأكلت كل شيء لهم. ولا تزال كذلك حتى ينشو في تلك القرى النمل، فيسلط الله ذلك النمل على تلك الأرضة، حتى تأتي على آخرها. وعلى أن النمل بعد ذلك سيكون له أذى، إلا أنه دون الأرضة تعديا. وما أكثر ما يذهب النمل أيضا من تلك القرى، حتى تتم لأهلها السلامة من النوعين جميعا وزعم بعضهم أن تلك الأرضة بأعيانها تستحيل نملا، وليس فناؤها لأكل النمل لها، ولكن الأرضة نفسها تستحيل نملا، فعلى قدر ما يستحيل منها يرى النقص في عددها ومضرتها مدى الأيام.*

Killing these pests sometimes becomes a necessity. The author himself tried to eliminate these ants, but in vain, and he did this by applying certain recipes[38]:

> *It is said that to kill ants, tar and sulfur must be poured at the entrance of their nests, and hair must be dropped on them. We have carried out this experiment but no results!*[39]
>
> *قالوا: وتقتل بأن يصب في أفواه بيوتها القطران والكبريت الأصفر، ويدس في أفواهها الشعر. وقد جربنا ذلك فوجدناه باطلا.*

The louse is another known and customary pest of the environment close to man. According to many medieval doctors, for lice to appear on a person, a set of external

[35] Ibid, vol. III, p. 319–320.

[36] Ibid, vol. IV, p. 15.

[37] Ibid., vol. IV, p. 34–35.

[38] Ibid., vol. IV, p. 36.

[39] It should be noted that Ǧāḥiẓ does not explicitly mention Aristotle, but this method for helping to get rid of ants is found in Aristotle's *Historia animalium* (IV, 8, 534b, 20).

and internal[40] causes had to be present. Dirt and a confined environment (hair, feathers, dirty clothing) were external elements that promoted the presence of lice. But our author observes that clean and perfumed people could be afflicted with lice. He thinks, indeed, that some bodies are likely to catch lice[41] despite refined external care and cleanliness. It is then necessary to work on the internal causes determined by the temperament of each person, the temperament that the medieval doctors tried to adapt and balance:

> *… it is important to prevent bodily ailments through fasting and food (food and drink), rest and activity, moments of awakening and sleep, bathing, maintenance of habitat and any other condition of life depending on man. Even when drug therapy is required, these precautions, taken as a whole, are part of the care required to help everyone maintain his balance and get rid of the disease.*[42]

In this context, Ǧāḥiẓ does not hesitate to give advice on food to maintain the balance of the body; he particularly discourages the consumption of figs,[43] considered as an element that modifies certain internal factors in the human body in such a way as to favor the production of lice. This opinion was shared by many doctors before and after Ǧāḥiẓ.

When all this care is insufficient, he offers advice on medication to get rid of pests like lice that feed directly on the human body[44]:

> *When infested with and suffering from lice, a person applies mercury to his head if the lice are there, or to his body if the lice are in his clothes and this kills them .*[45]
> وإذا قمل إنسان وأفرط عليه ذلك، زأبق رأسه إن كن في رأسه أو جسده، وإن كن في ثيابه، فموتن.

3.4.2 *Examples of Venomous and Impressive Animals*

As the Arabs lived on lands where venomous animals such as snakes and scorpions abound, venom fascinated them.[46] These people were very interested in toxic substances, both as a way to protect themselves and to use them. Ǧāḥiẓ also admired these animals, which have the ability to frighten humans; he contemplates them and describes them at length. In his *Kitāb al-Ḥayawān*, he explains why the Arabs and especially the nomads were greatly interested in wild animals and particularly in venomous species. He writes:

---

40 Katouzian-Safadi and Chebbi-Lamouchi 2015.

41 Ǧāḥiẓ 1945, vol. V, p. 372.

42 Katouzian-Safadi and Chebbi-Lamouchi 2015, p. 172.

43 Ǧāḥiẓ 1945, vol. V, p. 371–372.

44 Ibid. vol. V, p. 371.

45 Ǧāḥiẓ 1945, vol. V, p. 371. (with good transliteration for Gahiz). Scientists and doctors were aware of the toxicity of some products. It was frequency and dilution that prevented mortal dangers. We find references to this kind of dangerous products in chapter 3 of *Al-Mansūrī fī al-ṭibb* of Rāzī (died in 925) concerning the simple medicines (*Al-Mansūrī al-ṭibb*, ed., Kuwait, 1987).

46 For more details, see Aarab et al. 2001.

The zoological knowledge of nomadic Arabs is not the result of a systematic study (العناية والفلاية) *of the animal world around them, or a way to animate discussions or to make profit. But as these people live in the wilderness among wild beasts, they have often faced dangers and been victims of fangs and claws, stings and bites, and sometimes have even been devoured. Thus, it was need, which led them to know in detail the nature and behavior of each of these aggressors, and by observing them; they were able to find various means to avoid them and to find a pharmacopoeia to cure themselves from their evils. These Bedouins have accumulated all this knowledge over time and they passed it on from one generation to the next….*[47]

وإنما أعتمد في مثل هذا على ما عند الأعراب، وإن كانوا لم يعرفوا شكل ما احتيج إليه منها من جهة العناية والفلاية، ولا من جهة التذاكر والتكسب، ولكن هذه الأجناس الكثيرة، وما كان منها سبعا أو بهيمة أو مشترك الخلق، فإنما هي مبثوثة في بلاد الوحش من صحراء أو واد ... وهي في منازلهم ومناشئهم ... وربما بل كثيرا ما يبتلون بالناب والمخلبن وباللدغ واللسع والعض والأكل، فخرجت بهم الحاجة إلى تعرف حال الجاني والجارح والقاتل، وحال المجني عليه والمجروح والمقتول، وكيف الطلب والهرب، وكيف الداء والدواء، لطول الحاجة، ولطول وقوع البصر، مع ما يتوارثون من المعرفة بالداء والدواء.

Let us begin by presenting a few explanations reported by Ğāḥiẓ, in connection with the venoms and their mode of action. These ideas were elaborated and discussed by the scientists and physicians of his time. Thus, Ğāḥiẓ indicates that for the Indians, the venom kills because it is a foreign body. Because, according to them, every foreign body which penetrates the body of an animal kills it. Moreover, Ğāḥiẓ reports that this assertion is unfounded. Our author also describes his master, Abū Isḥāq al-Naẓẓam's[48] interpretation according to which the venom of the serpent alone is not mortal. It becomes so according to the following scheme: the body of the victim possesses its own venom whose action is inhibited. When the venom of the animal penetrates the human body, it lifts this inhibition and releases the venom of the body and the victim dies.[49]

The method of injection was also a topic of reflection for the scholars. Baḫtīšū' Ibn Ğibrīl[50] thought, as we are told by Ğāḥiẓ, that the venom was injected by the fang (in the case of the snake) or by the sting (in the case of the scorpion) which are hollow. He himself claims to have observed the hollow part in the sting of the scorpion. As far as the viper is concerned, this physician thought that during the bite, it was only when the animal rolled over itself that it injected its venom.[51] Today the sequence of events and the mechanism of the venom injection are known.[52]

---

[47] Ğāḥiẓ 1945, vol. VI, p. 29. Bedouin means nomadic Arabs.

[48] Abū Isḥāq Ibrāhīm Ibn Sayyār Ibn Hāni' al-Naẓẓam (775–846) Arab mu'tazilite poet and theologian; see table n° 1.

[49] Ğāḥiẓ 1945, Vol. V. p. 21.

[50] Great doctor at the Abbasid court and contemporary of Ğāḥiẓ. See table n° 1.

[51] Ğāḥiẓ 1945, Vol. IV, p. 124.

[52] The viper belongs to the group of solenoglyphous, whose venomous apparatus is equipped with an improved device characterized by a pair of fangs traversed by a canaliculus, completely closed, associated with the venom gland. The mechanism of the bite in the viper is currently described as follows: to bite, the viper stands and, with its mouth open, strikes a head blow to reach its prey with its fangs. The contraction of a complex set of muscles allows the mandible to be raised, the fangs to be pulled onto the pulpits, and the poisonous gland compressed: the venom is thus violently injected (Platel et al. 1991). This set of contractions results in a behavior described as a reversal of the animal.

At the time of Ǧāḥiẓ, it was impossible to precisely follow the sequence of movements preceding the attack of the snake and to determine the mechanisms allowing the injection of the venom. Some did not accept Ǧāḥiẓ's explanations. The author cites as example the case of an expert on snakes[53] who does not agree with Baḫtīšū regarding the relationship between the injection of the venom and the overturning of the body of the snake at the time of the bite. According to him, given the shape of the fang of the noxious organ, this behavior does not serve for injection but is intended to facilitate the removal of the fang from the body of the victim.[54]

After a long presentation of various observers and scholars, Ǧāḥiẓ presents the results of his own observations. He believes that mortal venoms have three characteristics:

- The first: quantity – a small quantity (kamm, كم) of venom is sufficient for this effect. According to our author, experiments of which he was informed, consisting of weighing the scorpion before and after the sting, had shown that it was so small that the weight of the scorpion remained practically the same.[55]
- The second: process (kayf, كيف) – in order for the venom to penetrate the body of the victim, an incision must first be made.
- The third: nature of the venom (*ğinss*, جنس) – our author presents the various questions and hypotheses concerning the nature of the venom and its mortal action on the body of the animal.[56]

This third point was of great concern to scholars in Ǧāḥiẓ's time. One of the questions raised was to know whether the sting or the fang was poisonous by themselves or whether they were tools allowing the passage of the venom to the body of the victim. According to our author, both cases could exist. He explains the first case by the fact that one can be painfully stung just by walking on the sting of a dead scorpion or on the skeleton of a snake.[57]

Moreover, with regard to the venom injected, the author points out that the adverse effect produced by the sting or fang is not due to the presence of the incision itself; the effect of the venom of snakes, scorpions, and other animals on the blood is that they coagulate or liquefy it; some venoms act on the nerves.[58] As to the mode of action of the venom, Ǧāḥiẓ refuses any explanation that attributes the cause of it being mortal to cold or heat, as proposed by Aristotle and later by Galen. He writes:

---

[53] Ibn Abī 'Ajūz.

[54] Ǧāḥiẓ 1945, Vol. IV, p. 124.

[55] Ibid. Vol. IV, p. 319.

[56] At present, it is known that the toxicity of a given species is based on three parameters (Grassé 1976; Platel et al. 1991): the efficacy of the mechanical venom injection device, the toxicity of the substance injected, and the dose. Indeed, a minimal dose of venom is needed to kill an animal.

[57] At present, it is confirmed that the nature of the sting or fang does not play a major role on the venomous effect (Stockmann 1990). However this organ plays the role of an indispensable device allowing the passage of the venom to the body of the prey. The role given by Ǧāḥiẓ to the sting or fang could be explained by the fact that generally there is a residue of venom in the organ and that even if it has dried up, the venom retains its toxic property.

[58] Ǧāḥiẓ 1945, Vol. IV. p. 126.

If this injected element acted owing to its high temperature, why then does fire not act in the same way? And if it was acting because of its very low temperature, why then does ice not cause the same effect? It is therefore clear that the venom does not kill because of its warmth or its coldness .[59]

فإن كان الشيء المقذوف من شكل الشيء الحار، فلم قصرت النار عن مبلغ عمله؟ وإن كان من شكل الشيء البارد فلم قصر الثلج عن مبلغ عمله؟ فقد وجب الآن أن السم ليس يقتل بالحرارة، ولا بالبرودة.

According to our author, the mode of action of the venom is explained by a theory, which he elaborated himself. It is the theory of properties, of specificities, and of the intrinsic nature of the elements. That is to say, according to Ǧāḥiẓ, it is the intrinsic nature (*الغرائز*) of the venom that possesses the property (*الخاصيات*) of causing death by a phenomenon called correspondence (*المقابلات*). In other words, the venom has the property of acting only on something that is specific to it. This theory was originally proposed to refute the theory which, since Aristotle, consisted in giving heat the primary role in several physiological mechanisms, especially that of digestion. While recognizing the need for a certain degree of heat, Ǧāḥiẓ asserts that in addition to this, other factors and other properties are required. These factors, he says, do not bear a name and cannot be expressed simply.[60] To explain his ideas, our author gives examples such as digestion in dogs and some herbivorous creatures. He writes:

I will prove to you that what I have just said about properties, specificities and the intrinsic nature of the elements is true. Don't you see that in the belly of the dog and the wolf the bones are dissolved but not the date? While these are much more tender than a bone? (...) Horses that can transform dry clover and straw into dung during digestion cannot digest barley grains. Camels that do the same with tough, thorny plants also cannot digest barley grains. All this can be explained only by properties and specificities.[61]

سأدلك على أن القول في الخاصيات والمقابلات والغرائز حق. ألا ترى أن جوف الكلب والذيب يذيبات العظام ولا يذيبان نوى التمر، ونوى التمر أرخى وألين وأضعف من العظام المصمتة ... والبراذين التي تحيل أجوافها القت والتبن روثا، ولا تستمري الشعير. والإبل تقبض بأسنانها على أغصان أم غيلان، وله شوك كصياصي البقر، والقضبان علكة يابسة جرد وصلاب متينة، فتستمرئها وتجعلها ثلطا ولا تقوى على هضم الشعير المنقع. وليس ذلك إلا بالخصائص والمقابلات.

According to Ǧāḥiẓ, the severity of venom's activity depends on several factors. It varies from one poisonous species to another. However, in the same species, it depends on the following aspects:

- The quantity injected, for even if it is very weak and no means were available at the time to weigh it, Ǧāḥiẓ specifies that the venomous animal does not always inoculate the same amount of venom. Moreover, the author emphasizes the fact that for the venom to act, a minimal dose[62] is required.

[59] Ibid., vol. IV, p, 318.
[60] Ibid. vol. IV, p, 318.
[61] Ibid. vol. IV, p. 313–314.
[62] Ibid., vol. IV, p., 319.

- The geographical region, since according to our author, the effect of the venom on a given species varies depending on whether it lives in one given region or another. Thus, he says, the action of the venom of the same species of scorpion called *gārrārah* varies depending on whether this species comes from the region of Šahrazor[63] or from another region,[64] which is ʿAskar Mokram.[65]
- The state of the animal is also important, because when speaking of scorpions, Ǧāḥiẓ specifies that the severity of the sting varies depending on whether or not the scorpion has gone out for the first time after its hibernation or whether it is in a state of gestation.[66]

Furthermore, Ǧāḥiẓ specifies that the damage caused by the venom varies from one victim to another,[67] but in the case of a single individual who has been stung or bitten, the effect depends on the part of the body where the victim was stung or bitten,[68] on his physiological state, and even on the meal taken by the victim and the state of his respiration/breathing.[69] These are not the only factors that come into play, because Ǧāḥiẓ lists a number of environmental factors that also make a contribution. He specifies that the action of the venom depends on whether it is day or night, that people feel much more pain at night than during the day.[70] This action also depends on the seasons. It would be more serious in summer than in winter.[71] The action of temperature cannot be neglected since Ǧāḥiẓ indicates that people think that the action of the scorpion sting is more harmful on people having just taken a bath. To explain this phenomenon, our author advances the explanation given by his master Abū Isḥāq al-Naẓẓām who thinks that the action of the venom is more effective when it is hot, because it is due to the opening of the pores of the body, the dilation of the vessels, and the heat of the body. This is why scorpion venom is more powerful in summer. The warmer the air, he says, the warmer the body and the more harmful the venom. Yet our author cannot explain why people feel much more pain at night than during the day since it is during the day that it is hotter! The only explanation given by Ǧāḥiẓ in this respect is that when it is cold the

[63] According to the classic concept, during the Sassanid era, the region of Šahrazor corresponded to the modern provinces of Kirkuk and parts of the Diyala province; see http://dictionary.sensagent.com/shahrazor/en-en/

[64] Ǧāḥiẓ 1945, vol. V, p., 363.

[65] ʿAskar Mokram is a town of the medieval Islamic province of Ahvāz (Khūzistān) and also the name of the district of which it was the administrative center; see http://www.iranicaonline.org/articles/askar-mokram-lit

[66] Ǧāḥiẓ 1945, p., 363.

[67] Ibid., vol. IV, p., 221–222.

[68] Ibid., vol. V, p., 363.

[69] Ibid.

[70] Ibid.

[71] Ibid., vol. IV, p., 222.

inside of the body is warmer than when the surrounding temperature is high.[72] The psychological state of the victim must also be taken into consideration. Thus, according to Ǧāḥiẓ, fear can play a decisive role in the survival of the victim. He specifies that there are snakes whose venom is mortal only if it is associated with the fright of the bitten person. Sometimes, he says, fright alone kills. To illustrate this phenomenon, Ǧāḥiẓ reports an anecdote of a man who was bitten on the head by a snake while sleeping under a tree. He was awakened by the bite, without realizing at once that it was due to a snake. He remained there for some time until a person who had witnessed the scene told him what had actually happened. Then, scared, he uttered a great scream and died on the spot.[73] According to our author, it is as if the effect of the venom on the vital organs had been inhibited, and the action of the fright lifted this inhibition. Ǧāḥiẓ tries to explain how the combined action of these two factors[74]occurs at the level of the body. He writes: "Either the fright leads the venom to vital places, or it combines the action of the venom to its own." [75]

Ǧāḥiẓ concludes that to be able to say that a venom is really mortal, one should see its effect on someone who is asleep or has fainted or on a naive child or on a madman who does not discern things.[76] On the other hand, Ǧāḥiẓ remarks that the venom might pass from blood to breast milk. He tells us in this context that a camel which was suckling her young was bitten by a poisonous snake. The calf which continued to suckle suddenly fell dead, long before its mother. Ǧāḥiẓ expresses his astonishment at the speed of the passage of the venom to the mother's milk.[77] He explains this, however, by using other observations made with humans.[78]

Finally, Ǧāḥiẓ reported that in his day, when a scorpion stung a victim, a specialist, who was also called a *ḥaǧǧām* (blood cupper), was brought to him to suck out the venom at the site of the sting before it spread to the whole body. Ǧāḥiẓ points out that this action could cause much harm in the *ḥaǧǧām*, his mouth having been in direct contact with the victim's poisoned blood. To avoid this problem, cotton was placed on the sting before the sucking operation. The cotton would thus prevent the passage of blood to the mouth.[79] We note here that for Ǧāḥiẓ the presence of snakes or scorpions is a pretext to engage in medical and philosophical debate on the nature of man, the venom, and the relationship between the two.

---

[72] Ibid.

[73] Ibid., vol. IV, p., 121–222.

[74] While repulsion against snakes is an acquired trait, some fears can induce deadly cardiac stimulation.

[75] Ǧāḥiẓ 1945, vol. IV, p. 123.

[76] Ibid.

[77] Ǧāḥiẓ remarks that a woman who drinks wine while breastfeeding intoxicates the baby; similarly, taking a purgative in a lactating woman can cause diarrhea in the baby (Ibid., Vol. V, p. 366–367).

[78] Although it is now known that wine and purgative medicines can have an effect on breast milk, we do not have data on the effect of the sting or bite on the mother.

[79] Ǧāḥiẓ, 1945, Vol. IV. p. 219–220. The verb "*hağama*" in Arabic, means "to suck" or "to carry out cupping." For cupping (*hağama*) as treatment, see http://cdn.intechopen.com/pdfs/26488/InTech-Chapter_download.pdf. For contemporary cupping, see https://en.wikipedia.org/wiki/Hijama

3.5 Conclusion

Ǧāḥiẓ was an author who described the human societies of the ninth century with much subtlety, from angles that today we would call sociological and psychological. It is in the sociological and ethnographic fields that his works are particularly recognized and studied, to the detriment of his zoological work. Ǧāḥiẓ analyzed the behavior of man in society among the powerful, the masters, the wretched, the slaves, the eunuchs, the handicapped,[80] and the able bodied. He examines the connections between these different classes in society. He deciphered all the levels of the human society of his time offering pertinent remarks and observations through a high-level literary style, meaningful, sometimes light-hearted, and even with an ironic or mocking tone. It is with this same scrupulous approach and the same lively and amusing style that he wrote his *Book on Animals*[81] as he indicates in his introduction.

The short passages discussed above describe animals in various ways: as the object of observation in nature, the object of admiration when in captivity, or as intruding and undesired, when disturbing the life of humans or as a source of problems of hygiene and health for them. In all these situations, our author observes these animals with the same view as an experienced, attentive, and curious zoologist. He is interested in their formation, in their physiology, and in the relationships that exist between these animals. Thus, he noted the place of the food chain between the different animal species and stressed the necessity of the existence of each one of them within this whole. He takes into account the psychological fragility of men and notes that a bite from a nonvenomous snake can kill simply by the fright it induces in some people.

It must be emphasized, however, that in Ǧāḥiẓ's *Kitāb al-Ḥayawān* or *The Book of Animals*, man has a central position in his conception of the world. It is therefore as an excellent zoologist that Ǧāḥiẓ considers the relationship of human health with its animal environment. It would be interesting to extend the same study to all the works by this author, in order to better apprehend the relation of man/animal in the different layers of the human society of his time.

Transliteration Adopted for Arabic in This Article

Usual characters in the Arabic alphabet	Arabic international transliteration
أ	ʼ
ا	ā
ة	a

[80] Katouzian-Safadi 2013.

[81] Ben Saad and Katouzian-Safadi 2013.

Table 3.1 Presentation of the authors cited

	Name	Geographical area	Date	Work	Work category
1	Al-Asmāʿī	Basra	740–828	*Book of Horses, Book of Camelins, Book of Insects …*	Zoological terminology
2	Abū Isḥāq al-Naẓẓām	Basra	775–846		Grand Master of Gahiz and Head of a Mu'tazilite branch
3	Ǧāḥiẓ	Between Basra and Bagdad	776–869	*Kitāb al-Ḥayawān Book of Animals*	Zoology
4	Baḫtīšūʿ Ibn Ǧibrīl	Baghdad	Ninth century Died in 870		Physician at the Abbasid court
5	Ibn Abī ʿAǧūz	Baghdad	Ninth century		Serpent breeder and specialist
6	Qazwīnī	Baghdad	1203/1283	*Aǧâʿib al-Maḫlūqāt The Wonders of Creation*	Cosmology, zoology, botany, etc.

Usual characters in the Arabic alphabet	Arabic international transliteration
ث	ṯ
ج	ǧ
ح	ḥ
خ	ḫ
ذ	ḏ
ش	š
ص	ṣ
ض	ḍ
ط	ṭ
ظ	ẓ
ع	ʿ
غ	ġ
ق	q
و	ū/w
ي	ī

Short Glossary

ṭabāiʿ, طبائع: nature
waḥšī, وحشي: wild
ahlī, أهلي: domestic

Waḥšah, وحشة:	attachment
kamm, كم:	quantity
kayf, كيف:	procedure
ğins, جنس:	nature (of the venom)
muqābilat (المقابلات):	correspondences (sing. *muqabila*)
ġarāyiz (الغرائز):	intrinsic natures (sing. *ġariza*)
ḫaṣiyyat (الخاصيات):	properties (sing. *ḫasīyya*)
ḥaǧǧām (حجام):	sucker using incision
ǧarrārah (جرارة):	kind of scorpion

Geographical Zones

Iraq
Byzantium
The region of Šahrazor
The region of 'Askar Mokram

References

Primary Source

Ǧāḥiẓ, *Kitāb-al-Ḥayawān*, Abū Ūṯman 'Amr Ibn Baḥr Ǧāḥiẓ, 1945, ed. 'Abd al-Salām Muḥammad Hārun, Cairo : Maktabāt al-Bābī al-Ḥalabī, 1938–1945. Reed. Cairo: al-ḫanǧī, 1964–1979. Beirut, Dār al-ǧil.

Secondary Sources

Aarab, Ahmed, Philippe Provençal, and Moḥamed Idaomar. 2001. The Mode of Action of Venom According to Jahiz. *Arabic Sciences and Philosophy* 11 (1): 79–89.

———. 2003. La méthodologie scientifique en matière zoologique de Ǧāḥiẓ dans la rédaction de son œuvre *Kitāb-al-Ḥayawān*. *Anaquel de estudios arabes* 14: 5–19.

Aarab, Ahmed, and Philippe Lehrminier. 2015. *Le livre des animaux de Ǧāḥiẓ*. Paris: L'Harmattan.

Ben Saad, Meyssa, and Mehrnaz Katouzian-Safadi. 2013. Raconter une Histoire Naturelle: lorsque le verbe du conteur trace les observations du naturaliste. *Arabic Biology & Medicine* 1. November. Online access: http://media.wix.com/ugd/4c7a74_7aad88cb0a544bd1a552478 69d727069.pdf.

Chauvin, Remy. 1967. *The World of Insects*. London: World University Library.

Gimaret, Daniel. 1981. Théories de l'acte humain en théologie musulmane. Vrin, collection Etudes Musulmanes, n° 24.

Grasse, Pierre Paul. 1976. *Précis de zoologie : Vertébrés, Agnathes, Poissons, Amphibiens et Reptiles*. Vol. 2. Paris: Masson.

Katouzian-Safadi, Mehrnaz. 2004. Salon ou *mağlis*, lieu d'échanges scientifiques et point de départ pour l'élaboration d'un traité. In *Les éléments paradigmatiques thématiques et stylistiques dans la pensée scientifique*, Facultés des Lettres et des Sciences Humaines, Rabat, série : Colloques & séminaires n° 116, 157–165.

———. 2013. Histoire de quelques infirmités visibles, les maladies de la peau: la lèpre, la variole, la rougeole. *Arabic Biology & Medicine* 1 (1): 39–55. http://media.wix.com/ugd/4c7a74_583b 6990514b43db85a1009b8f8f15aa.pdf.

Katouzian-Safadi, Mehrnaz, and Kaouthar Chebbi-Lamouchi. 2015. Pluralité des regards face aux nuisibles, Textes médiévaux arabes et persans. In *Poux, Puces, Punaises, La Vermine de l'Homme, Découverte, description et traitements, Antiquité, Moyen Âge, Epoque moderne*, ed. Franck Collard and E. Samama, 165–178. Paris: L'Harmattan.

Lewicki, Tadeusz. 1997. Kazwînî. In *The Encyclopaedia of Islam*. ed. E. van Donzel, B. Lewis, and Ch. Pellat, 3rd ed., vol. IV, 865–867 Leiden: Brill.

Platel, Roland, François J. Meunier, Jean-Marc Ridet, and Hélène Vieillot. 1991. *Zoologie des Cordés*, vol. 1. Paris: Ellipses.

Stockmann, Roland. 1990. Les Scorpions. In *Encyclopédie Larousse des Animaux*, Fascicule no. 41, 1–20. Paris: Société des périodiques Larousse.

Wiberg-Larsen, Peter. 1998. "Light Trapping of Trichoptera Near the Coast of NW Zealand", Denmark. *Natura Jutlandica* 23 (6): 69–77.

Chapter 4
Landscaped Environment and Health in Han China (208 BCE–220)

Catherine Despeux

Abstract Medical and Taoist sources written or compiled during the Han dynasty (208 BCE–220) provide the first accounts, reflections, and theories on the self, on disease, and on the relationships between humans and the world in which they live. This chapter focuses on this particular period of time which, in fact, lays important foundations for Chinese society and culture. Relying mainly on medical and Taoist sources, it firstly sheds light on how the self was thought of and represented at this time and examines how the links between human beings and their environment were conceived of. It then focuses on what, for humans and in the environment, were thought of as potential threats to health and how the image of an ideal landscape was thus constructed, that is, a landscape into which humans could fit perfectly and live healthy lives. Finally, we analyze how this ideal landscape was supposed to work to preserve good health in humans.

Keywords Exchange · Link · Unity · Mobility · Correlation · Nature · Breath (*qi*) · Resonance (*ying*) · Climate · Symbolic · Spirits

4.1 Introduction

The founding principles of theories relating to the environment and health were elaborated around the first centuries of the Common Era.[1] In ancient China, health-care was the responsibility of several different actors, from the doctors who treated illnesses to the "masters of methods" (*fangshi* 方士) who used a variety of techniques to ward off illnesses, maintain the life-force and "nourish life" (*yangsheng*

[1] I would like to warmly thank the two anonymous reviewers for their pertinent remarks and very stimulating questions that allowed me to improve this article.

C. Despeux (✉)
INALCO, Sorbonne Paris Cité, Paris, France
e-mail: Catherine.despeux@wanadoo.fr

F. Bretelle-Establet et al. (eds.), *Making Sense of Health, Disease, and the Environment in Cross-Cultural History: The Arabic-Islamic World, China, Europe, and North America*, Boston Studies in the Philosophy and History of Science 333, https://doi.org/10.1007/978-3-030-19082-8_4

養生). The conceptions of health and the means proposed to preserve it were developed by scholars who were connected to proto-Taoist thinkers, like Laozi 老子 (fifth century BCE), Zhuangzi 莊子 (fourth century BCE), and Huainanzi 淮南子 (second century BCE). They were summarized in the text of reference for medical theory in antiquity, the *Inner Classic of the Yellow Thearch* (*Huangdi Neijing* 黃帝內經),[2] compiled in the first centuries before the Common Era, under the Han dynasty. These founding principles have endured almost to the present day, despite the odd insubstantial controversy and minor modifications that have been made to them.

Correlative thinking developed and became established under the Han dynasty (206 BCE–220). Things are classified in categories according to their affinities and are linked to one another, forming patterns for the organization and functioning of the world. While the best known of these systems is that of the 5 agents based on 5 classifications, there are others, such as the system of the 9 sectors, that of the 8 trigrams, and that of the 6 or 12 classes, associated with the weather. These systems express a theory of nature, of the human body, of society, and of empire that are applied to various areas of life and to techniques; they give rise to changes, notably in state rituals, in the interpretation of canonical texts, and in medical theory and practice. Considering that there is an interaction between the different classes and the elements of the categories of these systems, we see a certain conception of the world gradually elaborated, from which the elements, connected to each other, obey a dynamic and set various processes in motion. If one of the elements moves, "is affected or creates a stimulation" (*gan* 感), another element responds and "resonates" (*ying* 應) with the first. This idea of resonance plays the role of a cosmological principal.[3]

It is for this reason, with regard to the question of the environment and health, that it is not so much the environment on one side and the individual, ill or in good health, on the other that is important, as the links between the two. There is then a strong conviction that the individual is not apart from nature and that the two resonate. In Chinese medicine, environment and the human body have been imagined as being on similar models which resonate with each other, the two maintaining various ties which call into question our usual conceptions of the boundaries and links between the environment and human beings.

The starting point for our reflection is a text which presents analogies and correspondences between the macrocosm where, according to the Chinese expression, "space-time" or "heaven-earth" (*tiandi* 天地) (heaven referring to time and the earth

[2] This book is made up of two parts: *Basic Questions* (*Suwen* 素問) and *Magic Pivot* (*Lingshu* 靈樞). In our contribution, we will rely for the *Basic Questions* on the translation by Paul Unschuld and Hermann Tessenow, in collaboration with Zheng Jinsheng (2011 and 2016).

[3] This notion of *ying* 應, "correspondence and resonance," well developed especially in the *Huainanzi*, is at the heart of this issue of the relationships between things and beings. It is because things are classified according to categories that if one thing within these categories begins to move, there is a principle of stimulus and response between two things of the same category due to their affinity. A special bond is then created between these two indivisible things, like sound and its echo. On this notion, see in particular Charles Le Blanc 1985.

to space) and the human body considered as a microcosm.[4] This presentation already provides us with a good picture of what the ideal environment is.

This short but essential passage includes an enumeration of the elements of nature and their correspondences in the human body, which we will present and from which a cartography of the ideal landscape that forms a device is drawn. This inventory both describes the beauty of the world and constitutes an appropriation and organization by man. The elements of the world and their device possess fundamental characteristics deduced from the way in which the maintenance of a balance of life has been observed and imagined. It was important to maintain constant harmony between nature and humans through the quality of the complex links and interactions between the two.

4.2 The Environment and Health

From the second century before the Common Era, texts on medical literature as well as documents of political literature establishing standards in the correlations described correspondences between the body and an organized nature. The system of correspondences arising from the five agents (wood, fire, earth, metal, water), while not unique, became one of the fundamental keys for understanding the world, which made it possible to classify things and beings. A proto-Taoist writing, the *Huainanzi* 淮南子, had, in the second century before the Common Era, already established correlations between nature and the human body.[5] The comparison is developed perfectly in the *Luxuriant Dew of the Spring and Autumn Annals* by Dong Zhongshu 董仲舒 (179–104), a culminating work for these theories of natural government and correlations between the cosmos, the state, and the human body, written by one of the most influential scholars in the court of Emperor Wu of Han (r.141–87).[6]

The three main orders of existence, namely, the human body, the cosmos, and the state, are composed of elements which play similar roles in each system; they operate on the same models, they are in correspondence and harmony with each other, and, moreover, they interact with one another. This ordering, *zhi* 治 in Chinese, applies equally well in social and political fields as the guarantor of the peace of the country, as for the individual as the guarantor of good health, and in nature as the preservation of a cosmic equilibrium.

[4] Our choice of the expression "space-time" is also intended to convey the fact that in Chinese thinking the two are interdependent; in other words, the body is not considered as something static but moving and in constant evolution.

[5] *Huainan zi zhushi* 淮南子註釋, Chap. 7, p. 100.

[6] For correspondence with the human body, see in particular the *Chunqiu fanlu* 春秋 繁 露 attributed to Dong Zhongshu 董仲舒 (175/179–115/104) but which would in fact be a compilation by multiple authors, *juan* 13, Chapters 56 and 60; other chapters in this work also address these correspondences. Cf. Nathan Sivin (1995), p. 5–37.

A human being is surrounded by nature, by his habitat, by the objects it contains, and by all the human constructions in a given society. We will focus here on the natural environment as it was conceived of in ancient Chinese medicine and also in Taoism which developed methods of disease prevention and healthcare.

This construct of an ideal environment implies, on one hand, that the individual himself has a harmonious body whose mechanisms are unhindered and has a balanced psyche bringing him joy and happiness and, on the other, that his environment, especially his natural environment, is beneficial to him. But even more important is the fact that the individual and his environment interact in such a way so as to establish that health, to maintain it, and to nourish it. The boundary between the subject and the object, between humans and nature, and between oneself and the environment is thus blurred and endowed with a certain plasticity.

The notion of nature is translated into modern Chinese as *da ziran* 大自然, literally "great spontaneity," coming from the old notion of *ziran* 自然, "that which is natural, spontaneous, that happens by itself." This latter notion, a return to untamed/wild nature, as put forward by the Taoists and the Han physicians creates a tension with the ideas of managing the land and taming nature conveyed by other scholars and in particular by the myth of the land being arranged in nine regions by Yu the Great who tamed the streams and canalized them.

The emblematic text on the notion of *ziran* in Taoism is Chap. 25 of Laozi's *Tao Te Ching*, which says:

> Man models himself on the Earth;
> The Earth models itself on Heaven;
> Heaven models itself on the Way;
> And the Way models itself on that which is so on its own (*ziran*).[7]

Thus the wise man is not perceived by the people as only human; paradoxically, he is himself a part of nature and its processes.[8] This chapter by Laozi, a text at the heart of the Taoist religion, which served and still serves as a reference to the point of being elevated to the rank of a canonical text, presents the three essential elements of the Chinese worldview: heaven, the earth, and man whose life is based on the Way, which is nature or spontaneity. The natural environment is above all "heaven-earth." It is constructed and organized intellectually and symbolically on the basis of systems of numerical correlations. Nature becomes a landscape; its elements are carefully selected and arranged. It is a cultural construct resulting from the way man has observed, imagined, and worked with nature, in which, moreover, gardens occupy a privileged place. All peoples transform nature and the Chinese people are no exception. The great civilizing acts have led it to transform this nature,[9] be it by digging waterways, by deforestation for the development of agriculture, or by the construction of grandiose palaces and equally gigantic tombs for high dignitaries. But the ideology defended by the Taoists and the doctors

---

[7] *Daode jing*, Chap. 25. (transl. From Robert G. Henricks, *Lao-Tzu Te-Tao Ching. A new translation based on the recently discovered Ma-wang-tui Texts*. New York, Ballantine Books, 1989, p. 77.).

[8] Cf. Christoph Harbsmeier (2010), p. 220–254.

[9] Cf. Heiner Roetz (2010), p. 198–219.

emphasized a management in compliance to the laws of nature, in a desire to penetrate the mysteries so as to conform to nature, to adapt to it, and to contribute to its evolution by modifications which conform to the laws of nature and above all by human behavior which follows these laws.

This question of the relationship between the environment and health cannot be considered without asking what the individual is within an environment, what his limits are, what the separation between the self and the nonself is, and what characterizes the individual, defines him, and distinguishes him from the environment.

According to Chinese conceptions under the Han, an individual is made up of a body and a number of different powers and souls; the material body is animated by them.[10] This originates in the energy seeded by the father and the mother, which are particular types of *qi* 氣 (breath, pneuma, energy), the cosmic principle which constitutes the lowest common denominator for all things in the world. In the medical and Taoist context of the first centuries of the Common Era, the world was conceived of as proceeding from a unique *qi* which divides, diversifies, and assumes particular qualities; this unique *qi* results in the multiplicity of the manifestation, a manifestation whose various parts are interdependent and interacting. This *qi* is also, along with other elements, what animates both the human body and nature.

As we have already mentioned, three main areas have been defined: the cosmos, the state, and the individual. They resonate with each other, they interact, and to do so they have to match. Thus, in Taoism and Chinese medicine, man is a landscape and a state; his body is a landscape and a state (see Hennesy 2001); in the same way, the state is a human body and nature is a giant. In other words, while humans extend into nature, nature extends into humans: humans are not categorically distinct from but yet not merged with nature. We must rethink the relationship humans have with the world, and (also) the boundaries between nature and culture. To look at a landscape is also to contemplate oneself (see Selin 2003).

The human body, a small world which is both closed and open, is a microcosm, just like the cosmos. Since it is closed, each individual is separated from everyone else and his environment by his body, a space-time circumscribed by the skin, which plays a multiple role. Indeed, while it is a barrier and a boundary, it also constitutes an interface and a link between the inside and the outside. But this barrier has gateways, through which communication and exchange with the outside can be effected, it is porous and can thus let elements from inside the body escape as it can let those from the outside enter. A similar role is played by the orifices of the face and the body's two lower orifices.

What are the factors that will then endanger health and cause disease? The first factor is the aggression of what today would be called pathogens but which at the time were referred to as an evil or demonic influence. Nature was in fact populated by entities that could exert an evil or beneficial influence and were known by a variety of terms (essences, breaths, demons, ghosts, gnomes, mountain spirits, and others). Any intrusion into the human body by these harmful forces was believed to cause disease.

[10] Cf. Catherine Despeux (2007), p. 71–94.

But in order for these forces to enter the body and do harm, it was first necessary that the skin and the orifices no longer played their role as a barrier and were weakened and also that the defensive function of the human body no longer functioned correctly, due to disorder that had established itself inside. The disturbance concerned the inner self, the outside, and the links between the two. There could be a loss of harmony between the individual's psychic forces, his body, and the environment; there could be an imbalance in the cosmic elements and the correlations between them; there could be an insufficient renewal of the individual's nourishing energies which caused channels to become clogged. While this could lead to the intrusion of harmful elements into the body, it could also result in the loss of certain elements from the person, be they physical or physiological elements, or the different souls that populate the body and which, if they escape, attack the integrity and the completeness of the being. Health was therefore perceived as an unstable state that also depended on the manner in which it was nourished and nurtured; a deficiency in the elements in the living was also an important cause of the appearance of pathological states.

In fact, if we summarize the main causes of illness briefly outlined above, it can be seen that they can again be reduced to two main elements, the environment and the behavior of the individual, or, in the words of Wang Bing 王冰 in the eighth century,[11] to the harmful *qi* of heaven and the earth and the desires of human beings.[12]

4.3 The Ideal Environment Described in the Medical Sources

The founding text of medical theory under the Han, the *Inner Classic of the Yellow Thearch*, provides us with a vision of the ideal landscape through a series of correspondences established between heaven-earth and the human body:

"Huang Di asked Bo Gao 伯 告: "I wish to be informed of how man's limbs and joints correspond (are in cosmic resonance *ying* 應) to heaven and earth?"

Bo Gao replied:

> Heaven is round, the earth is rectangular. Man's head is round, his feet are rectangular, thereby corresponding to the [measures of heaven].
> Heaven has sun and moon; man has two eyes. The earth has nine regions; man has nine orifices.
> Heaven has wind and rain; man has joy and rage. Heaven has thunder and lightning; man has the sounds of his voice.

[11] Wang Bing, who had a very senior position in the Imperial administration, is the author of the oldest commentary on the *Basic Questions* (*Suwen* 素問), part of the *Inner Classic of the Yellow Thearch (Huangdi neijing* 黃帝內經) to have reached us; the *Basic Questions* was submitted to the throne in 762. This author not only commented on the text but redesigned it, changing the order of the chapters and increasing the length of some. From the Tang (618–907), the text became a study document for the official instruction given in medical schools.

[12] See the translation by Paul Unschuld and Hermann Tessenow, in collaboration with Zheng Jinsheng (2011 p. 218).

> Heaven has the four seasons; man has the four limbs. Heaven has the five tones; man has the five long-term depots.
> Heaven has the six pairs of flutes; man has the six short-term repositories.
> Heaven has winter and summer; man has cold and heat sensations.
> Heaven has ten [stems counting the] days; man has ten fingers. [The earth branches beginning with] *chen* are twelve; man has ten toes, the stalk, and the pendant to correspond to them. Women lack [the latter] two parts. This way they can carry [an additional] human body within them.
> Heaven has yin and yang; man has husband and wife. A year has 365 days; man has 365 [bone] sections.
> The earth has high mountains; man has shoulders and knees. The earth has deep valleys; man has armpits and the hollows of the knees.
> The earth has twelve stream waters; man has twelve stream/conduit vessels. The earth has spring vessels; man has guard *qi*.[13]
> The earth has growth of herbs; man has hair on his body and head. Heaven has day and night; man has [times of] resting and rising.
> Heaven has the constellations of stars; man has teeth. The earth has small hills; man has small [body] sections.
> The earth has mountains of stone; man has elevations of bones. The earth has forests of trees; man has gatherings of sinews.
> The earth has human settlements; man has bulgings of flesh. A year has twelve months; man has twelve joints. On earth it may happen that during all four seasons no herbs grow; among men there are those who have no son.
> That are the correspondences of man to heaven and earth.[14]

The correspondences between the human body and the macrocosm had varied before what we have quoted above became established in medicine.[15] This text, which lays the founding principles for the image of the body/microcosm in medicine, constructs and systematizes the anatomical and physiological knowledge of the time, as well as providing the essential landmarks for the life of man in his natural environment, an environment that becomes a chosen landscape.

4.4 Structure and Composition of the Ideal Landscape

Starting from the aforementioned text of the *Inner Classic of the Yellow Thearch*, which describes the basic elements of the construct of the landscape and thus of the human body, we will see what the structure and composition of the ideal landscape are, what the elements are, and how they are arranged.

---

[13] Traditional Chinese medicine distinguishes two main types of breath: nourishing breath, which circulates in the deeper layers of the body and is carried by the blood, and defensive (guard) breath, more on the surface of the body, which protects it from external aggressions.

[14] *Huangdi neijing Lingshu* 黄帝内經靈樞, Chap. 71 (translation by P. Unschuld, 2016, p.634–636).

[15] While some comparisons seem to be explicit, such as the heaven and the head, round, or the Earth and the feet, flat and square surfaces, others are less so, such as the correspondence between the 12 earthly branches and the toes plus the penis and the testicles which seem stranger to us and the relation established between the 4 seasons and the 4 limbs and between the stars and teeth.

This construct of space is structured around a center, man, according to a first bipartite classification between the heavens above (time) and earth below (space) and then a division of terrestrial and celestial space in oriented sectors, numbering 5, 8, 9, or 12, which correspond to the numerological bases of the systems of correlations.

4.4.1 The Sun, the Moon, and the Stars

In the heavens, the sun and the moon are the two main elements, which diffuse yin and yang lights. Hence, the balance of yin and yang is the basis of health; it is conceivable that the way in which the individual is subject to the influence of these two lights is primordial. The rays of the sun bring a tempered contentment; places exposed to excessive cold or heat are obviously not an ideal environment. The North and long cold spells are associated in the Chinese imagination with the underworld and with inhospitable regions. The South with its heat waves and its humidity was a little-known and frightening land, full of miasmas and poisons of all kinds. In the "Discourse on Different [Therapeutic] Patterns Suitable [for Use in Different] Cardinal Points" chapter of the *Inner Classic of the Yellow Thearch*, which is undoubtedly one of the oldest in the book, the importance of the earth and its climate in the production of diseases is highlighted:

> Huang Di asked:
> When physicians treat diseases,
> One identical disease may be treated differently in each case,
> And is always healed. How is that?
> Qi Bo responded: Physical features of the earth let it be the way[...]
> The centre,
> Its land is flat and damp.
> It is here that heaven and earth generate the myriad beings in large numbers.
> Its people eat diverse [foods] and are never fatigued....[16]

Although this chapter associates specific diseases to each of the four cardinal regions and to the center, the center nevertheless appears the most favorable region for the growth and development of living beings.

As far as the moon is concerned, there are periods during each lunar month in which it is harmful and establishes an environment to be avoided, whereas at other times it brings energy through its moonbeams and benefits to human beings and especially to women. Healthcare practices consist in soaking up the moonbeams, in certain circumstances and at certain moments of time. A moonless landscape is frightening, while the round moon is sung of by poets and is the occasion for celebrations. The full moon is also a preferred time for certain healthcare practices.

---

[16] "Yifa fangyi lun" 異法方宜論, *Huangdi neijing Suwen, juan* 4, Chap. 12, (translation by P. U. Unschuld and Hermann Tessenow, in collaboration with Zheng Jinsheng, 2011, vol. 1, p. 210, 216).

Three types of light were differentiated: the sun, the moon, and the stars. The landscape taken into account in health is also that of the night and the beneficial influence of some of the stars that shine in the firmament. This is all the more evident in Taoism, where practices for absorbing the essence and light of the moon and stars are at the heart of well-being and techniques for longevity and where stellar deities inhabit and are visualized in certain parts of the body with which they are in correspondence, as the body is imagined both as an earthly space and as a celestial vault. In medicine, one of the roles of the meridians associated with acupuncture points is precisely to divide the surface of the skin into sectors which correspond with sectors of the celestial vault. Some Taoist writings explain, in fact, that the individual is a set of lights or rays associated with the components of the body (flesh, blood, muscles, bones, viscera), points of light whose external counterparts are the stars. In the Taoist school of High Purity (Shangqing 上清), formed in the fourth century of the Common Era, someone adept in the practices of longevity should absorb light, so that he transforms himself, as was thought, into a luminous and transparent being, having assimilated the light of the stars.[17]

These three principal sources of luminosity, the sun, the moon, and the stars, are indispensable, and their quality is carefully studied, as it varies with time. The first chapter of the *Inner Classic of the Yellow Thearch* explicitly states:

> Next, there were the exemplary men.
> They took heaven as law and the earth as rule;
> Their appearance resembled sun and moon.
> They distinguished among and arranged the stars and the northern star
> [on the basis of their]movements contrary to or following [the movement of] yin and yang.
> They distinguished among the four seasons.[18]

Among the stars, the pole star and the stars of the Big Dipper (the Northern Bushel, in Chinese) are specifically referred to in the medical sources. The heart is compared to the pole star, whose immobility it must imitate, leaving the celestial vault or its body, with its meridians to revolve around it. The seven stars of the Big Dipper are associated with the face and its sense organs and sometimes also with the teeth.

> [Huang]Di:
> I have heard [the following]:
> The nine needles correspond above to heaven and earth, to the four seasons and to yin and yang.
> I should like to hear the methods [underlying] their [use] so that they may be transmitted to later generations to serve as rules.
> Qi Bo:
> Now,
> One [is] heaven,
> Two [is] the earth,

[17] On this theme, cf. Isabelle Robinet (1986), p. 87–105.

[18] *Huangdi neijing Suwen*, *juan* 1, Chap. 1 "*Shanggu tianzhen lun*" 上古天真論. (transl. by P. U. Unschuld, 2011, vol. 1, p. 44).

Three [is] man,
Four [are the] seasons,
Five [are the] tones,
Six [are the] pitch-pipe tones,
Seven [are the] stars,
Eight [are the] winds,
Nine [are the] fields.
The body corresponds to the [numbers], too.
Each of the needles has what it is appropriate for.
Hence, it is said: nine needles.
Man's skin corresponds to heaven.
Man's flesh corresponds to the earth.
Man's vessels correspond to man.
Man's sinews correspond to the seasons.
Man's voice corresponds to the tones.
When yin and yang in man merge [their] *qi*, this corresponds to the tones of the pitch pipes.
Man's teeth, face, and eyes correspond to the stars.
The *qi* leaving and entering man corresponds to the winds.
Man's nine orifices and 365 network [vessels] correspond to the fields.[19]

It should also be noted that the stars contribute to Chinese medicine through the names given to 16 acupuncture points which are the names of stars.[20] These are located mainly in the region of the heart and on the abdomen around the umbilicus.[21]

4.4.2 The Climate and the Seasons

The normative evolution of time is that of the four seasons, during which the modalities of yin and yang are balanced in a four-phase movement that also makes up the progression of human life: birth, growth, decline, and death:

Man comes to life through the *qi* of heaven and earth ;
He matures in accordance with the laws of the four seasons.[22]

This description first of the balance between yin and yang, then of the alternation of the four seasons, implies that regions that are not subject to the four seasons do not form an ideal environment. The ideal landscape is thus not only made up of ele-

[19] *Huangdi neijing Suwen, juan* 14, Chap. 54 "*Zhenjie*" 鍼解. Translated by Unschuld, 2011, vol. 2, p. 21–22.

[20] Cf. Catherine Despeux (2011).

[21] For example, Xuanji 旋 機 (21 VC), "armillary sphere," a point on the median line of the front of the body, not far from the heart, is the name of four stars of the "Northern Bushel," the Chinese name of the Big Dipper; the seven stars of the Big Dipper revolve round the motionless pole star; they evoke the form of a measure of capacity used in China, the bushel, similar to a large parallelepiped ladle.

[22] *Huangdi neijing Suwen, juan 8,* Chap. 25 Treatise on the Preservation of Life and the Integrity of the Body (*Baoming quanxing lun* 保命全形論). Transl. by Unschuld, 2011, vol. 1, p. 419.

ments that are positioned in well-defined places, which would have certain qualities or characteristics, like a cartography. That alone is not enough. This landscape must be considered in its dynamic, in other words, also in its temporal dimension.

The ideal environment is therefore that of a temperate and regulated climate, in which the six climatic elements taken into account in Chinese medicine, namely, heat, summer heat, cold, wind, humidity, and drought, are well-balanced. Of these six elements, special attention is paid to the wind: "Wind designates that which makes things grow."[23] Chapter 77 of the "Magic Pivot" in the *Inner Classic of the Yellow Thearch*, entitled "The Nine Palaces and the Eight Winds" (*Jiugong bafeng* 九宫 八 風), is devoted to eight main types of climatic wind, classified according to nine sectors, to the qualities of these winds, and to the diseases they bring when they do not come from their usual direction or when they carry harmful elements.[24] The winds, and this also means air, must be pure; their quality varies according to the time of day or night and the direction from which they come. Pure airs must be light, dynamic, harmonious, and not too weak or too violent, especially those you breathe. Very early in its history, ancient China developed methods of prolonging life and nurturing it regularly: healthy living, gymnastic, dietary, respiratory, and contemplative methods. One of the first authors to mention these methods is Zhuangzi 莊子 in the fourth century before the Common Era.[25] In techniques for a long life, respiratory practices are essential, even replacing ordinary food with food made of pneuma. The aspirant to immortality "eats pneuma" (*fuqi* 服氣). These practices privilege certain periods of the day or night and certain orientations of the body toward different cardinal points according to the seasons.[26]

The wind also plays an essential role as it is related to sounds. Numerous musical references evoke the sound of the wind in pine trees or in bamboo, for example. Indeed, the ideal landscape is made up of lights and sounds that come into harmony with all the elements of the being. In ancient China there were five principal sounds correlated with the world's five fundamental agents, and with the body's five principal organs, sound had six pitches corresponding to the six organ-stores and eight types of sonority or eight classes of musical instruments. Hence, in this idea of interaction of things in resonance with each other, sounds have an influence on health. Our organs are our musical instruments, through which the air passes and each plays its own melody, notably through the harmonious expression of the emo-

[23] *Baihutong yi* 白虎通義, *juan* 2, Chap. "Eight Winds" (*Bafeng* 八風).

[24] *Huangdi neijing Lingshu, juan* 11, Chap. 77 "Nine palaces and Eight Winds" (*Jiugong bafeng* 九宮八風), p. 117–119.

[25] Zhuangzi criticizes these methods which, according to him, are not worth a spiritual attitude of calm and simplicity which he calls the "fasting of the heart." He says, "Exhale the sounds *chui* 吹 and *xu* 呴, spit out the old breath and absorb the new, imitating the bear hanging on the tree or the bird taking flight is the ideal of those who want to maintain their bodies through gymnastic practices (*daoyin* 導引) and are in the same vein as Pengzu 彭祖, who lived more than eight hundred years" (translation by Burton Watson). See *Zhuangzi jijie* 莊子 集解 by Wang Xianqian 王先謙, ed. *Zhuzi jicheng*, Shanghai, Shanghai shudian, 1986, volume 3, Chap. 15 "Carving out Creative Thinking," p. 96.

[26] Cf. Catherine Despeux (2006), p. 37–67.

tions which in Chinese medicine are associated with the five viscera. The harmonious music of nature, which Zhuangzi calls "celestial flutes" (*tianlai* 天籟),[27] is part of the ideal environment; it emerges from the landscape, thanks to the sound of the wind, but also water and all the other constituents of nature.

4.4.3 The Land

As far as the ideal terrestrial environment is concerned, the conceptions have been developed in a technique known in Chinese as "water and wind" (*fengshui* 風水), which has been commonly translated as "geomancy": "grasp the breath of the mountains, identify the best marriage of water and wind so that the living and the dead can inhabit the earth with happiness, or at least without too much inconvenience, this is the goal of the art of geomancy in China."[28] Both in geomancy and in the art of garden design, the ground must be a delimited area, both closed and open, avoiding panoramas that are too open or too closed; one never puts oneself at the top of a hill or a mountain or at the bottom of a valley. Walls and enclosures must exist, but they are effaced by artifices, particularly in the art of landscaping the garden. They become porous and transparent in some way; they protect but do not separate, like the skin on the human body. In front of those entrance gateways that are not always closed, screens are placed so as to avoid leaving a complete opening.[29]

This area is centered through the choice of an anchor point in harmony with the telluric currents, from a point that gives the best possible device to a whole. It is the notion of *shi* 勢 "propensity of things," "device," which makes it possible to obtain the maximum efficiency in a network of correlations.[30] The ideal environment therefore assumes that the individual finds his ideal anchor point and orientation. The latter depends both on the general configuration of the terrain and on the particular properties of the individual therein, which is called his *ming* 命, "destiny," which is revealed among other things in the eight spatiotemporal markers defined by the date and place of birth.[31] In other words, while there is a theoretical model of an ideal environment for man, each human being adapts it according to his own characteristics: he chooses and constructs his own ideal environment.

[27] Cf. *Zhuangzi jijie*, Chap. 2 "The Equality of Things" (*Qiwu lun* 齊物論).

[28] Cf. Frédéric Obringer (2009), p. 7.

[29] Cf. Ronald G. Knapp (2006).

[30] Cf. François Jullien (1992).

[31] A first series of spatiotemporal markers, 10 in number for a series based on the 5 agents, the 5 orients (4 cardinal points plus the center), and the 5 periods of the year (4 seasons plus an intermediate period), is combined with a series of 12 markers qualifying the 12 2-hour periods in 1 day or the 12 months in a year, to give a cycle of 60 which serves to qualify the days, months, and years. Thus, for the birth date, two markers (one from the quinary series and one from the duodenary series) indicate the time, two markers indicate the hour, two markers indicate the day, two others indicate the month, and two others indicate the year.

Each region of the earth, according to its cardinal direction, has its specificities and imbalances that cause particular diseases and consequently require different treatments. This is a major theme of a chapter in the *Inner Classic of the Yellow Thearch*, which details the specific diseases of each region and the therapies employed. Here is, for example, how the eastern region is presented:

> Huang Di asked:
> When physicians treat diseases, one identical disease may be treated differently in each case, and is always healed. How is that[32]?
> Qi Bo responded:
> Physical features of the earth let it be this way.[33]
> The facts is,
> The region of the East,
> This is where heaven and earth give life first.
> [This is] the land of fish and salt;
> Beaches border on the water.
> Its people eat fish and crave for salty [meals].
> All of them consider their place of residence as comfortable and they regard their food as delicious.
> As for the fish, it lets one develop a heated centre;
> As for the salty [meals], they defeat the blood.
> Hence,
> Its people all have a black complexion and open [skin] structures.
> Their diseases are always *yong* – abscesses and ulcers.
> For their treatment, pointed stones are suitable
> Hence,
> It is for sure that [the therapy with] pointed stones originated in the East.[34]

There are therefore basic elements in the terrestrial world, which are listed and invested with a precise symbolism. They must be part of the landscape. Two elements emerge from the whole: the mountain (*shan* 山) and water (*shui* 水), which have given the word *shanshui*, one of the terms used for "landscape."[35]

4.4.4 Water

The qualities of water are due to its purity and its movement: all stagnant water is to be avoided, which means that even a pond will have some of its water flowing, if only artificially, in a garden. Water must be drained. Let us recall the myth of Yu the

---

[32] Wang Bing's eighth-century commentary specifies that the different therapies are acupuncture and stone needles, moxa and cauterizations, toxic remedies, gymnastic movements, and massages.

[33] Wang Bing's commentary says: by the configuration of the land, we mean that we model ourselves on the devices of birth, growth, contemplation, burial, on the top and bottom, and dry and damp.

[34] *Huangdi neijing Suwen, juan* 2, chap. 12 "*Yifa fang yi lun* 異法方宜論." (transl. Unschuld, 2011, vol. 1, p. 211–212.).

[35] Cf. Christoph Harbsmeier, (2010), p. 224.

Great, the demiurge who organized the waters of China and thus divided the land into nine terrestrial regions resonating with nine regions of the heavens. In the human body, the drainage of liquids, which ensures good circulation and the regulation of flows, particularly by means of water-shedding zones, are part of the maintenance of health.[36]

Purity, lightness, and luminosity are water's essential qualities. That which is pure is nearer to heaven. The conservation/preservation of water's purity and respect for nature were, therefore, preoccupations of the time in China, which are to be found in the precepts of the oldest Taoist community, that of the Celestial Masters, which appeared at the end of the second century. Of the 108 precepts that were to be observed by the members of this community, no fewer than 20 were concerned with the preservation of the natural environment. It was recommended not to defecate and urinate on plants or in drinking water, not to dig deep holes and destroy the land, not to fell trees unnecessarily, not to put toxic substances in lakes and rivers, not to put dirty things in wells, etc.[37]

Mountains are equally indispensable to an ideal environment. They evoke strength, political power, verticality, the link between the world below and the celestial world, and the link between the heavens and the earth but also a place apart, hollow, with its cave-wombs in which man regenerates, reconstructs, and finds himself closest to his origins. A totally flat landscape is therefore not ideal, so that the scholars who had the means had ponds dug and formed a hill with the ground that was removed so as to be surrounded, even artificially, by mountains and water.[38]

4.4.5 Plants and Minerals

In addition to these essential elements, a good environment must include vegetation, trees, and rocks. Trees and plants are essential for the life of man who eats many plants, treats himself with certain medicinal mineral or vegetable substances, and for buildings uses trees and earth in the form of cob or fired or unfired bricks. Rocks are concentrations, nodes of *qi*, and therefore concentrations of power. They are the bones of the earth and are indispensable in the ideal Chinese landscape. They contain minerals which are among the most prized substances for self-transformation and immortality; the practice of outer alchemy made extensive use of cinnabar, orpiment, realgar, lead, and other minerals. If there are no rocks in the surroundings, get some, or buy some - certain stones are worth a fortune.

---

[36] This is one of the roles of a typically Chinese functional organ called the "triple burner" (*sanjiao* 三焦). According to the *Inner Classic of the Yellow Thearch*, "The triple burner has the official function as opener of channels. The paths of water originate in it" (*Huangdi neijing Suwen, juan 3*, Chap. 8, "Secrets Documents from the Library of Orchids and Spirits" (*Linglan bidian* 靈蘭秘典) (Discourse on the Hidden Canons in the Numinous Orchid [Chambers], p. 49–51)).

[37] Cf. Kristofer Schipper (2001), p. 80–86.

[38] Cf. Baud Berthier et al. (2004).

4.4.6 Nature's Invisible Powers

We realize that this inventory does not include the animal kingdom, which is absent from this device, in which soil, water, and vegetation predominate. This presentation of the principal geographical features of the heavens and the earth which form the ideal landscape as given in the text of the *Inner Classic of the Yellow Thearch* quoted above only posits the structural and visible part of the whole. It lacks the essential: the invisible forces that animate nature.

The quality of man's environment lies above all in invisible and sacred forces, powers of *qi*, "essences of plants and objects" (*jing* 精), "souls or spirits" (*shen* 神), "demons" (*yao* 妖), and "manes" (*gui* 鬼), to mention only the main ones, that is, forces invisible to the common man, which may be beneficial or evil. Man is not unique in having souls: nature too. These forces are supported by various elements from the mineral world, vegetation such as trees and plants, rocks, and different types of water. As is written in the *Book of Rites*[39]: "Since mountains, forests, rivers, valleys, and hills have the power to produce clouds, to bring wind and rain, to manifest tremendous phenomena, we say that there are spirits (*shen*). The sovereign of the whole empire offered sacrifices to the hundred spirits; the feudal princes did so only to the spirits in their own lands."[40] These spirits take the form of animals, humans, or half-human half-animals: the body of a bird and a dragon's head, a dragon's body and a human head, a horse's body and a human head, a human body and a panther's tail, the body of a sheep and a human head, a snake's body and a human head, etc.[41]

There were texts of omens taken from the observation of strange phenomena caused by "essences" (*jing* 精) and "evil spirits" (*guai* 怪) in the environment, in the form of the body of an animal, objects, or unusual apparitions in the domestic environment, causing the cauldron to murmur, making people feel scared, and upsetting peoples' relationships.[42] On the other hand, some good omens were observed from various natural elements, such as the "Drawings of the Happy Portents of Resonance," a manuscript which presents 23 drawings of fabulous animals such as the tortoise, the dragon, and the phoenix, whose appearance is a good omen.[43]

It was important to know these invisible forces, to describe them, and to depict them. The "Hall of Numinous Brilliance in Lu" (*Lu lingguang dian fu* 魯靈光殿府) mentions the habit of painting them:

[39] A work attributed to Confucius or his disciples but in reality a collection of ancient texts compiled under Emperor Wu (r. 156–87) of the Han. It describes the social forms, the administrative system, and the ceremonial rites of the Zhou dynasty (1046–256 BCE). This book had a great influence on Chinese culture and social behavior throughout history.

[40] *Liji jishuo* 禮記集說, *juan* 8, Chap. 23, p. 253.

[41] Cf. *Shanhai jing*, transl. Rémi Mathieu 1983.

[42] Cf. *Baize jingguai tu* 百澤精怪圖, Text of the Six Dynasties. See Catherine Despeux (2003), p. 436–443.

[43] Cf. *Ruiying tu* 瑞應圖. See Catherine Despeux (2003); Lipiello (2001), p. 432–436.

The Hall of Numinous Brilliance in Lu
Here they have painted Heaven and Earth,
Multiform beings of every type and kind:
Various creatures wondrous and strange,
Mountain demons, sea spirits.
They have sketched and preserved their forms,
Conferring them to vermeil and blue.
A thousand changes, ten thousand transformations,
Each thing distinctly described.
Following set colors to image each kind,
They perfectly capture their essence.[44]

These spirits could enter the human body, itself endowed with souls which could leave the body and wander around, causing diseases. According to the first chapter of the *Inner Classic of the Yellow Thearch*, the soul is a major component of the person, along with the body:

The people of high antiquity,
Those who knew the Way,
They modelled [their behaviour] on yin and yang and
They complied with the arts and the calculations.
[Their] eating and drinking was moderate.
[Their] rising and resting had regularity.
They did not tax [themselves] with meaningless work.
Hence,
They were able to keep physical appearance and spirit together
And to exhaust the years [allotted by] heaven.
Their life span exceeded one hundred years before they departed.[45]

The physician's role consisted also in controlling these forces, as evidenced in particular by the title of a chapter in the *Inner Classic of the Yellow Thearch*, "Discourse on moving the essence and changing the *qi*."[46] This was characteristic of the Taoist master who created a sacred area from his own breath and divinities; his body was no longer an ordinary body but a cosmic body.[47]

[44] A poem by Wang Yanshou 王延壽 from the Eastern Han, in *Wenxuan zhu* 文選註 (Anthology of Commented Literature) (circa 520) by Xiao Tong 蕭統, commentary by Li Shan 李善 of the Tang, *juan* 11, p. 23, ed. SKQS. (translation by David R. Knechtges, *Wen Xuan or Selections of Refined Literature*. Volume two "Rhapsodies on Sacrifices, Hunting, Travel, Sightseeing, Palaces and Halls, Rivers and Seas," Princeton, Princeton University Press, 1987, vol. 2, p. 273).

[45] *Huangdi neijing Suwen, juan* 1, Chap. 1, p. 1. Transl. Unschuld, 2011, vol. 1, p. 31–32. Cf. Catherine Despeux (2007), p. 71–94.

[46] *Huangdi neijing Suwen, juan* 4, Chap. 13, "Moving the essences and transforming the breaths" (*Yijing bianqi lun* 易精變氣論).

[47] Cf. John Lagerwey (1994), p. 323–331.

4.5 Characteristics of the Ideal Landscape

Having laid the foundations necessary for the composition of an ideal landscape, let us see what characterizes it. Bear in mind that the environment is an integral part of the human being: nature and the human being not only have the same origin, but they feed from each other, they interfere with one another, and moreover, they are one and the same thing. Knowing how to keep in harmony with nature, to fit into this environment by keeping ones rightful place, is key to the attainment of good health and longevity. We do not do harm to nature with impunity. Before digging the ground for a building, for example, special rituals were performed; offerings and sacrifices were made to the god of the soil. While medical texts do not mention this because it is not within their remit, it goes without saying in the context of Antiquity, historical and literary writings abound with examples of this type. By virtue of these links, be they observed or imagined, between the human being and his environment, the individual tried to remove himself from it or, on the contrary, to expose himself to it, to adapt to it, or to restructure it.

Nature, like the Way nourishes the human being and enables him to transform himself. She is a beneficial mother, the womb that provides refuge and protects him. Laozi's book says that "The Way is the mother of all under heaven."[48] This means that, in a certain respect, the heavens and the earth are the womb for the human being: the individual grows in the landscape and constantly reconstructs himself. Nature is therefore characterized by a leading role in the transformation of the human being, playing a role that is both nurturing and protective. Relationships with nature will therefore be diverse: we scrutinize it, we seek the secrets of life and eternity within it, we feed on its essence and its pneuma, we melt into it, we incorporate it in ourselves, and we become nature.

4.5.1 From Nature to a Constructed Landscape

This nature becomes a shaped, constructed landscape, and constituted to contain – even symbolically – all the elements essential to life. The categorization of natural objects establishes a hierarchy in nature, as well as in the human body. Space is occupied by the human being who appropriates it: if on one hand he is conceived of as an integral part of nature, of the 10,000 beings, while being considered as the most intelligent, the most efficient (*ling* 靈) of all, on the other hand, he anthropomorphizes nature. According to a widespread creation myth, nature is a cosmic

[48] *Daode jing*, Chap. 25.

giant named Pangu: "When Pangu, the first born, was about to die, he transformed his body: his pneuma became the winds and clouds, his voice the claps of thunder, his left eye the sun, his right eye the moon, his four limbs and the five (parts of his body) the four extremes and five sacred mountains, his blood and his moods the Blue River and the Yellow River, his nerves and arteries, the veins of the earth, his muscles the earth of the fields, his hair and beard the stars and sidereal reference points, the hairs on his skin the vegetation, his teeth and bones the metals and stones, his semen and marrow the pearls and the jades, his sweat and his excreta the rains and the marshes."[49]

This anthropomorphism is taken up later in Chinese painting. Guo Xi (ca. 1000–1090), a Song painter, writes: "Water is the blood of the mountains, the plants and trees are its hair, the mist and clouds are its divine colors. This is why if the mountain is bestowed with water, it is alive, if it contains vegetation and trees, it thrives, if it is in the mists and clouds, it is full of elegance."[50]

4.5.2 A Space That Is Both Closed and Open

The environment is a space that is both closed and open, because, like the womb, it serves as protection and allows the being's elements to move in a circular manner; open, for exchanges with its environment, this double aspect of opening and closing promotes a free flow of things but also responds to a certain suppleness and malleability of the individual. Depending on needs or circumstances, the individual opens or closes these gateways to the outside while always maintaining the link and the possibility of the exchange between the inside and the outside.

An ideal environment has a certain dynamic: the wind must blow, but harmoniously, air that does not move is harmful. Air that is not renewed is bad. Similarly, stagnant water is bad, whether it is the water in a well that is not being drawn or that of the marshes. Only this renewal guarantees the purification of the nature in which we bathe, just as it purifies the interior of the body in which our organs bathe.

The closed character of the ideal environment makes it possible to place there all the things considered as necessary, which must be present for the life of the human being. Moreover, the history of the world and the human being, in other words the memory of the individual, his family, and his society, is contained within it. Natural objects are invested with the history and representations of humans and their experiences and feelings. They participate in the reconstruction and reinterpretation of cultural heritage.[51]

[49] Transl. Rémi Mathieu (1989) p. 28–29.

[50] Guo Si 郭思 (ca. 1000–1090), "*Lingquan gaozhi* 林泉高致," part "Instructions on the mountain and water" (*Shanshui xun* 山水訓), ed. SKQS.

[51] Cf. Dorofeeva-Lichtmann, (1995), p. 57.

4.5.3 A Harmonious Place

From this whole is born a harmony, which is also a harmony of the five senses: sounds and odors must respond to the theories of correspondences, the basis of the theoretical foundations of traditional Chinese medicine. Some Confucian writings emphasize two cardinal virtues that the wise must develop, the center and harmony,[52] virtues which correspond to qualities that they have attributed to nature. While the term harmony initially evokes for us music and sounds, in the Chinese context it also refers to luminosity. Hence, the choice of the location, of its altitude, of its orientation, and of its qualities at different times of the day is studied learnedly; the luminosity and transparency of the air are highly valued qualities. As for music, it is that of the five sounds and the six pitch pipes[53] which are expressed through nature's various instruments.

Maintaining harmony means respecting the environment's powers, and doing anything to the environment involves taking these powers into account. Thus, before digging the ground, there are rituals for the spirits of the ground, likewise for the trees that are felled and for other elements of the environment.

4.6 Importance of Links Between an Individual and His Environment

Being a moving space-time, the environment is constantly changing, as are the human beings who live there. Not only is it moving, but it weaves links with individuals according to how the individual positions himself and interacts with it, due to the thinking that there is no separation, but mutual interaction between the two. In other words, an ideal landscape can be transformed into a harmful landscape by the human beings who reside there and vice versa: there is a constant back and forth between the two wherein, in particular, lies the art of geomancy, on one hand, taking advantage of good environmental conditions and, on the other, circumventing those which are unfavorable. Building a fish pond in a certain position in front of his house can transform the destiny and health of the inhabitant. And the way these links are woven provides certain characteristics to the mobility and the dynamic of both.

The qualities of the links between the human being and his environment are therefore as essential for health as the environment itself. It is primarily the capacity for communication and exchange (*jiao* 交) through the passageways between the

[52] This is the central theme of one of the *Four Books, The Doctrine of the Mean* (*Zhongyong* 中庸), elaborating on "the center and harmony" (*zhonghe* 中和).

[53] Musical theory in China is based on 5 principal sounds corresponding to the 5 agents and thus enter into the system of correlations of the 5 agents; it is also based on 6 pitch pipes (in fact 12) which fixed the pitch of the notes and inserted them into the system of correlations of the 12 periods of time.

two, the human being and his environment, via the gateways and channels of communication in space, and via the orifices in the human body. This implies a certain permeability (*tong* 通), an interpenetration of one and the other. Cosmic pneuma enters the human body through the pores of the skin and all its orifices. The body and the being become transparent, the human being lets the landscape and all its elements penetrate him, and he becomes the landscape. The openings, the passageways, and the channels of communication, arranged by the human being, are indispensable in an ideal environment.

These qualities of exchange and interpenetration lead to the oneness and the unification of man and his environment (*yi* 一): there is a basic unity behind all this diversity; recognizing and embodying it are the guarantor of health, and all that unifies is the bearer of life and health.

Like the environment, individuals are in a state of constant mobility (*xing* 行): movement, whether external or internal, guarantees good health. Through codified movement the temporal dimension reappears, be it the gymnastic movements of the body whose integrative function for the human being within his environment and regulation of internal movements is essential or be it the internal movements within the body. These two types of movements contribute to the transformations and changes (*hua* 化) of life and to the maintenance and restoration of health: things, the individual, his body, and nature change with time and the conditions, everything is transformed but in an ideal of mutual service. Animal and human excreta have their utility in fertilizing the soil and were even transformed into remedies. Waste has thus been integrated into a cycle of transformation, making it beneficial for the human being and his environment.

In this relationship between man and his environment, the connection must seek equilibrium (*ping* 平): just as the five agents must be in equilibrium in the body, the elements of the environment or the landscape must be so too. This bond maintains (*yang* 養) the vital power of one and the other, of the human being, and of his environment. Life and the body maintain each other: nurturing, preserving, and taking care. To look after the landscape is to look after oneself, in a narcissistic extension. While the pneuma, the *qi*, is often put forward as the element that unifies and creates the link between things, it is not alone. Equally important are the invisible powers, *jing* 精, which inhabit nature and the body. However, there was a cardinal notion under the Han, which cannot be forgotten, that of imaginary thought, of intention (*yi* 意), and of mental representation, which surely played a major role in techniques for maintaining life but also in medical techniques such as acupuncture.[54] This symbolic and imaginary construct has a certain effectiveness. The ideal environment is equally that of our imagination.

[54] Cf., for example, *Huangdi neijing Lingshu, juan 1*, Chap. 1 "The Twelve Sources of Acupuncture" (*Jiuzhen shier yuan* 九針十二原), on the manipulation of the needle: "We welcome it (the pneuma, *qi*); we follow it in imagination to harmonize it; like that, the way of acupuncture is accomplished." 迎之隨之以意和之箴道畢也.

4.7 Conclusion

The Han period, with the *Inner Classic of the Yellow Thearch*, gives us an elaborate system of correspondences between man, the body, nature, and empire. This was not specific to the medical environment and responded to the formation of dominant notions and conceptual models of thought, enunciated and presented by scholars close to power such as Lü Buwei 呂不韋 (?–235), a minister under the Qin and the presumed author of the *Spring and Autumn Annals* (ca. 239 BCE), a book that lays the foundation for the theory of correspondences between the cosmos and government; and Dong Zhongshu (179–104), who describes ethical and political principles in terms of yin and yang; and also the authors of the *Comprehensive Discussions in the White Tiger Hall* (*Baihu tong* 白虎通) which established, in 79, the orthodoxy in terms of correlations. The ideal landscape is made up of elements arranged according to a device which constitutes the best possible framework for the preservation of health and the maintenance of life. However, outside this collective framework, each individual must look for his own ideal environment and construct it according to his own characteristics, in a search for harmony and correspondence between these elements and the environment, from which he is not separated and with which he interacts.

While there are favorable and unfavorable elements in the environment, it is not possible to reduce the relationship between the landscape and health to these. For example, a toxic element may be beneficial, as evidenced by the use of highly toxic plants and minerals in medicine. Much more important is the ability of the individual to create a connection with that environment, to situate himself in relation to it, and to maintain a malleable relationship with it.

The elements of the enumerated landscape include not only those that would come under what is called physical geography but also those that are rather in the domain of the invisible and do not relate to terrestrial configurations. They are firstly all the invisible forces that belonged thus to the sacred world, which animate nature and which had to be learned and mastered. But it is also all the elements that nourish the organs of the senses and not just the vision. It is therefore necessary to introduce the qualities of luminosity into this ideal landscape and also sound, olfactory, and other qualities.

The elements are arranged according to devices which give them the maximum power. First they form an enclosed space, such as the womb in the mother's body, which defines the first space linked to the exploration of the subject; the human being's environment is ideally the maternal womb, a place closed on itself, which protects and nurtures while being open for other reasons. This symbolism can be vested in a natural element such as caves or constructed like the artificial lakes, hills, and canals, the displaced rocks, or trees planted to create a landscape having the expected qualities.

The human being must consider the possibilities of integration and the creation of networks so that the cyclical and correlative functioning work. He observes and accompanies the evolutions of his environment; he transforms it without doing violence to it and even participates in its transformation. There is an efficiency in the symbolic representation. The environment is a route, a path and a track, and a cartography of the self, which is constructed according to our own individual environment and experience.

How does the ideal landscape contribute to disease? We should appropriate or reappropriate space-time itself and all its elements and conform to the rhythms of time through an observance of the calendar which dictates the often ritualized behaviors and choice of rhythms of life. We need to recreate the link within and outside of ourselves. When there is a loss of meaning, it is useful to reactivate the symbolism and the imaginary, whose effectiveness is real, while being different from that of the real landscape. The forces of the path, of the route, of the track, of the image, and of the symbol are such that they can serve to restore a disorder in the link between oneself and the environment.

References

Baihutong yi 白虎通義 (Directions on the understanding of the White Tiger Hall) by Ban Gu 班固, ed. *SKQS* (*Wenyuange Siku quanshu* 文淵閣四庫全書).

Baud Berthier, Gilles, Sophie Couëtoux, and Che Bing Chiu. 2004. *Le jardin du lettré : synthèse des arts en Chine*. Besançon: éditions de l'Imprimeur.

Chunqiu fanlu 春秋繁露 (Luxuriant Dew of the Spring and Autumn Annals) by Dong Zhongshu 董仲舒 (175/179–115/106), ed *SKQS*.

Daode jing 道德經 (Tao Te Ching), attributed to Laozi 老子.

Despeux, Catherine. 2003. Auguromancie. In *Divination et société dans la Chine médiévale*, ed. Marc Kalinowski, 431–470. Paris: Bibliothèque Nationale de France.

———. 2006. The Six Healing Breaths. In *Daoist Body Cultivation*, ed. Livia Kohn, 37–67. Magdalena: Three Pines Trees.

———. 2011. Toponymie des points d'acupuncture. In *Dictionnaire des noms de points*, ed. Elisabeth Rochat de la Vallée. Paris: Institut Ricci.

———. 2007. Âmes et animation du corps. La notion de *shen* dans la médecine chinoise antique. In *De l'esprit aux esprits. Enquête sur la notion de shen. Of Self and Spirits. Exploring shen in China, Extrême-Orient-Extrême-Occident*, n° 29, 71–94. Paris.

Dorofeeva-Lichtmann, Vera V. 1995. Conception of Terrestrial Organization in the *Shanhai jing*. *Bulletin de l'École française d'Extrême-Orient*, tome 82: 57–110.

Harbsmeier, Christoph. 2010. Toward a Conceptual History of some Concepts of Nature in Classical Chinese: *Ziran* 自然 and *Ziran zhi li* 自然之理. In *Concepts of Nature, a Chinese-European Cross-Cultural Perspective*, ed. Hans Ulrich Vogel and Günter Dux, 220–254. Leiden/Boston: Brill.

Hennesy, Anna Madelyn. 2001. *Chinese Images of Body and Landscape Visualization and Representation in the Religious Experience of Medieval China*, University of California, Santa Barbara, UMI 3473750 (thesis).

Huainan zi zhushi 淮南子注釋 (Book of the Master of Huainan) by Liu An 劉安, commented by Gao You 高誘 of the Han Dynasty [1973], Taibei: Hualian chubanshe.

Huangdi neijing Suwen Lingshu 黄帝内經素問靈樞 (Magic Pivot and Basic Questions two parts of the Inner Classic of the Yellow Thearch), commentaries by Wang Bing 王冰, edition *Sibu congkan* 四部叢刊. Taipei: Xuanfeng chubanshe, 1974. (Translated by P.U. Unschuld & H. Tessenow).

Jullien, François. 1992. *La propension des choses. Pour une histoire de l'efficacité en Chine*. Paris: Seuil.

Knapp, Ronald G. 2006. *Chinese Houses: The Architectural Heritage of a Nation*. Boston: Tuttle Publishing.

Knechtges, David R. 1987. *Wen Xuan or Selections of Refined Literature*. Volume two "Rhapsodies on Sacrifices, Hunting, Travel, Sightseeing, Palaces and Halls, Rivers and Seas", 2 vol. Princeton: Princeton University Press.

Le Blanc, Charles. 1985. *Huai-nan Tzu: Philosophical Synthesis in Early Han Thought: The Idea of Resonance (Kan-Ying) With a Translation and Analysis of Chapter Six*. Hong Kong: Hong Kong University Press.

Lagerwey, John. 1994. L'espace sacré taoïste. In *Aménager l'espace*, ed. Flora Blanchon, coll. Asie, vol. 2, 323–331. CREOPS, Paris: Presses universitaires de Paris-Sorbonne.

Liji jishuo 禮記集說 (Collection of Commentaries on the Book of Rites). 1973. Taipei: Xuesheng shuju.

"Lingquan gaozhi" 林泉高致 by Guo Si 郭思 (son of Guo Xi 郭熙), part "Instructions on the Landscape/Landscape Painting" (Shanshui xun 山水訓), ed *SKQS*.

Lipiello, Tiziana. 2001. *Auspicious Omens and Miracles in Ancient China. Han, Three Kingdoms and Six Dynasties*, Sankt Augustin: Institut Monumenta Serica, Monograph Series 39.

Mathieu, Rémi. 1983. *Études sur la mythologie et l'ethnologie de la Chine ancienne*. Tome 1: *Traduction annotée du* Shanhai jing. Tome 2: *Index du* Shanhai jing. Paris: Collège de France, Institut des hautes études chinoises.

———. 1989. *Anthologie des mythes et légendes de la Chine ancienne*. Paris: Gallimard.

Obringer, Frédéric. 2009. *Fengshui, l'art d'habiter la terre: une poétique de l'espace et du temps*. Paris: Philippe Picquier.

Robinet, Isabelle. 1986. The Taoist Immortal: Jesters of Light and Shadow, Heaven and Earth. *Journal of Chinese Religions* 13–14: 87–105.

Schipper, Kristofer. 2001. Daoist Ecology. The Inner Transformation. A Study of the Precepts of the Early Daoist Ecclesia. In *Daoism and Ecology*, ed. Norman Girardot, James Miller and Gang Liu, 80–86. Cambridge, MA: Harvard University Center for the Study of World Religions.

Selin, Helaine. 2003. *Nature across Culture: Views of Nature and the Environment in Non-western Cultures*. Dordrecht: Kluwer Academic Publishers.

Sivin, Nathan. 1995. State, Cosmos, and Body in The Last Three Centuries B.C. *Harvard Journal of Asiatic Studies* 55 (1, June): 5–37.

Unschuld, Paul U., and Hermann Tessenow, in collaboration with Zheng Jinsheng. 2011. *Huang Di nei jing su wen. An Annotated Translation of Huang Di's Inner Classic-Basic Questions*, volume I, chapters 1 through 52, vol. II, chapters 53 through 71, and 74 through 81, Berkeley/Los Angeles: University of California Press.

Unschuld, Paul U. 2016. *Huang Di Nei Jing Ling Shu. The Ancient Classic on Needle Therapy. The complete Chinese text with an annotated English translation*. Berkeley/Los Angeles: University of California Press.

Wenxuan zhu 文選註 (Anthology of Commented Literature) (circa 520) by Xiao Tong 蕭統, commentary by Li Shan 李善 of the Tang, ed *SKQS*.

Zhuangzi jijie 莊子集解 (Explanations and compilation of commentaries by Zhuangzi) by Wang Xianqian 王先謙, ed *Zhuzi jicheng* 諸子集成, Shanghai, Shanghai shudian 1986.

Part II
Observations, Definitions, and Theories About Environment, Disease, and the Body: Questioning the Meaning of the Environment

Chapter 5
The Construction of Thinking on the Environment: The Words, Their Meanings, and Their Uses from 1790 to 1970

Valérie Chansigaud

> *La plupart des mots de notre vocabulaire représentent plutôt une zone de 'suggestion' qu'un objet clairement défini.*
> *André Maurois (1932, 179) (Most words in our vocabulary represent a zone of "suggestion" rather than a clearly defined meaning).*

Abstract Between approximately 1800 and 1830, many words appear in French and English ("habitat", "station", "milieu", "environment") in relation to the same idea: living beings are influenced by and dependent on what surrounds them. However, the analysis of scholarly texts written by naturalists or by scholars pertaining to religious circles stresses that this idea was popular before this time. In fact, the birth of this vocabulary points to a new phenomenon: the transformation by sciences (mainly botany and zoology) of a common but indistinct idea into an object which can be analysed through scientific methods. The reshaping of the idea of environment by scientists led to the leaving aside of spiritual or magical elements (only material manifestations are considered as real) and to the retention of that which could be measured, thereby narrowing the number of elements to be studied. During the nineteenth century, a second set of words (chorology, ecology, mesology) appears, translating the expanding autonomy of this field of research, even if only ecology becomes an autonomous discipline. Finally, in the 1960s, we observe a new evolution of the idea of milieu among a lay culture, attested by the adoption in French of the word "environnement". From this time on, the precision in the vocabulary becomes blurred, and the word "environment" conveys a more general

---

V. Chansigaud (✉)
Université de Paris, SPHERE, UMR 7219, CNRS-Université de Paris, Paris, France
e-mail: valerie.chansigaud@noos.fr

F. Bretelle-Establet et al. (eds.), *Making Sense of Health, Disease, and the Environment in Cross-Cultural History: The Arabic-Islamic World, China, Europe, and North America*, Boston Studies in the Philosophy and History of Science 333, https://doi.org/10.1007/978-3-030-19082-8_5

and vague meaning, becoming, at times, synonymous with the negative impact from human activities and including, sometimes, a spiritual dimension.

Keywords Environmentalism · Ecological history · Ecology history · Subject terms · Vocabulary development

5.1 Introduction

In Western Europe at the end of the eighteenth century and the beginning of the nineteenth century, we witnessed the reformulation of an old vocabulary which allowed the environment to be named *environment* in English, *milieu* and *ambiance* in French, and *station* and *habitat* in both languages. These terms had already existed, but they took on additional meaning. In the middle of the nineteenth century, new terms were created to designate new sciences: two of them, *mesology* and especially *ecology*, gave rise to active disciplines. To understand the meaning and the scope of these two successive waves of vocabulary reformulation, we have to step back to a longer perspective as many authors (doctors, naturalists, philosophers, and poets) affirmed two inseparable ideas long before the nineteenth century: the influence of what surrounds a living human or non-human being and the interdependence of that living being with its milieu.[1]

5.2 Notions of the Influence of Milieu and Interdependence Before the Nineteenth Century

Foreign travel undertaken by Europeans from the fifteenth century onwards considerably increased the number of species and natural substances (mainly plant-based) used in pharmacopoeia and in food. This rapidly led to a debate on the consequences of the consumption of foreign products, often coming from far away countries. To their supporters, these foreign substances had a higher efficacy than local plants, while their opponents challenged this idea completely.[2] The opposition to exotic products was based on a set of economic (their higher costs would ruin countries) and health (it is not possible to guarantee the quality of these products) arguments leading sometimes to an attitude that could be described as patriotic and contributed to the valorization of inventories of European flora. The idea that there is a correspondence between the human body and the plants and animals which flourish in

---

[1] The word "milieu" is used here as it was the term used in French at that time. It is an almost perfect synonym for the English word "environment". Since this contribution focuses on the words used in Western Europe to talk about the environment, we have paid attention to present quotations in their original language, followed by an English translation. When published translations exist and are faithful to the original text, we chose them; when there is no published English translation or when it appears to be too far from the original text, we produced our own translation.

[2] See in particular works by Alix Cooper (2003, 2007).

the same region was also expressed in this controversy, the supposed harmfulness of the exotic products coming from their foreign origin.

Jan van Beverwyck (1594–1647) was one of the opponents of the use of exotic plants and other foreign products. He formulated thus the correspondence between the milieu and the organisms which live in it: "La nature aurait été très injufte de nous envoyer des maladies qui proviennent du climat du païs, fans y avoir fait naître des fimples & de remedes pour les guérir" (Nature would have been very unjust to send us illnesses which result from the climate of the country, without having created simple remedies to cure them).[3] This was not purely theoretical reasoning, as Beverwyck presented acclimatization as proof of the influence of the milieu: "On voit de même que les perfonnes qui font tranfportées dans des Païs éloignez de leur lieu natal, changent infenfiblemant par l'air nouveau qu'ils y refpirent" (it is seen equally that persons who are transported to regions far from their native land, change imperceptibly through the new air which they breathe there).[4] Jan van Beverwyck gives other examples to demonstrate the dependence of organisms on their environment of origin. Firstly, plants transplanted from one place to another end up dying: "Et fuivant cet air naturel, il eft impoffible que toute forte de plantes & les arbres de ce païs là puiffent être tranfplantez ici ; il y en a quelques-unes qui y viennent, mais elles n'ont plus les mêmes vertus : il y en a même beaucoup qui ne produifent ni fleurs ni fruits, & meurent à la fin" (And according to this natural air, it is impossible that all the plants and trees of this region may be transplanted here; there are some which come here, but they no longer have the same virtues: there are even many which do not produce flowers or fruit, and die in the end).[5] This kind of degeneration also occurs, according to Beverwyck, when animals and human beings change countries. Thus, a human being perfectly adapted to his/her original environment remains healthy if he/she is not exposed to foreign countries or substances.

This idea of adaptation is not unique to physicians nor even to naturalists. It is found in the *Ideen zur Philosophie der Geschichte der Menschheit* (1784–1791) by Johann Gottfried von Herder (1744–1803), a poet, theologian, philosopher, and mentor of Goethe (translated in English and French, respectively, in 1800 and in 1827–1828). Herder recognized, like Beverwyck 150 years earlier, the influence of the surrounding milieu and adaptation to its region of origin:

> Alle Pflanzen wachsen hin und wieder wild in der Welt; auch unsre Kunstgewächse sind aus dem Schoß der freien Natur, wo sie in ihrem Himmelsstrich in größester Vollkommenheit wachsen. Mit den Tieren und Menschen ist's nicht anders; denn jede Menschenart organisiert sich in ihrem Erdstrich zu der ihr natürlichsten Weise (All plants grow wild in some part or other of the World. Those, which we cultivate with art, spring from the free lap of Nature, and arrive at much greater perfection, in their proper climes. With animals and with man it is the same: for every race of men, in its proper region is organised in a manner most natural to it.)[6]

[3] Beverwyck (1730): 62 [translation by the editors].

[4] Beverwyck (1730): 75 [translation by the editors].

[5] Beverwyck (1730): 73 [translation by the editors].

[6] Herder (1800): 32. For the original text, see http://www.zeno.org/Literatur/M/Herder,+Johann+Gottfried/Theoretische+Schriften/Ideen+zur+Philosophie+der+Geschichte+der+Menschheit/Erster+Teil/Zweites+Buch/2.+Das+Pflanzenreich+unserer+Erde, last access 12122017.

In addition to the dependence of living organisms on their milieu of origin, Herder was also interested in the interrelationship of animals and plants which he describes as a permanent struggle "da die ganze Schöpfung in einem Kriege ist und die entgegengesetztesten Kräfte einander so nahe liegen ("since the whole creation is but a war, and that the most opposite powers are found so close to each other")".[7] This reasoning presents the world, certainly in a very general way, as an interweaving of influences that in our contemporary vocabulary we would call "biotic" and "abiotic". The interrelation between living organisms and the influence of the milieu is often supported by a theological reading: the world created by God is perfect, everything is in its proper place, and everyone assumes his proper role. This vision leads scholars and philosophers to describe nature as a system that is both stable (it is immutable) and effective (each organism plays a role ensuring the proper functioning of nature).

It is in this double perspective of stability and efficiency that one speaks of the "economy of nature". Carl Linnaeus (1707–1778) contributed significantly to the popularization of this expression. Thus, we read in the 12th edition of Linnaeus' *Systema Naturae*:

> Oeconomia Naturae versatur circa Rerum Generationem, Conservationem, Destructionem, ut opus Creatoris illibatum perfittat, ad quod tota conspirat Natura (The economy of Nature revolves around the Generation of things, their Preservation, their Destruction, so that the Creator's work remains complete, which is what all Nature tends towards).[8]

Linnaeus' *Systema Naturae* was translated into vernacular languages with many variations. In the French translation by Jean Emmanuel Gilibert (1741–1814) published in 1805, we read:

> L'économie de la nature roule sur la génération, la conservation la destruction des individus, pour que l'œuvre du Créateur reste toujours complète ; car toutes les opérations de la nature tendent à sa conservation (The economy of nature revolves around the propagation, the preservation, and the destruction of individuals, so that the work of the Creator remains always complete; because all nature's operations tend towards its preservation.)[9]

Similarly, J.H. Moreau, in his letter "Considérations philosophiques sur l'Histoire naturelle en général, et en particulier celle des Insectes" ("Philosophical considerations on natural history in general, and in particular that of insects"), sets living organisms in an extensive chain of interdependence: some being eaten by others, death generating life, and the proliferation of some is limited by the consumption of

---

[7] Herder (1800): 35. For the original text, see http://www.zeno.org/Literatur/M/Herder,+Johann+Gottfried/Theoretische+Schriften/Ideen+zur+Philosophie+der+Geschichte+der+Menschheit/Erster+Teil/Zweites+Buch/3.+Das+Reich+der+Tiere, last access 12122017.

[8] Linnaeus (1793): 15 [translation by the editors].

[9] Gilibert (1805): 27 [translation by the editors] https://books.google.fr/books?id=gS68KfAAmOoC&pg=PA18&dq=%22oeconomia+naturae+verfatur%22&hl=fr&sa=X&ved=0ahUKEwjfzu_aq4TYAhULLMAKHYwAAeYQ6AEISDAF#v=onepage&q=%22oeconomia%20naturae%20verfatur%22&f=false

others. For Moreau, this interdependence makes all animals and plants useful for the conservation of nature:

> La Nature a fait tous les êtres les uns pour les autres, et n'a point voulu qu'aucun d'eux fût isolé ou inutile. (Nature has made all beings for each other, and does not want any one of them to be isolated nor useless)[10]

Man is never excluded from this view of the world:

> L'espèce humaine, que des écrivains, moins philosophes que poëtes, veulent regarder comme le terme et le centre de toutes les productions de ce globe, cette espèce prétendue impériale n'est également qu'un point, un degré de l'échelle des êtres auxquels elle se lie par des rapports nombreux d'utilité et de dépendance réciproques. Périssable comme les victimes de sa puissance, éphémère comme l'objet de ses observations, l'homme, après quelques actes plus ou moins nécessaires, quelques scènes de mouvemens plus ou moins prolongés, s'évanouit et meurt, et ses débris, rentrés dans le torrent de la circulation universelle, deviennent la source où de nouveaux êtres doivent puiser la vie. (The human species, which writers, less philosophers than poets, wish to regard as the end and centre of all this globe's productions, this purportedly imperial species is also but a point, a degree on the scale of the beings to which it is bound by numerous relationships of mutual utility and dependence. Perishable like the victims of his power, ephemeral as the object of his observations, man, after a few more or less necessary acts, some scenes of more or less prolonged motion, vanishes and dies, and his debris returns to the torrent of universal circulation, and become the source where new beings must draw life.)[11]

The interdependence of man with other living organisms and his environment is not limited to the field of medicine nor to that of natural history, as this idea also permeates a plethora of literature, especially poetic. In his argument at the beginning of his *Essay on Man* (1734), the British poet Alexander Pope (1688–1744) remarks:

> That throughout the whole visible world, an universal order and gradation in the sensual and mental faculties is observed, which causes a subordination of creature to creature, and of all creatures to man. The gradations of sense, instinct, thought, reflection, reason; that reason alone counterwails all the other faculties.
>
> How much farther this order and subordination of living creatures may extend, above and below us; were any part of which broken, not that part only, but the whole connected creation must be destroyed.[12]

The idea that man is one element in the world from which he cannot escape is constantly taken up, reworked, and reformulated by various authors; it is a complex intellectual construct but not very precise, without a dedicated vocabulary; and with these authors, we find neither the terms milieu nor environment. While poets and

---

[10] Moreau (1798): 10 [translation by the editors]. This letter is published in 1798 in the *Magasin encyclopédique, ou journal des sciences, des lettres et des arts*, under the general heading "Histoire Naturelle".

[11] Moreau (1798): 11–12 [translation by the editors].

[12] The translation kept by the National Library of France and available on the site Gallica is used here. Notice n: FRBNF31136121, http://gallica.bnf.fr/ark:/12148/bpt6k9618705h

theologians evoke the interrelationship of living organisms, this idea constitutes one of the principal subjects in natural history, as is evident from the program of the French École centrale de la Côte-d'or (central school of the Cote d'Or) in 1801:

> La science des faits de la Nature est très étendue : on l'a séparée en deux branches, la Physique et l'Histoire naturelle. Cette dernière envisage les êtres sous le point de vue de leur existence individuelle, et de leurs rapports mutuels. (The science of the facts of Nature is very extensive; it has been divided into two branches, Physics and Natural History. The latter considers beings from the point of view of their individual existence, and of their mutual relationships)[13]

These mutual relationships always involve the adaptation of organisms to a given region and are not limited only to the interactions of living organisms. The author of this program emphasizes the importance of the study of the distribution of animals, which brings the idea of adaptation to a particular place:

> Les animaux ne se trouvent point indifféremment dans tous les lieux de la terre ; chacun a un climat qui lui est propre ; et certaines espèces sont confinées à des latitudes fixes. (Animals are not found everywhere indiscriminately on the Earth; each has its own climate; and some species are confined to fixed latitudes).[14]

The discourses of travellers, doctors, naturalists, philosophers, and poets before the end of the eighteenth century testify that the question of the links between plants, insects, animals, and humans and what surrounds them is not a new field of investigation. The emergence at the end of the eighteenth and early nineteenth centuries of a new vocabulary to describe what surrounds plants, animals, and man, and to evoke or study these links, indicates rather the scientists' application of a scientific method to old and intuitive reflections.

5.3 The Emergence of a Vocabulary Around the Notion of Environment, an Epistemological Break?

During a relatively short period, mainly from 1800 to 1830, we observe the emergence of a complex vocabulary relating to the environment in French and in English. Here is a quick inventory based on historical dictionaries[15]:

- *Environment* is an old word in the English language and refers to the state of being surrounded or the action of surrounding: in 1828 it also referred to the conditions under which an individual or things live or develop, as well as all the influences on the development of the life of a character.[16] The circulation of this

[13] Source: South Kensington General Library, 44NHM ALMA [translation by the editors].

[14] The text is attributed to Jean Nicolas Vallot (1771–1860). [translation by the editors].

[15] Oxford English Dictionary (2009) and Rey (1994).

[16] The first use of this word is attributed to Thomas Carlyle (1795–1881) in his *Critical and Miscellaneous Essays* (1904). See Jessop (2012).

word is difficult to trace, but its use by the British philosopher, biologist, and sociologist Herbert Spencer (1820–1903) in 1855 in *The Principles of Psychology* testifies to its popularity[17] outside the biological sciences.

- *Habitatio* was used in Latin texts in the mid-eighteenth century[18] to refer to the precise location of a plant or animal. It appears in the form of *habitatio* in the section defining botanical vocabulary in the flora that the British physician and naturalist William Withering (1741–1799) published in 1796. The definition stipulates that *habitatio* designates the place where a particular plant grows spontaneously. Withering also specifies: "This is now generally expressed by the word Habitat".[19]
- The word *station* was used by the British botanist Richard Bradley (1688–1732) in 1721 to designate the localities occupied by animals and plants, just like the term "habitation".

In French, the chronology is almost identical and changes only in the details. The words *habitat* and *station* are defined by the Swiss botanist Augustin Pyramus de Candolle (1778–1841) in his series of definitions of "physiological terms" published in his 1813 *Théorie élémentaire de la botanique*:

> §. 350. Station (*Statio*), lieu dans lequel un végétal croît naturellement, considéré, quant à sa nature physique, et non quant à sa position géographique, qu'on nomme son Habitation ou sa Patrie (*Habitatio, Patria*). Ainsi, l'habitation du Riz est dans l'Inde, sa station dans les Marais (Station (*Statio*), a place in which a plant grows naturally, considered, as to its physical nature, and not as to its geographical position, which is called its Habitation or its Homeland (*Habitatio, Patria*). Thus, the habitation of Rice is in India, its station in Marshes).[20]

Candolle then lists the different types of stations, and these three examples make it possible to understand his typology:

> Salines (salinæ, salsæ, salsuginosæ), qui croissent dans les terrains saumâtres ou salés.
> Maritimes (maritimæ), qui croissent au bord de la mer.
> Littorales (littorales, ripariæ), qui croissent sur les rivages de la mer, et surtout des lacs et des fleuves. (Saline (salinæ, salsæ, salsuginosæ), which grow in brackish or salty areas.
> Maritime (maritimæ), which grow by the sea.
> Littoral (littorales, ripariæ), which grow on the shores of the sea, and especially lakes and rivers)[21]

We see that the examples of stations described by Candolle make it possible to understand the meaning of his scientific project: it is a question of providing a

[17] Pearce (2010), Oxford English Dictionary (2009) and Rey (1994).

[18] The Oxford English Dictionary gives the date as 1762 without it being possible to link it to a particular publication.

[19] Withering (1796): 62. Pearce (2010).

[20] Candolle (1813): 423.

[21] *Op. cit.* [translation by the editors].

vocabulary that allows nature to be described in a standardized way. The terms "station" and "patrie (homeland)" are used in the same way as terms in anatomy (describing leaf position, the shape of flowers or fruit), physiology (describing the number of fruit borne by a flower, time of flowering or fruiting), or chemical composition (tannin, starch, gum). The precision of all this vocabulary makes it possible to ensure the validity of the work of determining species according to the method proposed by Candolle. These definitions were reused by other naturalists and introduced in English by the British geologist Sir Charles Lyell (1797–1875) in the first edition of his *Principles of Geology* (1830).

The word *milieu* appeared from the seventeenth century onwards, firstly to refer to a spatial situation and then in the fifteenth century to a temporal situation.[22] In the seventeenth century, it began to designate the physical world in which bodies are placed. It was only at the beginning of the nineteenth century that *milieu* designated both the conditions external to a being and the influences of these conditions on that being. This usage is generally attributed to the French naturalist Jean-Baptiste de Monet de Lamarck (1744–1829), while Candolle uses the expression "milieux environnants" (or "surrounding milieus") in a short note,[23] *Mémoire sur la vegetation du guy*, read at the Société philomatique in Frimaire year IX (November–December 1800). These old uses of *milieu* are ignored, willingly or not, by the French philosopher Auguste Comte (1798–1857), as he appropriates the authorship of this word in the third volume of his *Cours de philosophie positive* published in 1835:

> Il serait superflu, j'espère, de motiver expressément l'usage fréquent que je ferai désormais, en biologie, du mot *milieu,* pour désigner spécialement, d'une manière nette et rapide, non-seulement le fluide où l'organisme est plongé, mais, en général, l'ensemble total des circonstances extérieures, d'un genre quelconque, nécessaires à l'existence de chaque organisme déterminé. Ceux qui auront suffisamment médité sur le rôle capital que doit remplir, dans toute biologie positive, l'idée correspondante, ne me reprocheront pas, sans doute, l'introduction de cette expression nouvelle. (It would be superfluous, I hope, to expressly justify the frequent use which I shall henceforth make, in biology, of the word *milieu*, to designate in a clear and rapid manner not only the fluid in which the organism is immersed, but, in general, the totality of external circumstances of any kind necessary for the existence of each determined organism. Those who have meditated sufficiently on the essential role which the corresponding idea must assume in any positive biology will doubtless not reproach me for the introduction of this new expression).[24]

After their initial creation, however, the reuse of these words remains very variable. Following the proposals of Candolle, the French naturalist Jean François Pierre Deterville (1766–1842) in his *Nouveau dictionnaire d'histoire naturelle appliquée aux arts, à l'agriculture, à l'économie rurale et domestique, à la médecine* (1816–1819) gathers together the words "habitation" and "station" in the same

---

[22] Rey (1994): 1244.

[23] Candolle uses this expression to describe the type of environment in which certain plants thrive.

[24] Comte (1838): note 1, page 301 [translation by the editors]. An English adaptation of Comte's positive philosophy was published in 1896.

entry. The first term referred to both the climate preferred by plants and animals and the place they occupy: "le lion choisit son habitation dans les climats ardens de l'Afrique et de l'Asie, et le renne dans les régions glacées du Nord" (the lion chooses its habitation in the fierce climates of Africa and Asia, and the reindeer in the icy regions of the North).[25] Deterville refined this notion with the second word, "station", since it allows him to specify with more precision the type of place occupied by living organisms, often restricted to particular sites, "la station de la loutre est près des rivières, et celle du lièvre dans les campagnes et les buissons du même pays" (the otter's station is near rivers, that of the hare in the countryside and the scrubland of the same region).[26]

The *Dictionnaire classique d'histoire naturelle* (1822–1831) by the French geographer and naturalist Jean Baptiste Bory de Saint-Vincent (1778–1846) does not offer a definition of milieu or station (the latter word is however used in some articles) but only that of habitat:

> HABITAT, zool. et bot. Ce mot latin est passé dans notre langue pour désigner la patrie d'un Animal ou d'une Plante (HABITAT, zool. and bot. This Latin word has passed into our language to designate the homeland of an animal or a plant).[27]

The word *habitation* was frequently replaced by the word *habitat*, as by the French entomologist Félix-Édouard Guérin-Méneville (1799–1874) in his *Dictionnaire pittoresque d'histoire naturelle et des phénomènes de la nature* (1835–1839) which only offers the definition of "*Habitations*, ou simplement *habitat des plantes*. (bot. et agr.)" (*habitations*, or simply *habitat of plants*. (bot. and agr.))[28] as well as a series of articles on physical, botanical, and zoological geography.

In his *Philosophie zoologique* (1809), Jean-Baptiste de Monet de Lamarck (1744–1829) almost never uses the word *milieu* alone; he associates it with the adjectives *surrounding* or *prevailing*. Lamarck describes the factors modifying living organisms as follows:

> Les principales [circonstances favorables aux organismes vivants] naissent de l'influence des climats, de celle des diverses températures de l'atmosphère et de tous les milieux environnants, de celle de la diversité des lieux et de leur situation, de celle des habitudes, des mouvements les plus ordinaires, des actions les plus fréquentes, enfin, de celle des moyens de se conserver, de la manière de vivre, de se défendre, de se multiplier, etc. (The [principal circumstances favourable to living organisms] consist in the influence of climate, of the varying temperatures of the atmosphere and the whole environment, of the variety of localities and their situation, of habits, the commonest movements, the most frequent activities, and, lastly, of the means of self-preservation, the mode of life and the methods of defence and multiplication.)[29]

---

[25] Deterville (1817): 109.

[26] *Op. cit.*

[27] Bory de Saint-Vincent (1825): 2 [translation by the editors].

[28] Guérin-Méneville (1835): 540 p.

[29] Lamarck (1873): 238. [translation by Hugh Samuel Roger Elliot (1914): 114].

Elsewhere, it uses a periphrase:

> [...] que conséquemment ces animaux doivent leur forme générale aux influences du milieu dans lequel ils habitent. (that consequently these animals owe their general shape to the influences of the medium in which they live.)[30]

And, later on, he writes:

> Quant aux circonstances qui ont tant de puissance pour modifier les organes des corps vivants, les plus influentes sont sans doute la diversité des milieux dans lesquels ils habitent, mais en outre il y en a beaucoup d'autres qui ensuite influent considérablement dans la production des effets dont il est question. (As to the conditions which have so much power in modifying the organs of living bodies, the most potent doubtless consist in the diversity of the places where they live, but there are many others as well which exercise considerable influence the production of the effects in question.)[31]

The use of this vocabulary (habitat, station, milieus) indicates the transformation by naturalists of an idea of milieu that was widespread but vague. We can observe several elements common to the various authors: the description of the milieus constitutes a simplification and a standardization, at least in the work of a given author. This transformation is often supported by the quest for standardization, or even quantitative analysis, hence the interest given to climatology, which allowed numerous measurements.[32] Similarly, the reference to divine creation loses its importance, mainly because it was above all a question of describing reality and not of referring to a particular religious thought.

The study of the place where living organisms live (whether it is called milieu, habitat, or station) is not only a simple intellectual pastime; it has a strong practical and economic utility:

> Cette étude des stations est peut-être la plus importante de toutes celles de l'histoire naturelle pour l'économie rurale et l'agriculture. Comment pourra-t-on acclimater une plante dans un autre pays, si l'on ignore quel terrain elle demande ? (This study of stations is perhaps the most important of all those in natural history for the rural economy and agriculture. How can a plant be acclimatized to another country, if it is not known what land it requires?)[33]

Acclimatization occupied a prominent position in the scientific debates of the nineteenth century and has attracted the attention of many contemporary scholars[34] because of its role in the project of colonial expansion (its purpose was to make the colonists more resistant to exotic climates and increase the number of domesticated species[35]). The development of the study of the influence of the milieu is very much underpinned by the economic project of acclimatization.

---

30 Lamarck (1873): 154. [translation by Hugh Samuel Roger Elliot (1914): 74].

31 Lamarck (1873): 232 [translation by Hugh Samuel Roger Elliot (1914): 111].

32 Alexander von Humboldt (1769–1859).

33 Deterville (1817): 120 [translation by the editors].

34 See, for example, the movement in favour of acclimatization in France, Osborne (1994), and the cultural perception of acclimatization in the English-speaking world, Dunlap (1999).

35 See, for example, the *General Report on questions relating to the domestication and naturalization of useful animals, addressed to the Minister of Agriculture and Trade* (1849) by Isidore Geoffroy Saint-Hilaire.

5.4 Evolution as an Explanation of the Influence of the Milieu

The different terms (*milieu*, *environment*, *habitat*, *station*), which were adopted and defined in the early nineteenth century to describe the environment, were used increasingly from the 1830s, but most importantly, they were endowed with an evolutionary dimension: adaptation, as a result of the influence of the milieus, makes it possible to explain the transformation of species. Therefore, the milieu is no longer considered as a simple geographical characteristic or as the setting where living organisms evolve; it is one of the main causes for explaining the structure of organisms. Thus in 1830, the French naturalist Étienne Geoffroy Saint-Hilaire (1772–1844) used a very explicit didactic example:

> La rose qui a conservé ses étamines intéresse le botaniste sous le rapport du maintien de ses faits de famille, et la rose qui les a perdues, par une transformation en pétales, n'en plaît que davantage au jardinier, dont elle embellit les parterres. Mais pour le philosophe qui échappe aux inductions de ces positions spéciales, ces deux sortes de roses sont un seul et même végétal, variable sous l'influence des milieux ambiants [...] (The rose which has kept its stamens interests the botanist because of the maintenance of its reproductive power, and the rose that has lost them, through a transformation into petals, simply gives more pleasure to the gardener, whose flowerbeds it embellishes. But for the philosopher who evades the conclusions of these special positions, these two kinds of rose are one and the same plant, variable under the influence of surrounding environments [...]).[36]

This use of domestication as a model for understanding evolutionary mechanisms evokes, of course, the famous example of animal domestication used by Charles Darwin (1809–1882) in his 1859 *On the Origin of Species*.

Natural history then becomes a true story: time is a factor explaining the multiplicity of living organisms, a diversity that until that time was justified only by divine power. The passage of time not only influences living beings but also the milieus themselves that change over time, contributing to the emergence of new influences. This is what Lamarck points out in his *Zoological Philosophy* (1809):

> On sait que des lieux différents changent de nature et de qualité à raison de leur position, de leur composition et de leur climat, ce que l'on aperçoit facilement en parcourant différents lieux distingués par des qualités particulières ; voilà déjà une cause de variation pour les animaux et les végétaux qui vivent dans ces divers lieux. Mais ce qu'on ne sait pas assez et même ce qu'en général on se refuse à croire, c'est que chaque lieu lui-même change, avec le temps, d'exposition, de climat, de nature et de qualité, quoique avec une lenteur si grande, par rapport à notre durée, que nous lui attribuons une *stabilité* parfaite. [37] (It is known that different localities differ as to their character and quality by reason of their position, construction and climate, as is readily perceived in passing through different places distinguished by special qualities; this is one living in these various places. But what is not known so well, and indeed what is not generally believed, is that every locality itself changes in time as to exposure, climate, character and quality, although with such extreme slowness according to our notions, that we ascribe to it complete stability.

[36] Geoffroy Saint-Hilaire (1830): 118-–119 in Le Guyader (2004): 170.

[37] Lamarck (1873): 232 [translation by Hugh Samuel Roger Elliot (1914): 111].

The different terms used here (*milieu*, *station*, *habitat*) were first defined in the context of the natural sciences, but they were gradually re-employed in other disciplines that today are part of the social sciences: the ideas that these words cover provide an effective tool to explain the differences between peoples and the transformation of societies over time. One of the most often quoted examples is that of Hippolyte Taine (1828–1893), who specifies in the first volume of his *Histoire de la littérature anglaise*, published in 1863, that three "forces primordiales"[38] (primordial forces)[39] determine the "état moral élémentaire"[40] (elemental moral state)[41] of a people: the race or "le ressort du dedans" (the internal mainsprings),[42] the milieu or "la pression du dehors" (the external pressure),[43] and the "moment" or "l'impulsion déjà acquise" (acquired momentum).[44] The passage of time is fundamental in this construction because for Taine the influence of the milieu accumulates over time and the transformations that it produces are transmitted by heredity: the "moment"[45] according to Taine is thus thought of as the encounter of the milieu and heredity. He takes up the old idea that man inhabits a crowded universe of which he is one of the elements: "l'homme n'est pas seul dans le monde ; la nature l'enveloppe et les autres hommes l'entourent" (For man is not alone in the world; nature surrounds him and his fellow-men surround him).[46] The influences of milieu that Taine quotes are climate, political circumstances, and social conditions. As can be seen, Taine cannot be distinguished from thinkers active in the field of natural history.

5.5 The Emergence of Disciplinary Terms

In the middle of the nineteenth century, a second series of words appeared, this time entirely new. They allow the description of the scientific discipline studying the milieus and their influence.

Some of these terms exist today only in rare echoes, such as *zooethics*, coined by the French zoologist and anatomist Henri-Marie Ducrotay de Blainville (1777–1850) in his *Cours de physiologie générale et comparée: professé a la Faculté des Sciences de Paris* (1833) and "qui étudie les actes extérieurs ou évidents que ces animaux, ces assemblages d'organes réagissant les uns sur les autres, et qui sont pour nous en état de mouvement vital, exercent sur le reste de l'univers" (which studies the external

[38] Taine (1863): xxxiii
[39] Taine (1895): 18.
[40] Taine (1863): xxii
[41] Taine (1895): 12.
[42] Taine (1895): 18.
[43] Taine (1895): 18.
[44] Taine (1895): 18.
[45] Taine uses this word to define a short period in which his analysis is carried out. He gives several examples of it, such as "La tragédie française sous Corneille et sous Voltaire, le théâtre grec sous Eschyle et sous Euripide, la poésie latine sous Lucrèce et sous Claudien…". Taine (1895): 16.
[46] Taine (1895): 14.

or obvious acts which these animals, these assemblies of organs, reacting one with another, and which are for us in a state of vital movement, exert on the rest of the universe).[47] Zooethics would be replaced by *ethology*, a term created by French zoologist Isidore Geoffroy Saint-Hilaire (1805–1861). This scholar structured the organization of his *Histoire naturelle générale des règnes organiques: principalement étudiée chez l'homme et les animaux* (1854) into several major parts: the first explains the state of natural history and the principal theories in his time (Schelling, Etienne Geoffroy Saint-Hilaire, and Cuvier), the second recalls the fundamental biological notions (the notion of kingdoms, filiation, heredity, notions of species, etc.), the third deals with organological laws (organic beings considered in themselves), the fourth deals with ethological laws (relating to "aux instincts, aux mœurs, et plus généralement aux manifestations vitales extérieures des êtres organisés" (instincts, morals, and more generally to the vital external manifestations of structured beings),[48] the fifth presents geonemic laws (relating to the successive and present distribution of structured beings on the surface of the terrestrial globe[49]), and the sixth and last deals with natural philosophy which allows the convergence of organic, ethological, and geonemic laws.

Blainville pursued an identical project to Geoffroy Saint-Hilaire: to define the fields of investigation of the various subdivisions of biology. Zooethics and ethology concern the study of behaviour (or the reaction of animals to external influences), which can be considered as one aspect of the study of milieus. This is why the word *ethology* is sometimes used instead of *ecology*, to the point where scholars wondered, at the beginning of the twentieth century, about the correct term to use[50]: biology, ecology, bionomy,[51] ethology, etc.

It is interesting to compare the definitions given to the words *mesology* and *ecology*. *Mesology* was proposed[52] by the French physician, statistician, and anthropologist Louis-Adolphe Bertillon (1821–1883) in the early 1860s: "Nous avons appelé mésologie cette science abstraite des milieux, dont les sciences appliquées corrélatives sont l'hygiène, l'acclimatation, la domestication" (We have called mesology this abstract science of milieus whose correlative applied sciences are hygiene, acclimatization, and domestication).[53] In 1873 Bertillon proposed a long article in the *Dictionnaire encyclopédique des sciences médicales* on this discipline which can be considered as a true research program. It emphasizes the importance of the

---

[47] Blainville (1833): 5 [translation by the editors].

[48] Isidore Geoffroy Saint-Hilaire (1854): XXII [translation by the editors].

[49] Ibid. [translation by the editors].

[50] Wheeler (1902).

[51] Littré defines as an "uncommon synonym of biology". Littré (1873): 348.

[52] Georges Canguilhem (1904–1995) asserts (Canguilhem, 1994: 71–72) that Charles Robin (1821–1885), in a speech given in 1848 at the time of the creation of the Society of Biology, proposed the term mesology to designate "The science that studies the influence of the milieu, or if we want, external agents on the living being" (Robin 1850: iv). But if the printed discourse evokes this project well, there is no reference to this word.

[53] Bertillon (1865): 932 [translation by the editors].

influence of milieus on animals and plants, but its field of study is much broader, since it extends to human beings and to human communities.[54] Bertillon deals with subjects as varied as slavery, "habitudes vicieuses" (vicious habits), and familial and professional milieus.

There is none of this in the definitions proposed by the German biologist and philosopher Ernst Haeckel (1834–1919) for the term of *Oecologie.*[55] Haeckel reproduced this definition in numerous publications, and it was translated into French in 1872[56]:

> Sous le nom d'écologie, nous comprenons l'étude de l'économie, de la vie de ménage pour ainsi dire des animaux. Cette étude a à s'occuper de l'ensemble des rapports de l'animal avec les objets tant inorganiques qu'organiques de son entourage ; avant tout, elle recherche les relations amicales ou hostiles de l'animal vis-à-vis des animaux ou des végétaux avec lesquels il entre en contact direct ou indirect : ou plutôt, en un mot, celle étude comprend les rapports compliqués d'échange, que Darwin désigne comme les conditions de la lutte pour l'existence (By œcologie, we understand the study of the economy, of the domestic life, so to speak, of animals. This study deals with the entirety of relations of the animal with the inorganic and organic objects in its surroundings; above all, it seeks the friendly or hostile relations of the animal with the animals or plants with which it comes into direct or indirect contact: or rather, in a word, this study comprises the complicated relations of exchange, which Darwin designated as the conditions of the struggle for existence).[57]

While Haeckel attaches Darwinian theory to his new discipline (unlike Bertillon), he makes no mention at all of the social dimensions. Although to the present day mesology maintains the ambition of addressing social themes that relate in particular to sociology and social psychology, it is ecology that has become a true and dynamic discipline for which the creation of the revues *Journal of Ecology* in 1913 and *Ecology* in 1920 are extremely significant chronological milestones. The influence and the profusion of the progress made by ecology have led other disciplines to take up the vocabulary of this science: this is the case for certain sociologists who positioned their research under the designation social ecology from the 1920s.[58]

---

[54] Bertillon (1873): 265.

[55] Haeckel (1866): 286–287.

[56] The translation of the word "œcology" appears in the introduction to the English translation of *Generelle morphologie der organismen* (1866) published in 1876.

[57] Haeckel (1872): 582 [English translation by the editors]. Shortly after, an English translation was published (Haeckel 1876: 354): "*The œcology of organisms*, the knowledge of the sum of the *relations of organisms to the surrounding outer world*, to organic and inorganic conditions of existence; the so-called '*economy of nature*', the correlations between all organisms living together in one and the same locality, their adaptation to their surroundings, their modification in the struggle for existence, especially the circumstances of parasitism, etc. It is just these phenomena in 'the economy of nature' which the unscientific, on a superficial consideration, are wont to regard as the wise arrangements of a Creator acting for a definite purpose, but which on a more attentive examination show themselves to be the necessary results of mechanical causes."

[58] See Alihan (1938). This movement should not be confused with the social ecology of Murray Bookchin (1921–2006).

5.6 Upheaval in the Post-War Period

The popularization of the words *environment* and *ecology* in the general culture of Western societies is mainly a reaction to post-war environmental problems, which have the characteristic of mixing the destruction of natural resources and social problems. Two examples help to clarify the meaning of this vocabulary when it permeated society in the 1950s.

Romuald Zaniewski, in the foreword of the *Théories des milieux et la pédagogie mésologique* (1952), justifies his work as follows: "l'accroissement de la population dans les villes et surtout la formation de grands centres industriels créent de nouveaux aspects du problème mésologique concernant le milieu social. [...] Les efforts spontanés et héroïques de certains philanthropes chrétiens n'étaient pas en mesure de combattre la grande marée du malheur. [...] Heureusement, à l'heure actuelle, l'organisation de la vie communautaire est entreprise de façon méthodique et appuyée sur les résultats des recherches approfondies assurées par de nombreuses sciences" (population growth in towns and especially the formation of large industrial centres creates new aspects of the mesologic problem with regard to the social milieu. [...] The spontaneous and heroic efforts of certain Christian philanthropists were not able to fight the great tide of misfortune. [...] Fortunately, at the present time, the organization of community life is undertaken in a methodical manner and is based on the results of in-depth research carried out by many sciences).[59] The sciences he cites are pedagogy and scientific mesology, a branch of biology. His book seeks to make teachers aware of both the influence of the milieu on children and methods to implement in order to confront them.

Despite this book and some other scholarly publications, the word *ecology* was largely unknown to the general public in France at the beginning of the 1950s.[60] The comparison of the words used in the several editions of the French novel *Les racines du Ciel* by Romain Gary (1914–1980) which paid a great attention to the protection of nature illustrates it. In this novel, Morel, the main character, engages in acts of violence (prefiguring ecoterrorism), in order to obtain legal protection for African elephants. The main theme of the book, however, is not the great creature threatened with extinction, as elephants are mostly a symbol of the loss of human dignity following the crimes of the Second World War (Morel was a resistant and survivor of German prison camps). In his last edition, in 1980, Gary uses the words *environment* and *ecology*. Their meaning is quite explicit: "Ça a l'air de toucher les gens, cette histoire, ce qui prouve que la misanthropie, ou si vous préférez l'amour des bêtes, c'est vraiment très répandu. L'environnement, l'écologie, tout ça : empêcher l'homme de nuire, les amis de la vie, ils appellent ça. " (It seems to touch people,

---

[59] Zaniewski (1952): 10–11 [translation by the editors].

[60] One thinks, as well, of the 1941 publication by the French geographer Max Sorre (1880–1962), *Les Fondements biologiques de la géographie humaine, essai d'une écologie de l'homme*. One thinks of the *Précis d'ecologie animale* by Friedrich Simon Bodenheimer (1857–1959) published in 1955.

this story, which proves that misanthropy, or if you prefer the love of animals, is very wide- spread. Environment, ecology, all that: preventing man from doing harm, the friends of life, they call it). Morel himself discovers the word *ecology* through a Swedish naturalist friend: "Il avait toujours eu une passion tenace pour toutes les formes de la vie – il avait appris de Peer Qvist le mot "écologie", qu'il ne connaissait pas – dont les ennemis l'avaient toujours rencontré en travers de leur chemin" (He had always had a relentless passion for all forms of life - he had learned the word "ecology", which he did not know, from Peer Qvist - whose enemies had always found him in their path). And farther on: "Moi, c'est bien simple, tout ce que je défends, c'est la nature […] Appelez cela comme vous voulez. Liberté, dignité, humanité, écologie […] Cela revient au même. Je fais tout ça pour les amis de l'homme. On nous l'a appris à l'école, ce que ça veut dire. Le reste, je m'en contrefous" (Me, it's very simple, all that I defend is nature […] Call it what you want. Freedom, dignity, humanity, ecology […] It comes down to the same thing. I do all this for the friends of man. We learned it at school, what it means. The rest, I do not fight). The success of this book—he won the Prix Goncourt in 1956, and it was adapted for the cinema by John Huston in 1958—clearly indicates the sensitivity of a broad public to the protection of nature. But the words "ecology" and "environment" were not usual in the French language in the 1950s and were not used in the 1956 edition of the novel. Romain Gary certainly used these words later relying on what he knew from English-speaking countries.[61] It was finally the creation on January 8, 1971, of the first Ministry of the Environment that allowed this word to be finally adopted by the French.[62] The creation of this ministry was also a response to foreign influence resulting from the building of Europe. Despite the fact that Gary's intellectual training was largely fuelled by foreign culture (he was a polyglot and was a diplomat in several countries), he did not use these words in his first edition and added them in the 1980's edition. The world has changed meanwhile.

The words *milieu*, *habitat*, *station*, and *environment*, adapted and transformed at the beginning of the nineteenth century, correspond to a specific intellectual project corresponding to the descriptive study of milieus and their influences. This project had no moral or spiritual dimension. During the post-war period, the use of *environment* and *environnement* underwent a profound transformation, losing the precision of the scientific approach and acquiring a new and specific dimension in the modern era by designating a world profoundly disrupted by the actions of man (a similar shift of usage can be observed with the term ecology). For example, the Ministry of the Environment, created in 1971, deals only with environmental problems such as noise pollution, loss of biodiversity, waste management, etc. In new uses of this old vocabulary, however, we find a perception of the world close to that of the authors of the seventeenth and eighteenth centuries: the world is perfect, everything is in its

[61] Chansigaud (2017).

[62] Charvolin (2003).

proper place,[63] and human beings cannot free themselves from the ties which unite them to their environment at the risk of provoking chaos, a chaos envisaged by Alexander Pope in 1734 in his *Essay on Man*.

References

Alihan, Milla Aïssa. 1938. *Social Ecology: A Critical Analysis*. New York: Columbia University Press.

Benyus, Janine M. 2011. *Biomimétisme : quand la nature inspire des innovations durables*. Paris: Rue de l'échiquier [translation by Céline Sefraoui].

Bertillon, Louis-Adolphe. 1865. Mésologie. In *Dictionnaire de médecine, de chirurgie, de pharmacie, des sciences accessoires et de l'art vétérinaire : d'après le plan suivi par Nysten,* ed. Émile Littré and Charles-Philippe Robin, 932. Paris: J.-B. Baillière et Fils.

———. 1873. Mésologie. In *Dictionnaire encyclopédique des sciences médicales. 2e série. Tome 7, Mer-Mil,* ed. Amédée Dechambre, 211–266. Paris: G. Masson.

Bodenheimer, Friedrich Simon. 1955. *Précis d'écologie animale*. Paris: Payot.

Bory de Saint-Vincent, Jean Baptiste Geneviève Marcellin.1825. *Dictionnaire classique d'histoire naturelle*. Tome huitième. H-Inv. Paris: Rey et Gravier.

Bradley, Richard. 1721. *A Philosophical Account of the Works of Nature: Endeavouring to Set Forth the Several Gradations*. London: Printed for W. Mears.

Canguilhem, Georges. 1994. *Études d'histoire et de philosophie des sciences*. Paris: Vrin.

Carlyle, Thomas. 1904. *The Works of Thomas Carlyle in Thirty Volumes. Vol. VI. Critical and Miscellaneous Essays. Vol. I*. New York: Charles Scribner's Sons.

Chansigaud, Valérie. 2017. *Les Français et la nature : pourquoi si peu d'amour ?* Arles: Actes Sud.

Charvolin, Florian. 2003. *L'Invention de l'environnement en France. Chroniques anthropologiques d'une institution*. Paris: La Découverte.

Comte, Auguste. 1838. *Cours de philosophie positive. Tome troisième, contenant la philosophie chimique et la philosophie biologique*. Paris: Bachelier, imprimeur-libraire.

Cooper, Alix. 2003. The Indigenous Versus the Exotic: Debating Natural Origins in Early Modern Europe. *Landscape Research* 28 (1): 51–60.

———. 2007. *Inventing the Indigenous: Local Knowledge and Natural History in Early Modern Europe*. Cambridge: Cambridge University Press.

de Blainville, Henri-Marie Ducrotay. 1833. *Cours de physiologie générale et comparée : professé a la Faculté des Sciences de Paris*. Tome premier. Paris: Germer Baillière.

de Candolle, Augustin-Pyramus. 1813. *Théorie élémentaire de la botanique ; ou, Exposition des principes de la classification naturelle et de l'art de décrire et d'étudier les végétaux*. Paris: Déterville.

Deterville, Jean François Pierre. 1817. *Nouveau dictionnaire d'histoire naturelle appliquée aux arts, à l'agriculture, à l'économie rurale et domestique, à la médecine. Tome XIV*. Paris: Déterville.

Dunlap, Thomas R. 1999. *Nature and the English Diaspora: Environment and History in the United States, Canada, Australia, and New Zealand*. Cambridge: Cambridge University Press.

Gary, Romain. 1956. *Les Racines du Ciel*. Paris: Gallimard.

[63] The clearest example of this way of thinking is that of biomimicry (Benyus 2011), which affirms that nature has already found the right solutions to the technical problems encountered by humanity and that it suffices to seek inspiration from it, these solutions being presented as environmentally friendly.

Geoffroy Saint-Hilaire, Étienne. 1830. *Principes de philosophie zoologique, discutés en mars 1830 au sein de l'Académie royale des sciences*. Paris: Pichon et Didier, libraires.
Geoffroy Saint-Hilaire, Isidore. 1849. *Rapport général sur les questions relatives à la domestication et à la naturalisation des animaux utiles, adressé à M. le Ministre de l'agriculture et du commerc*. Paris: Imprimerie nationale.
———. 1854. *Histoire naturelle générale des règnes organiques : principalement étudiée chez l'homme et les animaux. Tome premier*. Paris: Librairie de Victor Masson.
Gilibert, Jean Emmamuel. 1805. *Abrégé du Système de la nature de Linné, histoire des mammaires, ou des quadrupèdes et cétacées. Contenant, 1. la traduction libre du texte de Linné et de Gmelin; 2. l'extrait des observations de Buffon, Brisson, Pallas, et autres célèbres zoologistes; 3. l'anatomie comparée des principales espèces: le tout relatif aux quadrupèdes et aux cétacées les plus curieux et les plus utiles*. Lyon: Chez l'éditeur.
Guérin-Méneville, Félix-Édouard. 1835. *Dictionnaire pittoresque d'histoire naturelle et des phénomènes de la nature. Tome troisième. Edredon-Hole*. Paris: Au Bureau de souscription.
Haeckel, Ernst. 1866. *Generelle morphologie der organismen. Allgemeine grundzüge der organischen formen-wissenschaft, mechanisch begründet durch die von Charles Darwin reformirte descendenztheorie* [volume 1], Berlin: George Reimer.
———1872. Des progrès et de l'objet de la zoologie. *Revue scientifique*, 2e série 2 (25): 577–584.
———. 1880. *The History of Creation, Or, the Development of the Earth and Its Inhabitants by the Action of Natural Causes: A Popular Exposition of the Doctrine of Evolution in General, and of That of Darwin, Goethe and Lamarck in Particular*. Vol. 2. New York: D. Appleton and Company.
Jessop, Ralph. 2012. Coinage of the Term Environment: A Word Without Authority and Carlyle's Displacement of the Mechanical Metaphor. *Literature Compass* 9 (11): 708–720.
Lamarck, Jean-Baptiste de Monet de. 1873. *Philosophie zoologique; ou, Exposition des considérations relatives à l'histoire naturelle des animaux. Tome premier* [Nouvelle édition, revue et précédée d'une introduction biographique par Charles Martins]. Paris: F. Savy.
———. 1914. *Zoological Philosophy: An Exposition with Regard to the Natural History of Animals*. London: Macmillan and Co., Limited [translation by Hugh Samuel Roger Elliot].
Le Guyader, Hervé. 2004. *Geoffroy Saint-Hilaire: A Visionary Naturalist*. Chicago: University of Chicago Press [Translated by Marjorie Glicksman Grene].
Littré, Émile. 1873. *Dictionnaire de la langue française. Tome premier : A-C*. Paris: Librairie Hachette et Cie.
Lyell, Charles. 1830. *Principles of Geology: Being an Attempt to Explain the Former Changes of the Earth's Surface, by Reference to Causes Now in Operation. Vol. The Second*. London: John Murray.
Maurois, André. 1932. *Mes songes que voici*. Paris: Grasset.
Moreau, J.H. 1798. Histoire naturelle. *Magasin encyclopédique, ou journal des sciences, des lettres et des arts* 4 (4): 7–23.
Osborne, Michael Andrew. 1994. *Nature, the Exotic, and the Science of French Colonialism*. Bloomington: Indiana University Press.
Oxford English Dictionary. 2009. *Oxford English Dictionary: Second Ed. on CD-ROM Version 4.0*. Oxford: Oxford University Press.
Pearce, Trevor. 2010. From 'Circumstances' to 'Environment': Herbert Spencer and the Origins of the Idea of Organism–Environment Interaction. *Studies in History and Philosophy of Biological and Biomedical Sciences* 41: 241–252.
Rey, Alain. 1994. *Dictionnaire historique de la langue française. 2 volumes*. Paris: Le Robert.
Robin, Charles. 1850. Sur la direction que se sont proposée en se réunissant les membres fondateurs de la société de biologie pour répondre au titre qu'ils ont choisi. *Comptes rendus des séances et mémoires de la société de biologie* 1–1849 : i–xi.
Sorre, Max. 1941. *Les Fondements biologiques de la géographie humaine, essai d'une écologie de l'homme*. Paris: Armand Colin.

Spencer, Herbert. 1855. *The Principles of Psychology*. London: Longman, Brown, Green, and Longmans.
Taine, Hippolyte. 1863. *Histoire de la littérature anglaise. Tome premier*. Paris: Librairie de L. Hachette et Cie.
——— 1895. *History of English Literature*. London: William L. Allison Company [translation by Henri van Laun].
van Beverwyck, Jan. 1730. *Éloge de la medecine et de la chirurgie : deffense de la medecine contre les calomnies de Montagne, en forme de dialogue, suivi des Lettre de la très fameuse Demoiselle Anne-Marie Schurmans, académicienne de la fameuse université d'Utrect.* Paris: Chez la V. Rebuffé, imprimeur-libraire.
von Herder, Johann Gottfried. 1786. *Ideen zur Philosophie der Geschichte der Menschheit.* ben Johann Friedrich Hartknoch. http://www.zeno.org/Literatur/M/Herder,+Johann+Gottfried/Theoretische+Schriften/Ideen+zur+Philosophie+der+Geschichte+der+Menschheit.
———. 1800. *Outlines of a Philosophy of the History of Man*. London: J. Johnson [translation by T. Churchill].
von Linné, Carl. 1793. *Système de la nature.. Classe 1re du règne animal, contenant les quadrupèdes, vivipares et cétacés. Traduction françoise.. d'après la 13e édition latine.. augmentée et corrigée par J. F. Gmelin.* Bruxelles: Chez Lemaire, imprimeur-libraire [translation by Joseph Van der Stegen de Putte].
Wheeler, William Morton. 1902. 'Natural History', 'Œcology' or 'Ethology'? *Science* 15 (390): 971–976.
Withering, William. 1796. *An Arrangement of British Plants, According to the Latest Improvements of the Linnaean System: Illustrated by Copper Plates ; to Which Is Prefixed an Easy Introduction to the Study of Botany. Vol. I,* [Third Edition]. Birmingham: Printed for the Author.
Zaniewski, Romuald. 1952. *Les Théories des milieux et la pédagogie mésologique*. Tournai: Éditions Casterman.

Chapter 6
Environment in Relation to Health, Wellbeing and Human Flourishing: The Contribution of Twentieth-Century Continental Philosophy of Life and of the Subject

Marie Gaille

Abstract Health, wellbeing and human flourishing are related in various ways to the emplaced dimension of human existence. In this paper, we would like to highlight how a network of thinkers who are part of twentieth-century continental philosophy gave specific content to this general and broad idea of a relationship between human health, wellbeing and human flourishing and environment, namely, K. Goldstein, G. Canguilhem and M. Merleau-Ponty. K. Goldstein aimed to determine an appropriate medical approach to brain injuries. Within this clinical context, he conceived of healthy and pathological states as essentially related to the patient's environment understood as "others and the world". G. Canguilhem focused on the human capacity to tailor their living environment and conceived of it as fundamental to their health. G. Canguilhem's contribution to the conception of the relationship between health and environment may indeed be considered as crucial. In its turn, M. Merleau-Ponty's conception of the "lived space" offers key tools for considering this issue, especially when one wants to take into account the embodied dimension of human life and the meanings of the "lived world". Taking K. Goldstein's inquiry as a starting point, one may consider the richness of the continental dialogue between him and his French readers, G. Canguilhem and M. Merleau-Ponty in conceiving the relationship between health, wellbeing and human flourishing: their three conceptions are original in regard to each other. However, they are deeply connected and propose complementary perspectives for thinking about this relationship.

M. Gaille (✉)
Université de Paris, SPHERE, UMR 7219, CNRS-Université de Paris, Paris, France
e-mail: marie.gaille@cnrs-dir.fr

F. Bretelle-Establet et al. (eds.), *Making Sense of Health, Disease, and the Environment in Cross-Cultural History: The Arabic-Islamic World, China, Europe, and North America*, Boston Studies in the Philosophy and History of Science 333, https://doi.org/10.1007/978-3-030-19082-8_6

Keywords Environment · Milieu · Lived space · Meaning · Medicine · Healthy and pathological states · Normalcy · Wellbeing · Human flourishing · Philosophy · Neuropsychiatry

6.1 Introduction

Health, wellbeing and human flourishing are related in various ways to the emplaced dimension of human existence. In this paper, we would like to highlight the contribution of a network of thinkers who are part of twentieth-century continental philosophy in giving specific content to this general and broad idea of a relationship between human health, wellbeing and human flourishing and environment, namely, K. Goldstein (1878–1965), G. Canguilhem (1904–1995) and M. Merleau-Ponty (1908–1961). Even though K. Goldstein was not a philosopher in the professional sense of the word – he practised as a neurologist and psychiatrist, he is part of this network, and his thinking is key for its constitution for two reasons: first of all, he himself drew on philosophical thinking in order to elaborate his "holistic" conception of medicine and second, his works were actively integrated into philosophical reflection by G. Canguilhem and M. Merleau-Ponty. The reception in France of K. Goldstein's major work, *The Organism: A Holistic Approach to Biology Derived from Pathological Data in Man* (1934), was extremely significant as early as the 1930s. This work is still well-known in French philosophy, thanks to its initial reception, in which G. Canguilhem played a prominent role, along with M. Merleau-Ponty. The translation of *The Organism: A Holistic Approach to Biology Derived from Pathological Data in Man* into French was published in 1951. G. Canguilhem mentioned M. Merleau-Ponty several times in his writing, usually along with K. Goldstein. He stressed M. Merleau-Ponty's role in the spreading of K. Goldstein's ideas in France.[1]

As a starting point, K. Goldstein had the aim of determining an appropriate medical approach to brain injuries. Within this clinical context, he elaborated a conception of pathological states as situations in which the patient cannot relate to others and the world in a confident and successful way because she fails in accomplishing tasks and acts she used to do easily and with ease before the injury occurred. In this line of thinking, healthy and pathological states cannot be conceived independently from the relationship to the patient's environment, understood as "others and the world" (I). Independently from any clinical context, G. Canguilhem confronted the dominant medical views of his time, according to which normal states were the standard from which to define pathological states. He promoted a conception of normalcy as relative to individual norms and individual capacity to accommodate to one's environment. He described this capacity as that of tailoring one's milieu – an issue about which he stood very close to K. Goldstein's conception (II). In his turn, M. Merleau-Ponty's elaborated a conception of the "lived space",

[1] Limoges, 2015, p. 133

with K. Goldstein's work as its background. It focuses on the lived and meaningful experience of the subject and the role played by the body in the relationship to one's environment. It thus offers innovating insights to the philosophy of health and disease that are complementary to G. Canguilhem's perspective (III).

6.2 Healthy and Pathological States as Modes of Relationship to the World and to Others in Kurt Goldstein's Conception

In this first part, we present K. Goldstein's conception of healthy and pathological states. We do not give a complete account, but focus on the central place that he granted to the relationship between human beings and their environment in order to define these healthy and pathological states.

K. Goldstein (1878–1965) himself explained how his professional practice both as a researcher and a physician led him to develop this conception of healthy and pathological states. He began studying medicine at university with a special concern for people with mental diseases, who seemed to him particularly in need of help. He did so at a time when such diseases "were considered the expression of abnormal brain conditions": "This study of the nervous system was taken for granted, and I became attracted by professors who were occupied with studies in this field".[2] After his studies, he started working in a university hospital dedicated to psychiatric pathologies. As a neurologist and psychiatrist, K. Goldstein took care of soldiers who had suffered brain injuries during the First World War.

At this time, he became critical of the dominant views on psychiatric patients and of the care provided to patients with brain injuries:

> While concentrating on investigation of organic neurological and psychiatric cases, I became aware that the usual procedures, following the method of natural sciences, studying carefully the outstanding symptoms and trying to base therapy on these results, revealed many interesting phenomena but was very unsatisfactory for purposes of therapy (...) The problem became particularly urgent during and after the First World War, when I was confronted with the task of treating a great number of young soldiers with brain injuries and defects in different mental capacities, particularly language. After surgical treatment, they were generally considered objects of charity and care, because it seemed that a real improvement could never be expected. Only a few neurologists, myself among them, protested that by adequate treatment these patients might be brought to a condition where life would again be worth living, despite some remaining defects.[3]

This experience thus led him to develop a reflection on the method through which the human organism is studied and understood. He affirmed that pathological phenomena played a key role in this study. His major work, *The Organism: A Holistic Approach to Biology Derived from Pathological Data in Man* (1934),

[2] Ibid., p. 5

[3] Ibid., p. 2

assumes that pathological states often tell us more about the organism than states of (good) health; hence, to understand the general conditions for (good) health, one is to take them as a starting point (and not the reverse). This thesis was contrary to the dominant medical thinking of his time, which made healthy states the norm from which pathological states could be defined:

> Our treatise differs from the usual one in that it proceeds from pathological rather than from normal phenomena. This approach requires some justification. The observation and analysis of pathological phenomena no doubt often yield greater insight into the processes of the organism than do those of the normal.[4]

The clinical examination of brain injuries also led K. Goldstein to stress the fact that their main effect was to make it difficult or impossible to carry out certain mental or bodily acts in a temporary or permanent way. As a result, the wounded persons' whole relationship to the world and to others was modified and became precarious. Their personalities could (totally) change.[5] K. Goldstein described this phenomenon as that of "the disintegration of function".[6] His intention was to describe it in terms of general laws and he stressed the anxiety generated by repeated failures to achieve such and such operation for the patient.

From this observation, he defined two main types of relationship to the world and to others. In his view, one may have "ordered"/"normal" or "disordered"/"catastrophic" behaviour:

> In an ordered situation, responses appear to be constant, correct, adequate to the organism to which they belong, and adequate to the species and to the individuality of the organism, as well as to the respective circumstances. The individual himself experiences them with a feeling of smooth functioning, unconstraint, wellbeing, adjustment to the world, and satisfaction (.) 'Catastrophic' reactions, on the other hand, are not only 'inadequate' but also disordered, inconstant, inconsistent, and embedded in physical and mental shock. In these situations, the individual feels himself unfree, buffeted, and vacillating. He experiences a shock affecting not only his own person, but the surrounding world as well.[7]

To K. Goldstein, both behaviours expressed a general state of living and were not related to specific objects or issues. In particular, catastrophic behaviour gave rise to a radical fear of danger.[8] This meant "the subjective experience of being in danger of losing 'existence'".[9] It thus affected each step in one's life and pervaded the whole atmosphere in which one lived. As K. Goldstein recalled in *The Organism: A Holistic Approach to Biology Derived from Pathological Data in Man*, he had searched for the proper word to express the difference between two experiences: that of being afraid of something and that of being generally anxious. He stressed the fact that patients felt an anxiety "without content". He saw it as the expression of a failure to be in a positive and ordered relationship with the world and others. In

---

[4] Goldstein, 2000, p. 29

[5] Goldstein, 1940, p. 38

[6] Goldstein, 2000, p. 33

[7] Ibid., p. 48–49

[8] Ibid., Chapter 7 and Goldstein, 1940, p. 92

[9] Goldstein, 1971, p. 6 for the quotation

order to qualify this feeling properly, he chose to use the term of anxiety ("angst" in German) so as to describe a feeling distinct from the fear of something (rather expressed by the word "furcht" in German).[10]

K. Goldstein did not consider that one person's relationship to the world and to others was a given. He rather saw it as partly established by the person and those who surrounded them, up to the society at large. To him, each one of us strives in order to design this relationship in which our goals may be achieved, much as a *tailor* that cuts a suit to measurements.[11] To come back to patients with brain injuries, he considered that their acceptation of the "restrictions"[12] (or some of them) generated by the brain injuries could be helpful. A person could feel "healthy" even if they could not do/carry out everything they did before, after redefining their sense of a life worth living.[13] In the case of his patients, K. Goldstein considered that it was the role of the physician to help them recreate a living environment in which they could relate to the world and to others in an ordered way.

This analysis had a decisive implication on K. Goldstein's conception of healthy and pathological states. The brain injuries could not be simply described as a brain disease. In fact, a brain injury could occur without engendering an actual pathological state as it could involve no alteration in the capacity of the person to achieve their specific goals in relation to the world and to others. In K. Goldstein's mind, it was legitimate to speak of a pathological state only under the following conditions: when the relationship between the organism and the world and others changed in such a way that the person entered a state of anxiety and remained overwhelmed by a feeling of constant danger.[14] K. Goldstein also assumed that this conception was not only relevant for patients with brain injuries, but in a more general way, for every patient that could not be cured totally.[15]

This conception led K. Goldstein to express a renewed view of medical practice. According to this view, medicine needed to embrace a "holistic approach" to human nature as disease endangered the whole relation of the person to the world and to others. This conception of medical practice was opposite to the way medical knowledge had improved, as K. Goldstein himself emphasized, that is, on the basis of an "atomistic"[16] methodology. However, both aspects were not incompatible with his view. Medical research could still "isolate" brain injuries to understand their nature and elaborate techniques or medication. However, with patients, a physician had to be aware that they had to deal with an overwhelming feeling of anxiety and help the patient to proceed from "catastrophic" to "ordered" behaviours.

---

[10] Goldstein, 2000, p. 230

[11] Ibid., p. 76

[12] Goldstein, 1971, p. 7

[13] Goldstein: "Thus, being well means to be capable of ordered behaviour that may prevail in spite of the impossibility of certain performances that were formerly possible. But the new state of health is not the same as the old one", 2000, p. 332.

[14] Ibid., p. 248

[15] Goldstein, 1971, p. 5

[16] Ibid., p. 3

K. Goldstein did not only consider that this conception led to a renewed view of medical practice. In *Human Nature in the Light of Psychopathology*, a later text with an explicit anthropological ambition, he assumed it could contribute to "a conception of the nature of man" in a decisive way.[17] This conception was to offer a "holistic approach"[18] to human life. K. Goldstein had a profound conviction in the capacity of medicine – understood both as a study of human organism and an art of curing the sick – to offer a fundamental insight into human nature.[19]

Finally, let us observe that this conception of medicine had political and social implications. In the 8th chapter of *The Organism: A Holistic Approach to Biology Derived from Pathological Data in Man*, he addressed the issue of "normalcy". Here, K. Goldstein emphasized that "normal states" should not be defined statistically or considered as "ideal" states. On the ground of his renewed conception of healthy and pathological states, he affirmed that each one of us determined our own state of normalcy. He also underlined the possibility of active, fulfilling and happy lives for people affected with anomalies or malformations. As a result, according to him, the vision of "different" people as "abnormal" was nothing but a dangerous social fantasy. In the context of German politics in 1934, he clearly stood against any justification of eugenics (and racist theories).

6.3 The "Encounter" Between G. Canguilhem's Philosophy of Life and Medicine and K. Goldstein's Holistic Approach to Human Nature

Let us now turn to G. Canguilhem's conception of normal and pathological states. His theory is both original – its specificity being anchored in a philosophy of life – *and* convergent with K. Goldstein's views. We will highlight here one key aspect of this convergence: both authors focus on the capacity of human beings to "tailor" (up to certain limits) their own specific and individual liveable space within the world – a space G. Canguilhem opted to describe with the world "milieu".

G. Canguilhem studied philosophy in the 1920s. Qualifying with a degree to teach this discipline (1927), he taught in high schools before deciding to study medicine in 1936. In his introduction to *The Normal and the Pathological*, G. Canguilhem reveals that his involvement into a medical curriculum was linked to his expectation that medicine would "provide" him with "an introduction to concrete human problems".[20] During the war, he actively participated in the *Resistance*. He also taught. In 1940, he had asked for leave from his teaching position, as he disagreed with the Vichy Regime. A turn in his career took place in 1941 when he

[17] Goldstein, 1940, preface, p. viii

[18] Ibid., p. 5

[19] Goldstein, 1940, p. 25 et p. 33

[20] Canguilhem, 1989, p. 134. On this matter, see Bing, Braunstein and Roudinesco, 1998, p. 121.

accepted a position in the University of Strasbourg that had settled in Clermont-Ferrand (because of the location of this city in the *zone libre*). Finally, he could concentrate fully on his medical studies and prepared his Ph.D. in medicine: *Essais sur quelques problèmes concernant le normal et le pathologique*.

This text, completed in 1943, is part of the essay we know today as *Le normal et le pathologique*. It was followed by another text 20 years later, *Nouvelles réflexions concernant le normal et le pathologique*, and published in 1966. In the reflection developed in this two-fold essay, G. Canguilhem focused on the notions of "normal" and "pathological". Like K. Goldstein, he faced a conception of medicine based on the primacy of the knowledge of normal states over that of the pathological states: in the 1930s, "it was obvious that one should learn what is normal before learning what is pathological. The pathological could only be understood from the normal. This was the doctrine".[21] G. Canguilhem rejected both this assumption and the conception of recovery as a "way back" to the normal state. In his view, it was simply wrong to think of the pathology as a disturbance of the normal and to define the cure as a way of returning the organism to how it was before this disturbance.

This criticism originated in his conviction that healthy and pathological states should be interpreted through the lens of a "philosophy of life". This perspective did not lead to neglecting the biological, chemical and physiological analysis of healthy and pathological states. Rather, it played the role of a frame of thought that highlighted an essential feature of life: that is, from the cell to the human being, the living being's own dynamics is to define its own norm of living. To G. Canguilhem, from the simplest single cell organism to the most complex multicellular being, every living being's dynamic is "to appreciate", to affirm some "values", to make some choices and to assess the environment according to one's needs and preferences. These needs and preferences guide the living being to opt for what fits it: "we can say of the universe of every living thing what Reininger says of the universe of man: 'Unser Weltbild ist immer zugleich ein Wertbild', our image of the world is always a display of values as well".[22] In the conclusion of the first part of *The Normal and the Pathological*, G. Canguilhem highlighted the spontaneous effort exerted by living beings in order "to dominate the environment and organize it according to values as a living being".[23]

On the basis of his philosophy of life, G. Canguilhem proposed a conception of healthy and pathological states that broke with the dominant conception in his era. In his view, a pathological state was not a disease in the common meaning of the term, but a weakened capacity to formulate or redefine some values that would make life not only possible but also easy and comfortable in one's environment. As a consequence, a "sick" person experienced a "direct and concrete feeling of suffering and powerlessness, the feeling of life gone wrong",[24] while "health"

[21] Péquignot, in: Bing, Braunstein and Roudinesco, 1998, p. 59–60 (my translation)

[22] Canguilhem, 1989, p. 179. In this instance, he referred to Reininger, *Wertphilosophie und Ethik*, p. 29, 1939, Vienna-Leipzig, Braumüller.

[23] Ibid., p. 229

[24] Ibid., p. 137

appeared as "an ensemble of securities and assurances (what the Germans call *Sicherungen*), securities in the present, assurance for the future".[25]

In *The Normal and the Pathological*, let us note that G. Canguilhem used several words to describe the creative and organizing activity of the living being: "to adapt" or "to accommodate" to one's environment and also "to modify" it in order to make it favourable to one's life. In his later paper "The Living and the Milieu" (1946), he also used the verbs 'to make" or "to compose <one's environment>",[26] and he pictured human beings as "creators" of their own environment: "Man, as a historical being, becomes the creator of a geographical configuration; he becomes a geographical factor".[27] With varying degrees, all these verbs convey the idea of an environment crafted by human beings, that of a "space" in which they arranged a liveable space for themselves.

This perspective is the key point of G. Canguilhem and K. Goldstein's convergence. After discovering K. Goldstein's work in a psychology course offered by Daniel Lagache in the University of Clermont-Ferrand, G. Canguilhem read an article written by one of his disciples, Aron Gurwitsch: "La science biologique d'après Goldstein" (1940).[28] His first mention of K. Goldstein is to be found in the *Cours de philosophie générale et de logique* (1942–1943).[29] He introduced this reference in the section dedicated to "normalcy" and "normativity". In this section, he questioned the meaning of the "variations" displayed by various forms of life in respect of what is biologically considered as normal/average. He developed the idea according to which there was no absolute normalcy: in the field of life and of medicine, "normalcy" was actually to be determined within the relationship between a living being and its environment; it had no meaning outside of this relationship; a variation could be considered as "normal" as far as it allowed life to develop, evolve and expand comfortably; a pathological state appeared when this dynamic became difficult or even impossible. In developing this argument, G. Canguilhem referred to K. Goldstein to emphasize the individual dimension of this dynamic.[30] Normalcy could not be generalized, but only systematically related to specific ways of dealing with one's environment:

> As far as an individual's biological dimension is concerned, the issue of the normal and the pathological lies within ordered and catastrophic behaviour - using Goldstein's words. An organism does not respond to the requests of the milieu on the basis of what it is able to

[25] Ibid., p. 198

[26] Canguilhem, 2008, p. 111

[27] Ibid., p. 109

[28] We follow C. Limoges' chronology of the discovery and the subsequent reading, in French or German, of K. Goldstein by G. Canguilhem. He established it from the documents available in the CAPHES Fonds Canguilhem and gave a detailed account of it in Limoges, 2015, p. 106, note 1.

[29] Canguilhem, 2015, p. 81–109

[30] In a text by K. Goldstein translated into French by G. Canguilhem and his wife, Simone, in 1949, one sentence perfectly summarized this view: "La Biologie a affaire à des individus qui existent et tendent à exister, c'est-à-dire à réaliser leurs capacités du mieux possible dans un environnement donné", in "Kurt Goldstein, Remarques sur le problème épistémologique de la biologie", in Canguilhem, 2015, p. 1162.

> accomplish. In order to bring this response some behaviours are preferred to others because, thanks to them, the organism better achieves its very nature, and experiences order and stability at their highest level. (…) A catastrophic reaction highlights the impossibility for a sick person to adapt quickly to new conditions in the milieu.[31]

In this present analysis, G. Canguilhem spoke of an "encounter" between his own conception and that of Goldstein. In his mind, both conceptions led to a representation of healthy people as beings able to live and adapt to the various demands of their environment.

A reference to K. Goldstein also appeared when G. Canguilhem introduced the fundamental distinction between the general idea of environment and that of "milieu". Following K. Goldstein's view, G. Canguilhem considered that living beings created their "milieu" within the world according to their own norms of living. As a result, in the broad sense, a living being's "milieu" does not equate to its environment. This distinction was also clearly expressed in a comment to Bergson's work, *L'évolution créatrice* (1943) and presented in relation to K. Goldstein:

> Here, one may underline Goldstein's ideas: 'The milieu (*Umwelt*) is tailored in the world according to the very nature of the organism – provided that an organism only exists as far as it finds an appropriate milieu for itself in the world.[32]

K. Goldstein was again mentioned in a later text, "The Living and Its Milieu" (1946), by G. Canguilhem. In this text, G. Canguilhem made K. Goldstein's images and words his own to describe the relationship between human beings and their environment in terms of healthy or pathological states. According to K. Goldstein, as we mentioned above, in an ordered situation, "the individual himself experiences them with a feeling of smooth functioning, unconstraint, well-being, adjustment to the world, and satisfaction", while in catastrophic reactions, "the individual feels himself unfree, buffeted, and vacillating. He experiences a shock affecting not only his own person, but the surrounding world as well".[33] G. Canguilhem's depiction of the normative dynamics of living beings, their successes and failures, is very close to this description:

> The relationship between the living being and the milieu establishes itself as a debate (*Auseinandersetzung*), to which the living being brings its own proper norms of appreciating situations, both dominating the milieu and accommodating itself to it. This relationship does not essentially consist (as one might think) in a struggle, in an opposition. That applies to the pathological state. A life that affirms itself against the milieu is a life already threatened. (…) A healthy life, a life confident in its existence, in its values, is a life of flexion, suppleness, almost softness. The situation of a living being commanded from the outside by the milieu is what Goldstein considers the archetype of a catastrophic situation.[34]

In "The Living and Its Milieu", G. Canguilhem also emphasizes the role of both J. Von Uexküll and that of K. Goldstein in the historical "reversal" in the under-

---

[31] Ibid., p. 107 (my transl.)

[32] Canguilhem, 2015, p. 133 (my transl.)

[33] Goldstein, 2000, p. 48–49

[34] Canguilhem, 2008, p. 113. The word "*Auseinandersetzung*" is borrowed from K. Goldstein, 1934.

standing of this relationship between living beings and their environment.[35] As far as J. Von Uexküll was concerned, his biology was first grounded on the use of the word "milieu". Later, he promoted a sharp distinction between what surrounds the living being, labelled as the "milieu" (in German, "Umgebung"), and the living environment created by each animal within the world, called "umwelt" (and less frequently "merkwelt"). J. Von Uexküll gave up the word "milieu" and never came back on this decision.[36] When G. Canguilhem developed his philosophy of life using the word "milieu", he actually associated it with a meaning similar to Von Uexküll's "umwelt": "The *Umwelt* is thus an elective extraction from the *Umgebung*, the geographical environment".[37] As G. Canguilhem stressed in a text dedicated to human work, this same pair composed of J. Von Uexküll and K. Goldstein opened a new door to the understanding of life: "With Von Uexküll and Goldstein, biologists finally understand that the very essence of living beings is to make one's own milieu".[38]

From this analysis of the relationship between G. Canguilhem and K. Goldstein, it appears that G. Canguilhem found in K. Goldstein a *retrospective* confirmation of his philosophy of life. This fact does not minimize the importance of their "encounter". It was a late but decisive reading: from then on, as G. Canguilhem stressed in a speech given in 1987, he had no doubt about his rejection of the common medical view on the primacy of the knowledge of normal states over the knowledge of pathological ones. As far as his Ph.D. was concerned, G. Canguilhem finally mentioned K. Goldstein in the last chapter and introduced references to him in earlier chapters. As C. Limoges indicates, these references were probably of significant help to G. Canguilhem in 1942–1943: "One must remember that Goldstein was already widely recognized. As a result, his works were a very significant support to heterodox views and gave them a certain legitimacy".[39]

We may consider this "encounter" as a crucial element of the conception of the relationship between "normal" and "pathological" states and the environment of living beings. Such a conception highlights the normative dynamics of living beings

[35] J. Von Uexküll (1864–1944) published *Umwelt und Innemwelt der Thiere* in 1909. In this book, he presented his influential conception of the "umwelt" as the living environment actively created by each organism in order to live. He depicted organisms as essentially characterized by their capacity of organization.

[36] Feuerhahn, 2009 and 2017

[37] Canguilhem, 2008, p. 112

[38] Canguilhem, 2015, p. 301. G. Canguilhem stressed this dimension within an intellectual context that dedicated significant attention to the relationship between human beings and their environment in various aspects of their lives. In the late nineteenth century and the first half of the twentieth century, several sciences contributed to this reflexion (geography, sociology, history, biology). They "exchanged" notions and conceptions. This distinction thus had implications that went far beyond the realm of medicine and it is no surprise to see it mentioned about human work in his reflection. For a detailed examination of this context, from the point of view of geography and/or sociology, see Robic, 1990; Muchielli and Robic, 1995; and Imbert, 2016.

[39] Limoges, 2015, p. 37 (my translation)

and emphasizes their capacity to create a “milieu”, that is, an individually tailored environment.

As one may observe, G. Canguilhem’s conception was a significant attempt to identify various meanings of the notion of “environment” and to examine their relationship. As a result, he proposed an original and integrative conception of “milieu”. In addition to this, it is to be noted that his attempt unfolded into two main instances, each one corresponding to the two parts of *On the Normal and the Pathological*: first, the *Essais sur quelques problèmes concernant le normal et le pathologique* (1943) and second, the *Nouvelles réflexions concernant le normal et le pathologique* (1966). In the first instance, the analysis of the normative dynamics of living beings was mostly intended to formulate a conception of life and of its healthy or pathological states as opposed to their understanding in terms of general laws. As we saw, G. Canguilhem considered this understanding as simplistic and was not convinced with a description of life only grounded on chemistry and physics.

Now, in the second instance, G. Canguilhem placed more emphasis on the social processes that imposed new norms of life on living beings. The individual normative dynamics that he had described previously also had to stand against these social processes. This dimension was not unknown in the previous text, and, as we saw earlier, K. Goldstein was quite aware of the impact of the social interpretations of “normalcy”. Now, the *Nouvelles réflexions concernant le normal et le pathologique* paid explicit and thorough attention to it. This text allowed G. Canguilhem to express his deep concern about the dense network of norms in which individuals are embedded as members of various institutions, schools, for example. According to him, their minds and their bodies are modelled by these norms. He especially questioned public health policies and their ambition to determine “appropriate” ways of life, regardless of individual norms: in so doing, they could lead to stigmatization and even generate pathological states.[40] G. Canguilhem was thus rather critical of societies which heighten individual’s difficulties in this respect, rather than helping the individual to actualize her own norms (and form) of life. Following him, the contemporary philosopher G. Le Blanc considered these pathological states as forms of “negative identities” that prevent the self from developing and relating to others properly.[41] According to him, individuals that are prisoners of these negative identities do not manage to tailor a “milieu” appropriate to themselves. They have to develop a strategy of resistance and struggle on a daily basis against them.

In this respect, G. Canguilhem’s definition of “normal” and “pathological” states also has important implications on the understanding of disability.[42] Within his frame of thought, many forms of deficiency or disability appear as relative. They are obstacles to the dynamics of life only in as much as the environment is radically

[40] Canguilhem, 2012, p. 48

[41] Le Blanc, 2004, p. 204. In the field of mental health or in that of colonial relationships, one may refer to Erving Goffman’s and Franz Fanon’s works to illustrate the effects of these negative identities.

[42] The contribution of Anne-Lyse Chabert in this book offers an alternative way to relate deficiency to one’s life environment.

incompatible with them. G. Canguilhem himself rarely mentioned this issue that has become central to contemporary disability studies. However, in the 1970s, he formulated a sharp criticism of medical decisions grounded on the conviction that one could establish an objective threshold between a life worth living and a life not worth living.[43] He stressed that society could greatly contribute to the creation of a favourable "milieu" for disabled people. Not surprisingly, in this line of thought, some of his commentators developed a criticism of predictive medicine and eugenics.[44]

6.4 The Lived World of Health and Disease

Let us now turn to M. Merleau-Ponty's reading of K. Goldstein. As we saw, for the latter, catastrophic or ordered behaviours are the two ways to relate to the world and to other being. While the first generates anxiety, the second is associated with ease and pleasure. And both experiences are anchored in the body. This point is the main aspect of M. Merleau-Ponty's interest in K. Goldstein's thinking. It may appear at first sight that his interpretation does not have a direct relationship with the conception of health and disease. But it does have one, in an explicit manner in his substantial comment on Schneider, one of K. Goldstein's patients.[45]

The reference to K. Goldstein appears in *The Structure of Behavior*, his first Ph.D. (completed in 1938). In this work, M. Merleau-Ponty intended to find an alternative theory to the conceptions of "behaviour" as an "idea" or as a "thing". In this perspective, M. Merleau-Ponty was particularly critical of behaviourism understood as a mechanistic type of explanation.[46] To him, one was to take into account the *meaning* of stimuli for an organism:

> In speaking here of knowledge and consequently of consciousness, we are not constructing a metaphysics of nature; we are limiting ourselves to denominating the relations of the milieu and the organism as science itself defines them as they should be denominated. In recognizing that behavior has a meaning and depends upon the vital significance of situations, biological science is prohibited from conceiving of it as a thing in-itself (*en soi)* which would exist, *partes extra partes, in* the nervous system or *in* the body; rather it sees in behavior an embodied dialectic which radiates over a milieu immanent to it. There is no question - as we have said often enough - of returning to any form whatsoever of vitalism or animism, but simply of recognizing that the object of biology cannot be grasped without the unities of signification which a consciousness finds and sees unfolding in it.[47]

M. Merleau-Ponty thus conceived of the world as a network of "atmospheres" and "milieus" – in German, as he mentioned himself, "umwelt", "merkwelt" and

[43] Canguilhem, 1976, p. 531; Gaille, 2010

[44] Gros, 1993, p. 104–109

[45] Merleau-Ponty, 2012

[46] Merleau-Ponty, 1963, p. 127

[47] Ibid., p. 161

"gegenwelt".[48] His translation into French of the German words may be best understood if we recall the uses of "umwelt" in German phenomenological thinking, be it that of Husserl or of Heidegger, to stress the subjective relationship to the world and its difference from "scientific" or "objective" visions, and in the conception of biology developed by J. Von Uexküll.[49]

In M. Merleau-Ponty's view, acts have no meaning in themselves: they express, at least for human beings, what he calls "the aims of life".[50] They are given meanings. When he returned to this issue in the late 1940s, he emphasized K. Goldstein's role in helping him understand the important distinction between the notion of "geographical setting" and that of "behavioural setting": the first describes what surrounds an organism, generally speaking, while the second designates the meaning given to these surroundings by the organism, the selection and the organization of these surroundings made by it according to its orientation and intention.[51]

From this analysis, we can perceive M. Merleau-Ponty's relationship to K. Goldstein, in some respects, close to G. Canguilhem's. They both share the same interest in the distinction between the idea of a (neutral) environment and that of a tailored "milieu". Like G. Canguilhem, M. Merleau-Ponty finally proposed a vision of the human relationship to the world based on the vision of each human being as a "centre of perspective", a phrase we will meet later on in his writings. In this line of thought, both authors converged toward a similar use of K. Goldstein's thinking.

However, the notion of "meaning" used by M. Merleau-Ponty certainly conveys something distinct from that of the "values", "needs" and "preferences" used by G. Canguilhem. Its implications become manifest when we turn to subsequent works by the French phenomenologist, in which the idea of the "lived world" is fully formulated. In his second Ph.D., *The Phenomenology of Perception*, completed during the Second World War, M. Merleau-Ponty related his ambition to formulate a theory of knowledge. Phenomenology was presented as a science with the aim of describing human experience, thus taking into account the "lived worlds". To M. Merleau-Ponty, this reference to phenomenology meant that this description should go beyond (or before) the scientific knowledge elaborated about the world. It implied returning to the world as it was seen prior to the elaboration of scientific knowledge, as "the world is there prior to every analysis that I could give of it".[52] In this perspective, perception is given the key role as the foremost and essential human access to the world:

> Perception is not a science of the world, nor even an act or a deliberate taking up of a stand; it is the background against which all acts stand out and is thus presupposed by them. The

[48] Ibid., p. 129–13

[49] Feuerhahn, 2017

[50] Ibid., p. 163

[51] Merleau-Ponty, 2010a, p. 343

[52] Merleau-Ponty, 2012, Preface, p. xxiii

> world is not an object whose law of constitution I have in my possession; it is the natural milieu and the field of all my thoughts and of all my explicit perceptions.[53]

In *The Phenomenology of Perception*, human beings' existence appears as nothing other than a presence in and a relationship to the world (including the relationship to other human beings).[54]

The Phenomenology of Perception also emphasized the embodied dimension of this relationship to the world, as later would *L'oeil et l'esprit*, through the example of the painter's embodied relationship to the world.[55] To M. Merleau-Ponty, the human experience of the world is an embodied one, and always relative, in that it is always anchored in specific individual bodies.[56] M. Merleau-Ponty gave a central place to this perspective in his teaching project elaborated for the Collège de France, when he presented the body (each body) as a "centre of perspective".[57] In his comment on the Schneider case, he highlighted the fact that the sick individual was unable to engage in "play acting" as the "normal subject" would, or the "actor" in living situations.[58] The patient struggles to replace "destroyed" functions with "substitute" ones but fails to offer "the direct image" of the "fundamental" functions.[59] It is a fact that:

> The reason is that all these operations require the same ability to mark out boundaries and directions in the given world, to establish lines of force, to keep perspectives in view, in a word to organize the given world in accordance with the projects of the present moment, to build into the geographical setting a behavioural one, a system of meanings outwardly expressive of the subject's internal activity. For these patients the world exists only as one readymade or congealed, whereas for the normal person his projects polarize the world, bringing magically to view a host of signs which guide action, as notices in a museum guide the visitor.[60]

This conception of human life (and of its embodied dimension) as a form of life that is essentially present in and to the world has significant implications for the understanding of the relationship between health, disease and environment. We first observe that it grounds the description of the experience of disease, especially when what is at stake is to stress the difference between the physician's point of view and that of the patient. A reflection by S. Kay Toombs, a philosopher and a psychologist who was diagnosed with multiple sclerosis in 1973, illustrates this role played by M. Merleau-Ponty's thinking. Taking as a starting point her observation that physicians and patients seem unable to communicate with each other,[61] she elaborated a phenomenological account of the lived experience of disease, in which

---

[53] Ibid., p. xiv

[54] See Marratto, 2012.

[55] Merleau-Ponty, 1993, p. 121–149

[56] Merleau-Ponty, 2001, p. 223

[57] Merleau-Ponty, 2010b, p. 1816–1817

[58] Merleau-Ponty, 2012, p. 104–105

[59] Ibid., p. 107–108

[60] Ibid., p. 112

[61] Toombs, 1992, p. xi

she referred to M. Merleau-Ponty. Then, she intended to show that the body is not an "object": To her, I am my body, or an embodied subject; and this body is the centre of perspective from which I set up my place in the world. The lived body is the channel of any relationship to the world. It precedes any process of reflection. Now, the disease entails a relationship to the world and to others that is distinct from any other, including at this pre-reflexive level.[62] Some actions become complicated, or even impossible. As an example, S. Kay Toombs described how she lost her capacity to walk by herself little by little and the successive phases of this process: moving autonomously, walking with a cane, walking with crutches, use of a walker and finally use of a wheelchair.[63] These changes were associated to negative emotions and concern about what she considered to have become. In this respect, she underscored the lack of "control" over her own body and the loss of an easy and obvious relationship with the world.[64] The contemporary enquiry led by Havi Carel on the role and meaning of illness for philosophy shares with S. Kay Toombs' reflection a reference to M. Merleau-Ponty.[65] For both of them, the issue at stake is to demonstrate that disease is not only an objective fact, but first and foremost an event that transforms the world of the ill person into a hostile environment.[66]

Besides this, M. Merleau-Ponty is also a reference used in the contemporary stream of theories that associate health and wellbeing, or pathological states to specific environments in the fields of environmental psychology, sociology and geography of health and wellbeing. Generally speaking, these theories are based on the study of spatial, environmental or geographical parameters of health and wellbeing – two terms that are sometimes loosely distinguished. These theories focus on urban planning, landscape planning and private and public space design and in the field of clinical practice, into some renewed architectural conceptions of hospitals and therapeutics.[67]

M. Merleau-Ponty offers significant insights to a sociological approach that focuses on "ambiances" – a word it borrows from the Frères Goncourt ("ambiance des milieu") that intended to characterize an evocative power of a "spiritual climate, or atmosphere, emanating from, hovering over a milieu or even a thing".[68] In the same line of thought, the sociology of "ambiances" proposed by J.-P. Thibaud is grounded on a vision of human life that owes much to phenomenology, particularly its contribution to thinking on aesthetics, and among various authors, to M. Merleau-

[62] Ibid., p. 63

[63] Toombs, 2001, p. 247

[64] Toombs, 1992, p. 91–97. For people with chronic conditions, the sociologist K. Charmaz noticed the same feeling of "loss of self"; see Charmaz 1983.

[65] Carel 2014 and 2015

[66] Perrier 2016

[67] See Williams, 1999; Sternberg, 2010; Dannenberg, Frumkin and Jackson, 2011; Halimi, 2013, Atkinson, Fuller and Painter, 2016. As far as green therapies are concerned, see Ulrich 1992 and for an overview of these theories, Rivasseau Jonveaux and Fescharek, 2013.

[68] Spitzer, 1942, p. 186 and p. 188 for the quotation

Ponty.[69] In a medical setting, such a sociology allows us to grasp, for example, the effects of architectural design and sounds on patients' and hospital employees' emotional state;[70] beyond this medical setting, this sociological approach highlights the capacity of human beings to feel aesthetic emotions and their effects on their wellbeing, and even health, and the emotional dimension of the relationship to one's environment, evidenced in the event of a natural disaster or a constrained move from a familiar place.[71] Such a sociological approach is very close to a M. Merleau-Ponty-based geography that aims to understand the human experience of environment and landscape.[72]

Finally, the reference to M. Merleau-Ponty and particularly to his conception of human existence as an embodied form of life and as a centre of perspective appears in the contemporary enquiry about the relationship human beings may have with the "Earth" or the "biosphere". This enquiry is led through various approaches, not all of them acknowledging the implications of M. Merleau-Ponty's thinking.[73] Among them, the search for an ecological self of A. Naess is worth noting.[74] Its ambition – to transform oneself and one's relationship to the world and enlarge one's perception so as to make oneself part of a whole of living beings – is shared by D. Abram. The latter has turned to M. Merleau-Ponty in order to elaborate his own portrait of human beings as "made for" the relationship with other human beings, other living species and their environment – the word "environment" being used to describe both the nature around them and the biosphere. In *The Spell of the Sensuous*, he proposes both a narrative and a theoretical type of discourse. The first is used to enhance the "sensuous" dimension of human life and to confer his theory a "personal" tone.[75] We may for example read the narrative of his nights spent under the starry Bali sky, in which a total sensory experience take place under the spell of a magical and majestic nature.[76] The theoretical part borrows its tools from M. Merleau-Ponty's phenomenology. On the basis of his conception of the perceptive and participatory mode of human presence in the world, D. Abram views this participation as both an intellectual and sensory, even sensuous one:

> The living, attentive body – which Merleau-Ponty called the 'body-subject' – is this very being that, pondering a moment ago, suddenly took up this pen and scribbled these thoughts. It is the very power I have to look and to see things, or to turn away and look elsewhere, the ability to cry and to laugh, or to howl at night with the wolves, to find and gather food whether in a forest or a market, the power to walk upon the ground and to imbibe the swirling air. Yet 'I' do not deploy these powers like a commander piloting a ship, for I am,

[69] Thibaud, 2004 and 2012

[70] Leroux and Bardyn, 2003 and Leroux 2004

[71] Centemeri, 2006 and 2015

[72] Yi-Fu Tuan 1974 and 2001

[73] As far as ecopsychology is concerned, see Roszak, 1992, and Thomas, 2014. In the present volume, the contribution of Laura Centemeri explores the conception associated to permaculture.

[74] Naess, 2017

[75] Abram, 1996, Preface and Acknowledgments, p. 9–11

[76] Ibid., p. 13

> in my depths, indistinguishable from them, as my sadness is indistinguishable from a certain heaviness of my bodily limbs, or as my delight is only artificially separable from the widening of my eyes, from the bounce in my step and the heightened sensitivity of my skin.[77]

In this work, M. Merleau-Ponty's phenomenology appears especially relevant to renewing our vision of the relationship between human beings and the biosphere. We may relate to it in a scientific way (to map and measure the biosphere, and explain the complex mechanisms of the planets and of their interactions). We may also relate to the biosphere in terms of a personal experience: through it, we become aware of being part of the world we have the experience of – an awareness that is, according to D. Abram, essential to our wellbeing and our flourishing.[78]

6.5 Conclusion

Between the two World Wars, K. Goldstein introduced the idea of environment into medical thinking in order to propose a renewed conception of both pathological states and medical action. According to him, such an approach was necessary to offer appropriate medical care to patients with brain injuries. It led him to define medicine on a new epistemological and anthropological basis. He thus promoted a vision of the human being as an emplaced and relational being. Such a bold and original conception attracted much attention from G. Canguilhem and M. Merleau-Ponty, who were early readers of K. Goldstein's works. Both were very interested in the idea according to which each one of us "tailors" a liveable, and if possible comfortable space, in which one lives at ease. We emphasized the fact that G. Canguilhem's reading was determined by his philosophy of life. To him, this tailoring process helped to define pathological states in a renewed way, but it had a wider scope: it was an essential feature of life as such. Despite their differences, we may spot a similar or analogous move in K. Goldstein's and G. Canguilhem's works: both found their starting point in medicine and went beyond the realm of medicine to propose an anthropological conception. As a philosopher of the subject, and one may add, of the embodied subject, M. Merleau-Ponty also based his own reflection on K. Goldstein's theory. He proposed a conception of the lived world in which the body, my body, is my centre of perspective, and in which the emotional and pre-reflexive dimension of life is taken into account. Today, it is very interesting to observe that while G. Canguilhem's "milieu" is a tool used in the field of disability studies in order to highlight the role played by one's living environment in the worsening of a disability, M. Merleau-Ponty's "lived world" helps thinkers in various disciplines to conceive of the role played by one's living environment in wellbeing and flourishing.

---

[77] Ibid., p. 38

[78] Ibid., p. 48

According to us, on the basis of their relationship to K. Goldstein, G. Canguilhem and M. Merleau-Ponty offer consistent[79] and complementary views one should take into consideration to face the various contemporary challenges raised by the impact of the environment of human health.

Bibliography

Abram, David. 1996. *The Spell of the Sensuous: Perception and Language in a More-than-Human World*. New-York: Vintage Books.

Atkinson, Sarah, Sara Fuller, and Joe Painter. 2016. *Wellbeing and Place*. London/New-York: Routledge.

Bing, François, Jean-François Braunstein, and Elisabeth Roudinesco. 1998. *Actualité de Georges Canguilhem, Le normal et le pathologique*, 59–69. Paris: Les empêcheurs de penser en rond.

Canguilhem, Georges. 1976. *Actes du colloque mondial "Biologie et devenir de l'homme" (1974)*. Paris: Université de Paris.

———. 1989. *The Normal and the Pathological* (1966), trans. C. R. Fawcett, in collaboration with R. S. Cohen. New-York: Zone Books.

———. 2008. "The Living and Its Milieu" (1946), In *Knowledge of life*, trans. S. Geroulanos and D. Ginsburg, ed. P. Marrati and T. Meyers. New-York: Fordham University Press.

———. 2012. "Health: popular concept and philosophical question" (1988), In *Writings on medicine*, trans. S. Geroulanos and T. Meyers. New-York: Fordham University Press.

———, 2015, in *Georges Canguilhem, Résistance, philosophie biologique et histoire des sciences, 1940–1965, Œuvres complètes*, tome IV, Paris, Vrin. This volume includes: - Cours de

[79] We only partially agree with the interpretation proposed by M. Foucault about twentieth-century French philosophy and its connection with German philosophy, particularly phenomenology – to which K. Goldstein referred. According to M. Foucault, two main schools of thought developed in twentieth-century French philosophy. The first was related to a philosophy that conferred a major role to experience. It developed a conception of the subject focused on their lived experience and her/his feelings. In this school of thought, knowledge is derived from the subject's lived experience. The second associated knowledge with rationality and concept. M. Merleau-Ponty belongs to the first, G. Canguilhem belongs to the second: "on the one hand, one network is that of Sartre and Merleau-Ponty; and then another is that of Cavaillès, Bachelard and Canguilhem" (Foucault, See Canguilhem, 1989, p. 8). In a second version of his analysis (1985), M. Foucault suggested that this distinction probably had a more distant origin. One may track it back to the nineteenth century: in the thinking of Bergson and Poincaré, Lachelier and Couturat, Maine de Biran and Comte (Foucault, 1985, p. 4). This text is a new version of the text written by M. Foucault in 1978 to introduce the English translation of G. Canguilhem's essay, *The Normal and the Pathological*. The two texts are quite the same. The 1985 version adds some complementary remarks here and there. However, the title of the 1985 version, "La vie: l'expérience et la science", clearly emphasizes the issue at stake in the distinction between the two trends in French philosophy. In any event, in Foucault's mind, this distinction was of particular relevance in interpreting the way German phenomenology came to be known in France in the 1930s and integrated into French philosophy according to two distinct frames of thought (Foucault, 1989, p. 23–24). However, G. Canguilhem indeed paid to the sick individual and to the experience of illness. As G. Le Blanc stresses, this is the very point of G. Canguilhem's criticism of R. Leriche's definition of health (Le Blanc 1998, p. 44). Even if the individual considered by G. Canguilhem is not similar to Merleau-Ponty's subject, it is not possible to associate the former exclusively to the philosophy of life approach.

philosophie générale et de logique, année 1942–1943 – Extraits, p. 81–109; - Commentaire au troisième chapitre de L'Évolution créatrice, p. 111–170; - Milieu et normes de l'homme au travail, p. 291–306.

Carel, Havi. 2014. The philosophical role of illness. *Metaphilosophy* 45 (1): 20–40.

———. 2015. Conspicuous, Obstrusive and Obstinate: A Phenomenology of the Ill Body. In *Medicine and Society,* éd. Darian Meacham, *New perspectives of Continental Philosophy,* 105–123. Dordrecht: Springer.

Centemeri, Laura. 2006. *Ritorno a Seveso. Il danno ambientale, il suo riconoscimento, la sua riparazione*. Milan: Bruno Mondadori.

———. 2015. L'apport d'une sociologie des attachements pour penser la catastrophe environnementale. *Raison publique*, https://www.raison-publique.fr/article768.html. Accessed 29 Sept 2016.

Charmaz, Kathy. 1983. Loss of self: A fundamental form of suffering in the chronically ill. *Sociology of Health and Illness* 5 (2): 168–195.

Dannenberg, Andrew L., Howard Frumkin, and Richard J. Jackson. 2011. *Making healthy places – Designing and building for health, Well-being, and sustainability*. Washington, DC: Island Press.

Feuerhahn, Wolf. 2009. Du milieu à l'umwelt: enjeux d'un changement terminologique. *Revue philosophique de la France et de l'étranger* 4 (tome 134): 419–438.

Feurhahn, Wolf. 2017. Les catégories de l'entendement écologique: milieu, Umwelt, environment, nature ... (dir.). In *Humanités environnementales – Enquêtes et contre-enquêtes*, ed. Guillaume Blanc, Elise Demeulenaere, and Wolf Feuerhahn, 19–41. Paris: Publications de la Sorbonne.

Foucault, Michel. 1985. La vie: l'expérience et la science. *Revue de Métaphysique et de Morale*, 90e Année, n° 1, Janv.-Mar., p. 3–14.

Foucault, Michel, Introduction (1978) to G. Canguilhem, *The Normal and the Pathological*, see above.

Gaille, Marie. 2010. *La valeur de la vie*. Paris: Les Belles Lettres.

Goldstein, Kurt. 1934. *Der Aufbau des Organismus*. The Hague: Springer Science + Business Media Dordrecht.

———. 1940. *Human nature in the light of psychopathology*. Harvard: Harvard University Press.

———. 1971. Notes on the developments of my concepts. *Selected papers/Ausgewählte Schriften*, ed. A. Gurwitch, E. M. Goldstein Haudek, W. E. Haudek, intr. by A. Gurwitch, La Haye, Martinus Nijhoff Phenomenologica, 43, p. 1–12 (initially published in the *Journal of Individual Psychology*, 15, 1959, p. 5–14).

———. 2000. *The Organism: A holistic approach to biology derived from pathological data in man (1934), with a foreword by Oliver Sacks*. New-York: Zone Book.

Gros, François. 1993. Hommage à Canguilhem. In *Georges Canguilhem, philosophe, historien des sciences*, 104–109. Paris: Albin Michel.

Halimi, Patrice. 2013. La nature en ville pour notre bien-être. In *Notre santé et la biodiversité – Tous ensemble pour préserver le vivant*, ed. S. Morand and G. Pipien, 183–122. Paris: Buchet Chastel.

Imbert, Christophe. 2016. *Retour sur les genres de vie: un détour par l'histoire lexicale et sémantique pour l'étude contemporaine de la relation homme-milieu* (Mémoire d'habilitation à diriger des recherches en géographie).

Le Blanc, Guillaume. 1998. *Canguilhem et les normes*. Paris: PUF.

———. 2004. *Les maladies de l'homme normal*. Bègles: Éditions du Passant.

Leroux, Martine. 2004. L'épreuve de l'hôpital La notion d'ambiance dans la méthodologie d'enquête. In *Ambiances en débats*, ed. Pascal Amphoux, 31–37. Bernin: A la Croisée.

Leroux, Martine and Jean-Luc Bardyn. 2003. *À l'écoute de l'hôpital*, enquête sociologique (unpublished fieldstudy).

Limoges, Camille. 2015. In Georges Canguilhem, *Résistance, philosophie biologique et histoire des sciences, 1940–1965, Œuvres complètes*, tome IV, see above.

Marratto, Scott L. 2012. *The Intercoporeal self: Merleau-Ponty on subjectivity*. Albany: State University of New-York Press.
Merleau-Ponty, Maurice. 1963. *The Structure of Behavior* (1942), trans. A. L. Fisher. Boston: Beacon Press.
———. 1993. *Eye and Mind* (1968), In *The Merleau-Ponty Aesthetics Reader*. trans. M. Smith, ed. G. Johnson. Evanston: Northwestern University Press.
———. 2001. *Husserl et la notion de nature*, In *Parcours 2, 1951–1961*. Lagrasse: Verdier.
———, 2010a, *Child Psychology and Pedagogy, The Sorbonne Lectures 1949–1952*, trans. T. Welsh. Evanston: Northwestern University Press.
———, 2010b, *Travaux et projet d'enseignement (1951)*, In *Oeuvres*, éd. Cl. Lefort. Paris: Quarto Gallimard.
———. 2012. *Phenomenology of Perception* (1945), trans. D. A. Landes. London/New-York: Routledge.
Muchielli, Laurent and Robic, Marie-Claire. 1995. Entre sociologie et géographie: la 'morphologie sociale', selon Durkheim, (dir.) M. Borlandi and L. Muchielli, *La sociologie et sa méthode. Les règles de Durkheim un siècle* après, 101–136. Paris: L'Harmattan.
Naess, Arne. 2017. *La réalisation de soi – Spinoza, le bouddhisme et l'écologie profonde*. Marseille: Wildproject.
Péquignot, Henri. Le normal et le pathologique en médicine. In *Actualité de Georges Canguilhem, Le normal et le pathologique,* eds. Bing François, Jean-François Braunstein and Elizabeth Roudinesco. see above.
Perrier, Jean-François. 2016. L'approche phénoménologique en medicine", *Phares*. Online access: http://revuephares.com/wp-content/uploads/2016/04/Phares-XVIa-07-Jean-Francois-Perrier.pdf. Accessed 3 June 2017.
Rivasseau Jonveaux, Thérèse, and Reinhard Fescharek. 2013. Les dimensions thérapeutiques de la nature. In *Notre santé et la biodiversité – Tous ensemble pour préserver le vivant*, ed. S. Morand and G. Pipien. Paris: Buchet Chastel.
Robic, Marie-Claire. 1990. "Milieu (géo.)", In (dir.) S. Auroux, *Les notions philosophiques. Dictionnaire, t. 2, Encyclopédie philosophique universelle,* 1632–1634. Paris: PUF.
Roszak, Theodore. 1992. *The voice of the earth*. New-York: Simon & Schuster.
Spitzer, Leo. 1942. Milieu and ambiance – An essay in historical semantics. *Philosophy and Phenomenology Research* 3 (2): 169–218.
Sternberg, Esther M. 2010. *Healing spaces – The science of place and Well-being*. Cambridge, MA/London, The Belknap Press of Harvard University Press.
Thibaud, Jean-Paul. 2004. Une approche pragmatique des ambiances urbaines. In *Ambiances en Débats*, ed. P. Amphoux, G. Celkoff, Grégoire, and J.-P. Thibaud, 145–158. Bernin: Éditions A la Croisée.
———. 2012. Petite archéologie de la notion d'ambiance. *Communications* 90: 155–174.
Thomas, Catherine. 2014. Le sacré dans la nature du point de vue de l'éco-psychologie. In *Y a-t-il du sacré dans la nature ?* ed. B. Hurand and C. Larrère, 79–89. Paris: Publications de la Sorbonne.
Toombs, S. Kay, ed. 1992. *The Meaning of Illness – A Phenomenological Account of the Different Perspectives of Physician and Patient, introduction*. Dordrecht/London: Kluwer Academic Publishers.
——— éd. 2001. Reflections on bodily change: The lived experience of disability. In Dans *Handbook of Phenomenology and Medicine* , 247–262. Dordrecht: Springer.
Tuan, Yi-Fu. 1974. *Topophilia: A study of environmental perception, attitudes and values*. Englewood Cliffs: Prentice Hall.
———. 2001. *Space and place, the perspective of experience*. Minneapolis/Londres: Minnesota University Press (1977).
Ulrich, Roger S. 1992. Influences of passive experiences with plants on individual Well-being and health. In *The role of horticulture in human wellbeing and social development: A National Symposium*, ed. D. Relf, 93–105. Portland: Timber Press.
Williams, Allison. 1999. *Therapeutic landscapes: The dynamic between place and wellness*. Lanham: University Press of America.

Chapter 7
Environment and Chagas Disease: An Elusive and Diverse Relationship

Gabriel Gachelin

Abstract We have chosen to address the complexity of the links between the environment and health by examining a particular case: the history of a major public health problem in Latin America, Chagas disease, a parasitic disease transmitted to humans by vector bugs such as *Triatoma*, from wild natural reservoirs. The historical analysis of its discovery and the socio-economic conditions of its propagation, shows that a multiplicity of different environments are in operation. Moreover, the reduction of this abundance of cause-and-effect relationships to a single dominant factor is at the origin of the considerable difficulties encountered in the attempts to control the disease. The history of Chagas disease, like that of most ecosystem-related diseases, demonstrates the illusion, or even the danger, of a simplified environmental approach in the management of these pathologies.

Keywords Environment · Latin America · Chagas'disease · Parasitic disease · Housing · Habitat · Adaptation

The meaning given in this text to the word "environnement" in French is close to that attributed to it by Vidal de La Blache, as a substitute for the word *milieu* and in correspondence with the English "environment" (*But if one reflects on everything the word "environment" implies in the English expression, to all the unsuspected threads which weave the fabric of that which enfolds us, what living organism could escape from it?* Vidal de La Blache, *Principles of geogr. hum.*, 1921, p. 7). Since the early 1970s, sociologists and historians of science have popularized the expressions of cultural, scientific, political, etc. environments, in which the word "environment" is obviously neither biological nor physical-chemical. Consequently, the word "environment" has become a kind of envelope used as a substitute for less "expressive" terms or older expressions (context, milieu, etc.). We will therefore consider the word "environment" to be merely a convenient term, while it is not definable, and one immediately seizes all or part of what it covers, only in the context of its use. It is therefore a convention which bears no temporality.
The author would like to thank A.-L. Chabert and V. Chansigaud for their very useful critical reading of the manuscript.

G. Gachelin (✉)
Université de Paris, SPHERE, UMR 7219, CNRS-Université de Paris, Paris, France
e-mail: gabriel.gachelin@gmail.com

F. Bretelle-Establet et al. (eds.), *Making Sense of Health, Disease, and the Environment in Cross-Cultural History: The Arabic-Islamic World, China, Europe, and North America*, Boston Studies in the Philosophy and History of Science 333, https://doi.org/10.1007/978-3-030-19082-8_7

7.1 Introduction

Discovered by Carlos Chagas (1879–1934) in Brazil in 1908 (Chagas 1909a, b, 1911), the pathology that bears his name is a human disease caused by the *Trypanosoma cruzi* blood parasite. The vectors of the parasite are large haematophagous forest bugs of the Reduviidae family, belonging to the genera *Triatoma*, *Rhodnius* and *Panstrongylus*. These insects are known in Brazil under the nickname of *barbeiros* due to the traces of blood left after a blood meal taken on the faces of their victims. They will be given the generic name of *barbeiros* in the present text. Chagas disease is a long-course disease. The acute initial phase following *T. cruzi* contamination is spectacular in children. It is characterized by a very high fever, sometimes fatal, and a hemilateral palpebral oedema at the preferred site of parasite penetration, the ocular mucosa (Romana's sign, seen from 1909 but only recognized as pathognomonic in 1935). At this stage, the trypanosome is generally detectable in the blood, which is the only evidence of infection by *T. cruzi*, the other clinical signs being unspecific in these areas with multiple endemic parasites. The trypanosome disappears rapidly from the blood, spreads to the internal organs, especially the heart, and becomes undetectable to direct observation. The disease becomes chronic and evolves over several decades in about half of those infected. There is a gradual weakening of the subject, while irreversible lesions affect the organs in which the parasite is located, with an evolution towards progressive cardiac insufficiency, often ending in a spectacular "sudden" death from the rupture of the myocardium with blood flowing into the pericardium. All these clinical aspects were brilliantly described by Chagas (Chagas 1909a, b, 1916, 1922). The heated debate that accompanied his work did not concern the clinic – irreproachable – as evidenced by examining the medical records of the period 1909–1915 (Gachelin and Bestetti 2017). It concerned the etiological hypothesis advanced by Chagas: according to him, the disease was a parasitic thyroiditis, a hypothesis soon proved inaccurate (Kraus et al. 1915). The contestation was also about the frequency and severity of the disease. Hence, the extent of the disease was underestimated for a long time, probably due to lack of unambiguous clinical evidence (Gachelin and Bestetti, 2017), and this until the end of the 1950s according to Simone Kropf, when the high prevalence of cardiac troubles in the population, was finally recognized (Kropf 2009).

Like all vector-borne diseases, the presence of Chagas disease depends on the vector-man relationship and therefore on the ecosystem(s)[1] in which the vector

[1] The term did not exist at the time. The Brazilian terms *ambiente* and *medio ambiente* were rarely used; preference was given to terms that select particular ecosystems or environments, such as *cafua*, which restricts the ecosystem of the disease to domestic space, for example. This is what Emile Brumpt does in his 1913 parasitology report, "The Conorhinus bite at night in natural conditions. They abound in dirty dwellings. The best thing to do is to destroy the old houses". In the 1922 edition of the same treatise (page 276), Brumpt describes the favourable "milieu" more broadly: he introduces the climate, the temperature, the seasons, the landscape and the nature of the country and the housing and that "the way of sleeping in beds fixed to the walls, usual in the

lives, reproduces, becomes infected and finally encounters and infects humans.[2] What would have been considered as an emerging disease in 1909 if the term had existed was quickly considered as the direct and singular consequence of the habitat where vectors and humans coexist: the geographical space of the disease was thus reduced to agricultural workers' slums in Brazil's intertropical zone. The Brazilian authors of the time defined houses with dry earth walls, *cafuas*, as the determining factor for the disease: according to their observations, the vector, infected outside in an animal reservoir, lives in the cracks of the walls and from there infects the humans who inhabit the house. In the absence of any treatment, the control of Chagas disease was thus historically concentrated on the destruction of the insect in situ in the *cafuas* and on the improvement of housing.[3]

Today the strategy remains the same (WHO 2007). Yet, despite considerable efforts, the disease still "recruits" hundreds of thousands of people every year, especially in Latin America. Seven to eight million people are infected chronically, and the disease is responsible for at least 15,000 deaths annually principally from heart failure. According to the WHO, between the Southern United States and Argentina, 50–100 million people are exposed to the vector. Thus, Chagas disease is not a minor public health problem. Moreover, it is not really on the decline, at least not consistently.

One might wonder about the origin of this semi-failure of the control of the disease. In recent years, epidemiological field work shows that environmental factors in Chagas disease are much more diverse and complex than was supposed and that they were very poorly defined by the decrees and practices set up to combat the disease only in the *cafuas*. The model outlined earlier is only partially valid, and Chagas' disease is better described by a multiplicity of situations, of which the *cafuas* are only one example. Each of these situations represents a facet of these physical, medical, socio-economic and even cultural environments that have contributed to and still contribute to the disease and its propagation.

The purpose of this article is therefore to identify at least some of these facets and to define the weight that each could have had, and can have, in the history of Chagas disease.

---

cottages of the Brazilian peasants, favors infestation". He adds a photograph of a cottage and one of the landscapes where the infection is endemic. Brumpt's environment follows the same conventions as the medical geography of the previous century.

[2] This was not a new finding in 1909. The dependence of certain diseases on the living conditions of their vectors was rapidly perceived between 1875 and 1900 as an essential mechanism for the singularities of the distribution and transmission of tropical diseases, malaria, yellow fever, leishmaniosis, etc. This finding led to a complete renewal of the content of the previous medical geography, which emphasized the direct effects of climate on health.

[3] A comprehensive bibliography of the whole of the introduction of this article can be found in Benchimol and Texeira (1990); Perleth (1997), Delaporte (1999) and Kropf (2009).

7.2 A Detrimental Scientific Environment

The discovery of Chagas disease occurs several years after the elucidation of the transmission of the main parasitic diseases and after the notion of vector-borne diseases had been established. Actually, following the discovery in 1877 by Patrick Manson of the role of biting dipterans in the transmission of *Wuchereria bancrofti*, the role of *Anopheles* in malaria was elucidated between 1895 and 1899 and that of *Glossina* in sleeping sickness between 1900 and 1903. The vector for yellow fever – which is not a parasite – was discovered in Cuba and then in Brazil at the beginning of the twentieth century (Löwy 2002). The notion of vectors renewed thinking about contagion and contamination. It would be a mistake to think that parasitology is a continuation of Pasteurian microbiology. Great bacteriologists like Koch were, of course, interested in the agent of malaria. But attempts to apply the methods and concepts of bacteriology to parasites have invariably failed. In fact, the dominant feature of parasitology was the place of the vector and its biology. In other words, parasitology is more of a spectacular transformation of chapters of medical zoology, which is evident in the shift from Raphael Blanchard's 1888 treatise to that of Emile Brumpt in 1910: the content is almost the same, but the titles change from "medical zoology" to "parasitology", with an important place given to medical entomology (Opinel 2008, Coluzzi et al. 2008).

Brazilian parasitology, which began in the early twentieth century around Adolpho Lutz and Oswaldo Cruz, was an immediate continuation of this new medical approach and was strongly influenced by the vectorial transmission models that had already emerged (Opinel and Gachelin 2005).

The issue of Chagas disease was rapidly completed by Carlos Chagas shortly after the founding of the Institute at Manguinhos near Rio de Janeiro (1907) and which would soon take the name of its founder, Oswaldo Cruz. The chance discovery (at the end of 1908) of a new species of trypanosome in the blood of Berenice, a young patient from Lassance in the state of Minas Gerais, was soon followed by the demonstration that this parasite, called *Trypanosoma cruzi*, is pathogenic in monkeys. The assertion that it is propagated by the bite of *barbeiros* was followed by the conclusion that the parasite reproduced through a sexual cycle and that it invaded the thyroid gland (acute cases) and finally that a thyroid defect was responsible for chronic forms such as progressive cardiopathies. This extraordinary series of claims, established in just a few months, defines a new human trypanosomiasis (the first being sleeping sickness in Africa) and the first infectious disease in which clinical signs appear to be caused by a thyroid gland defect induced by the parasite and the development of a parasitic goitre (Chagas 1909a, b, 1911). Goitre and hypothyroidism were major problems in Brazil. For Delaporte (1999), Chagas' hypothesis made it possible to bring together what was so far separated: the scourge of endemic goitre in Brazil is the consequence of infection by *T. cruzi*, which was also endemic. Thus, *doença de Chagas* could legitimately become *doença do Brasil*, the title of Simone Kropf's seminal book (2009).

Thus, in his first article in 1909, and particularly in 1911 (Chagas 1911), Chagas describes all the clinical signs of the various forms of the new disease and their aetiology. Several clinical articles followed, particularly in 1916 (acute cases) and 1922 (cardiac forms). While, as we ourselves have noted (Gachelin and Bestetti 2017), the clinical description of the disease was irreproachable, Chagas amassed publications from the laboratory that would prove incorrect almost immediately. Very soon, from 1912, European and Argentine researchers demonstrated that the data on the life cycle of the parasite were essentially inaccurate, that contamination did not take place through biting and that the goitre had to be separated from the infection (Perleth 1997, Gachelin and Opinel 2009, Delaporte 1999).

This catalogue of errors, which could be viewed as "commonplace" in the construction of knowledge, casts doubt on the whole work, scientific doubts amplified by violent antagonism within the Institute.

The period of uncertainty which extended from about 1914 to the end of the 1920s and which went as far as to deny the very existence of the disease was scientifically difficult and painful for Chagas but is directly linked to Chagas' scientific environment and to the environment linked to a science policy chosen by the Oswaldo Cruz Institute. The will to manifest the power of Brazilian medicine and research, tainted with nationalism, played a significant role.

Several reviews of the early stages of the case (Perleth 1997, Delaporte, 1999, Romero Sà, 2005, Kropf 2009) highlight in the succession of scientific errors that marks the period, the predominant role of the scientific environment and the influence of German parasitology. Actually, from 1907, the Institute sought to establish itself as the major research institution in Brazil, much like the Institut Pasteur in France in the 1890s (Gachelin 2013). In this context, the fight against yellow fever which started at the beginning of the century (Löwy 2002) and the discovery of Chagas disease are arguments that weighed heavily. They are brought to the fore in the film screened at the Dresden International Hygiene Exhibition in 1911 (Fiocruz 2011). Thus, soon after its discovery, Chagas disease becomes emblematic of Brazilian science. Oswaldo Cruz relied on the presence of renowned German parasitologists in Rio and engaged them in the study of parasitic diseases in Brazil. François Delaporte (1999) and Magali Romero Sà (2005) show that the errors by Chagas and his collaborators on the biology of the parasite and its mode of transmission are related to the recourse without critical distance to the scientific environment created in Brazil by the invited German parasitologists. Their inadequate scientific models, notably the "malaria cycle" paradigm, Hartmann's theory of pansexualism and finally Schaudinn and Prowazek's parasitology which takes some liberties in its observations, largely contributed to Chagas' errors. Under these conditions, the theory of parasite sexuality collapsed in 1912. At the same moment, the contamination by biting was proved wrong by Brumpt's work on monkeys (1912): the parasite no longer entered the body through biting (the vector's salivary glands do not contain trypanosomes) but indirectly from the droppings of *barbeiros* and from which the parasite passes through the ocular or buccal mucous membrane. Thus, *T. cruzi* does not resemble the malaria agent or the sleeping sickness trypanosome.

Brazilian specificities also played a significant role; this contributes to explaining why Chagas persisted in his belief that the disease was a parasitic thyroiditis, despite doubts expressed in 1914 by Brumpt (Gachelin 2009) and Kraus' demonstration in 1915 (Kraus et al. 1915). Thus, outside the Oswaldo Cruz Institute, the endemic goitre in Brazil has been separated from the *T. cruzi* infection since 1915. For Chagas, damage to the thyroid remained the cause of the disease; if he held to this idea until 1926, it was because addressing the aetiology of goitre in Brazil through infection settled a real national health problem: Could it be envisaged to abandon this infectious aetiology? Chagas was a national hero; he was nominated for the Nobel Prize. It was difficult to recognize his mistakes. This procrastination lasted 15–20 years during which an increasing number of researchers in Rio emancipated themselves from official thinking, but this is to depart from this research topic, while the sick and their disease are left on the sidelines (Kropf 2009).

Thus, with regard to the link between environment and health, Chagas' scientific milieu constituted a singular environment that has heavily influenced the history of the disease. This scientific and political environment undoubtedly delayed research on the disease in Brazil by rendering most of the researchers at the Oswaldo Cruz Institute if not blind, at least partially sighted, faced with contradictory results obtained in France, in England and, after the First World War, in Argentina and Paraguay. It may also have excluded researchers from the study of goitre, a disease that was known to be highly endemic in Brazil since the eighteenth century (Sigaud 1844). Finally, it certainly delayed the recognition of the burden of Chagas disease on public health in Brazil until about 1950.

It would, however, be inaccurate to see Brazilian medicine through the sole prism of the Oswaldo Cruz Institute and its passions. The Sao-Paulo Faculty of Medicine, which opened in 1913, recruited from abroad, and in 1913 the chair of parasitology was entrusted to the Frenchman Emile Brumpt, an opponent of the German theses on parasitology, an expert on trypanosomes and from the outset a scientific dissenter of Chagas' work on the trypanosome cycle and transmission (Opinel and Gachelin 2005). His teaching left a deep impression in Brazil, without however affecting Chagas' thinking. Maybe we can see in all that an additional manifestation of the well-known rivalry between the two cities is another form of the political environment of research.

7.3 An Ideal Environmental Culprit: The Adobe Wall of the *Cafuas*

Chagas' papers on the disease gave a social position to the disease. The type of environment defined by Chagas, which we might call the primary environment, positions itself in an immediate causal relationship by defining the conditions that allow an infected bug to infect humans. A coherent scenario took shape in 1915, based on the environment and the biology of the vector: the *barbeiro* lives in the cracks in adobe walls of *cafuas*; at night, it falls and climbs onto the face of a sleeper to take a blood meal. At the same time, it bites (Chagas) or rather empties its posterior intestine (Brumpt) which contains the infecting trypanosomes. By rubbing his eyes, the sleeper puts the parasite in contact with the ocular mucous membrane which the trypanosome easily traverses. This little drama is played out at night without the knowledge of the sleeper. The point of penetration of the parasite explains the palpebral oedema which later extends to the whole face. The initial step of the disease's transmission is rooted in this narrow zone of contact. In this model, the sufferers have to sleep in an enclosed space, in houses, in *cafuas* infested with *barbeiros*. As for the *barbeiro*, it can easily become infected as and where it wants but especially outdoors according to many modalities that Torres enumerated in 1915 (Torres 1915). According to the most frequently held, in the wild, *barbeiros* live in crevices in rocks, and as early as 1909 Chagas noted that armadillos infested by *T. cruzi* share this environment with them: armadillos thus became the "right", or main, reservoir. Under these conditions, it is of no importance that *T. cruzi* is found in most domesticated animals; the armadillo remained for long the main wild culprit. The environmental scenario of Chagas disease is written, and a consequence is that the disease affects the poor.

That was true that in 1909, Chagas disease is a disease of the rural poor: in a study of 80 such patients referred to hospital between 1909 and 1915, we note that all lived in rural areas; most lived in *cafuas* in the livestock-raising region that stretches from Belo Horizonte to Lassance and Rio San Francisco in the state of Minas (Gachelin and Bestetti, 2017). This is still true today: the disease remains largely a disease of poor people, rural and urban, living in urban peripheries, products of the rural exodus. This applies to the endemic areas of Brazil and most of the Andean States as well as to the present reinfestations of favelas and shanty towns.

The notion of the environment of the disease[4] has thus immediately been reduced to that of poor housing, to *cafuas*. However, this strong correlation existing between poverty and disease is not proof in itself: in 1909, it was an observation made in a given context, that of Lassance in Minas. The claim that contamination by *T. cruzi* is domestic is reminiscent of the domiciliary malaria theory proposed by Chagas in 1906 (Chagas 1906), and does not mean that contamination does not occur elsewhere; the observation that the vector is an anthropophilic bug that lives and reproduces in the cracks of walls has not been questioned since 1909. The model proposed by Chagas is domiciliary in nature, but the *barbeiro* is infected by a wild animal such as the armadillo. The landscape in which the armadillo lives consequently becomes a pathogenic environment. In Brazil, it is the landscape of Lassance, a kind of medium altitude savannah dotted with rocky ridges, and occasional treed. It was soon popularized by photography and associated with the disease (Brumpt 1913).

But is this landscape "natural" or has it been created for and by livestock-raising for at least two centuries, as evidenced by the immense fazendas that structure it (Dean 1996, Bergad and Knight 2006)? The possible anthropic origin of the land-

[4] See note 1.

scape is indeed important because human infection by *T. cruzi* is situated at the intersection of a domestic cycle and a sylvatic cycle of the *barbeiros*. That was not perceived at the time. The emphasis on the fact that the intrusion of man created the domestic cycle from scratch was the only one retained by scientists. According to Neiva, an entomologist from the Oswaldo Cruz Institute, Lassance had few inhabitants before the arrival of the railway, and *barbeiros* were rare. In a few months, the bugs invaded the newcomers' houses. When the houses were abandoned, the *barbeiros* returned to their natural environment. The disease exists only through the intrusion of human presence (Neiva 1910). It did not exist before. It was "created" by poor rural housing.

We are thus entitled to examine whether the disease already existed in the context of the agricultural activity which preceded railway building and the activities associated with it. If it did not exist, or rather was not detectable, it can be proposed that the 1909 publication revealed a consequence of colonial expansion, the move to the west of the "new frontiers" of Brazil: the arrival of humans, who built the railway and organized a small town around the station and created the very environmental conditions for the disease by offering accommodation and shelter to the infected *barbeiros*. This situation would then be quite similar to that of sleeping sickness in Africa, where the number of cases exploded with the construction of channels of communication, or that of forest leishmaniosis in coffee plantations in Brazil (Gachelin 2005; Benchimol and Da Silva 2008). What has been observed is the passage from a number of cases that were too low and therefore not observable, linked to a small-sized population, to a high and therefore observable level, linked to the influx of people. Moreover, it can also be concluded that the discovery of Chagas disease is due to another facet of modernity, the technical environment of medicine constituted by the microscope and the laboratory: the microscope reveals the cause of an illness previously unseen because of the ordinary character of the clinical signs (Gachelin and Bestetti 2017). Incidentally, the introduction of laboratory techniques was the basis of the rural medical policy of the Oswaldo Cruz Institute at that time. Medical expeditions carried out between 1911 and 1913 within Brazil completed this dimension of the resumption of the conquest of the territory under the auspices of science which followed the penetration of the communication channels (Benchimol and Da Silva 2008). Thus, the arrival of the railway at Lassance simultaneously created the conditions for the disease and the conditions for its observation, another form of environment of discovery, in a context of colonial expansion, another form of socio-economic environment. In other words, science reveals what has been rendered possible by the economy and colonial expansion.

Under these conditions, designating the cracks in the walls as the place out of which the disease organizes itself has an essential consequence: that of directing prevention exclusively towards "curing" poor houses. Unable to destroy contaminated wildlife and to treat the sick (which remains true today, there is no vaccine and treatment remains dangerous and disappointing), we can try to interrupt the chain of contamination by destroying the bugs in human housing. This can be done by sealing cracks, using various insecticides (sulphur oxides, toxic plant powders, toxic fumigants), and by manual destruction. All these procedures are identical to those used by Chagas to fight domiciliary malaria in Brazil (Chagas, 1906) and by Oswaldo Cruz in 1903 against the vector of yellow fever (Löwy 2002). In fact, the model of prevention established in 1909 used all the procedures known at the time to kill insects. The reduction of the pathogenic environment to the walls of adobe houses integrates prophylaxis of Chagas disease in the larger frame of anti-vector strategies. This is perfectly described in the film presented by Brazil at the International Hygiene Exhibition in Dresden in 1911, one of the very first Brazilian scientific films: the prevention of yellow fever and Chagas disease is described in the same film; the diseases are different, but their prevention uses the same strategies – kill the vector(s).

7.4 A Rapidly Evolving Disease

The nature of the insecticides used in Chagas disease control has changed, but prevention remains based on the same environmental approach. To the extent that the so-called traditional housing is declining in Latin America and that systematic prophylaxis is implemented seriously by the States, a large reduction in the disease is expected. This certainly is the case in some regions of Brazil, but is not universal, far from it.

The reasons are that the disease is elusive and everything happens as if a logic of reinfestation on the one hand and co-evolution of the parasite and its vectors on the other were at work in a recurring and rapid way. New vectors, hitherto considered to be of little danger, now appear as threats. This has been the case for *Triatoma infestans* for about 20 years. In other words, the landscape offered by parasites and vectors is mobile. *Barbeiros* need blood meals and a suitable site for egg-laying. Only is this environmental minimum common to the many species of the three Reduviidae genera capable of becoming infected in the wild and transmitting the trypanosome to humans.

The point is that by using a weapon of extermination such as large-spectrum insecticides, by targeting the cracks in a wall and the interstices under the roof, one cares little about the life and behaviour of the *barbeiros* and knowing if there are differences from one group of insects to another which explain endemicity. In fact, the persistence of the disease and the fear of its geographical extension well beyond the locations of its discovery make the biology and behaviour of Reduviidae and their preferred environments an active research subject for entomologists from the United States to Chile (e.g. Wozniak et al. 2015 to quote one of the most recent texts at the time of writing this article). These various bugs, while invariably robust and accommodating, nevertheless have different habitats and behaviours. Their taste for the blood of this or that animal, including man, varies greatly. Their rate of defecation at the time of the blood meal, an important criterion for their infectivity, varies from a few seconds to tens of minutes depending on the species. Depending on the location and species, the main animal reservoir is not the same. In the Andes, it may be a small rodent and, in Amazonia, it can be a bird. Their ability to fly varies according to species and altitude, which allows a reinfestation in one place, and not in another. These distinctions are not absolute; habitats overlap at least partially, and the trypanosome can go from point to point and from reservoir to reservoir in a period of near-selective adaptation, as these differences are based on genetic differences. It is also true that some species of *barbeiros* do not host trypanosomes, an apparent enigma that brings to mind the old debate on anophelism without malaria in the 1920s (Fantini 1994): how to understand that geographical places are free from malaria, when the vector is present and malaria exists all around it? We know now that these differences are genetic. On the side of the victims, humans do not live in the same way in Bolivia and Arizona; the poverty of the sufferers is the only common denominator. Thus, the *barbeiro*-trypanosome-human interface occurs in an astonishing diversity of biological, behavioural and material environments with equally diverse consequences on the capacity to infest man and thus the endemicity of the disease. This explains the variability of the geographical distribution of human infection by *T. cruzi*. A posteriori, this also explains the erratic results of the epidemiology of the disease between 1920 and 1935 (Kropf 2009). This makes it especially difficult to envisage the fight against the disease from a single angle. Finally, this points to the conclusion that *cafuas* are only one particular environment among possible other modes of contamination.

7.5 Chagas Disease and the Debate on the Nakedness of the Indians

Such a conclusion is correct except if this type of poor housing is precisely the main environment through which the disease could effectively pass to man? This is what has been suggested a little earlier by mentioning the transformation of the region of Lassance to a savanna favourable for raising livestock. This transformation took place as early as the arrival of the European settlers during the middle of the seventeenth century. Colonization was associated with a profound transformation in the status of Indians. Was it also associated with transformation of their dwellings? This question is not asked in the histories of Chagas disease until about 1970 with a work by Guerra (Guerra, 1970). It moves the line of sight towards the history of housing in Latin America. The cornerstone is the type of housing introduced by the colonizer in Brazil from the beginning of the sixteenth century. On the Atlantic slopes, this housing is the result of the cross-fertilization of Indian housing, light, mobile, made of branches and leaves, and the Mediterranean practice of building walls with mud and adobe. Insects would have quickly colonized the crevices in the walls, and the nuisance they brought to the inhabitants is attested from the sixteenth century according to Delaporte (1999) mentioning Guerra (1970). The conditions of the emergence would then have arisen with the initiative of building walls. This modification of the housing, always according to the same sources, which results in a new environment both for the insect and for man, is due to the conjunction of the religious struggle against the Indians' real nakedness and supposed shamelessness (See de Lery 1576, Thevet, 1557 for the French texts in which the debate for and against, and the origin of the idea of "savage" are rooted), with the security and economic necessity of settling the Indian population to meet the manpower needs of the fazendas and mines (Bergard 2006). The environmental conditions of the emergence of Chagas disease, at least in Brazil, since as we shall see, the situation is different in the Andes, would thus be directly linked to the colonizers' beliefs, practices and material needs. Assuming that the disease "appeared" or rather emerged at this time and under these conditions, it would have been induced or at least promoted by a set of social behaviours imposed by the Europeans: its *primum movens* would have to be found in the culture of the European colonizer. Its discovery in 1909 is, in this context, simply linked to the amplification of colonization in another form, this time associated with the conquest of the interior of Brazil.

Is such an "environmental story" convincing? There is no doubt that adobe housing currently plays a significant role in maintaining infection and reinfestation with *Trypanosoma cruzi* on the Atlantic seaboard of the continent and in the plains of Bolivia, Argentina and Paraguay.

However, dwellings with dry clay walls have existed in all the Andean cultures and those on the west coast generally, for at least 3–4000 years. Their existence is not related to Portuguese and Hispanic colonization. In addition, a few thousand years ago, human populations lived in areas where the disease is currently endemic. This raises the question of whether, before the introduction of a colonial environment, the disease existed in ancient Indian populations? However, due to the prevalence of *barbeiros* and the geographical distribution of their infestation by *T. cruzi*, it is hard to believe that trypanosomiasis did not exist outside *cafuas*, in the traditional wood and foliage dwellings, under the trees and in the caves, places where many bugs live.

Historical sources of Brazilian medicine say nothing. Since no obvious clinical symptoms reveal Chagas disease, it is not really surprising that there are no clues before the description given by Chagas. Perhaps there is mention of sudden death and weakness in reports by Jesuits' *missionnes*[5] (who contributed largely to the adoption of a sedentary lifestyle by the Indians) who settled in these regions where the two Argentine researchers, Mazza and Romana, who "reinvented" Chagas disease, worked. Perhaps, but in the classic texts of Brazilian medicine from the early nineteenth century, like that of Jean-François-Xavier Sigaud[6] (1844), nothing is apparent, and the notion of "sudden death" is absent. It is correct to say that medicine was then reduced to that of the populations of the coast and showed little inter-

[5] *Missiones*, especially the Jesuit *Missiones*, installed in what is now southern Brazil and Paraguay, played an essential role in the links between European and Indian communities. Members of these missions were required to send detailed reports on a regular basis on the life and practices of Indians. Peruvian Jesuits, for example, are at the origin of the use of Quina-quina bark against malaria.

[6] Jean-François-Xavier Sigaud (1796–1848), a French physician living in Brazil at the time of the Restoration, played a major role in Brazilian medicine (he was Emperor Pedro II's physician) and in the diffusion of Brazilian and European medical texts.

est in the inhabitants of the interior of the country. On the other hand, the presence of goitre is always mentioned as an endemic scourge with the characteristic of very rapid development (Sigaud 1844). Thus for physicians, it was therefore necessary to wait for the scholarly construct elaborated by Chagas from 1909 for the disease to become visible.

7.6 A Disease as Old as the Human Populations

The clear, unambiguous answer comes from studies on the prehistory of the South American world and particularly the study of the mummies of populations that lived in environments entirely different from those discussed so far. The results obtained by paleopathology and paleo-parasitology make it possible to conclude that the disease existed in the Andean populations settled in a postglacial desert zone in northern Chile about 9000 years ago, belonging to the Chinchorro culture.[7] This date coincides with the first known settlements on the Chilean coast as far South as Antofagasta.[8] As soon as he arrived, man could have been infected with Chagas disease.

Several hundred mummies, perfectly preserved due to climatic conditions, have been discovered in the region of Arica in Chile and studied for their parasites and physical anomalies and pathologies. We keep hereto human remains preserved by natural mummifications, without the internal organs removed. The tissue of 43% of the mummified bodies contains *T. cruzi* DNA and was therefore infected during their lifetime. There is no ambiguity; the DNA of kinetoplasts of *T. cruzi* is easily detectable through molecular biology techniques. The internal organs of certain mummies are sufficiently well preserved, as are their contents. The size and structure of coprolites show that some subjects suffered from megacolon, a non-lethal clinical sign of Chagas disease in a large fraction of infected individuals. In the articles that describe the anatomopathology of these mummies, we have not found indications concerning the heart and reporting on an abnormal size (increased in the case of Chagas disease) or position (heart point pointing upwards). Chagasic infection was anyway frequent and the chronic disease existed 9000 years ago.

The dwellings of the Chinchorro culture are known. These hunter-gatherers (Chinchorro means fishing boat) lived between the seashore and the Altiplano. Except at the bottom of the valleys, the region is desert. Rain is very rare. During the so-called pre-ceramic period, the Chinchorro lived most frequently in light, mobile shacks made of wood, straw and leaves (Rivera and Forthhammer 1991). Rock shelters were also used. The use of rock shelters is interesting as genomic DNA studies

---

[7] The general reviews of, for example, Rivera and Rothhammer (1991) and Silverman and Isbel (2008) can be consulted.

[8] The results of paleopathological and paleoparasitological studies are summarized in several books including Cockburn (1998) and Steverding (2014). The authors of the numerous scientific studies are not cited.

strongly suggest that *T. cruzi* is derived from a trypanosome of a bat species found in the area in question. One can imagine a scenario of passage from bats to bugs then to man in a context of human's living in caves, equally conducive to the life of the bats and the proliferation of *barbeiros*. It should also be noted that the Chinchorro civilization practised weaving: the mummies are wrapped in blankets in which we have found remains of *barbeiros*. Jumping some 10,000 years, it should be noted that François Noireau (Cochabamba IRD, Bolivia), in the course of studies on the reinfestation of the Bolivian Indian populations by *T. cruzi*, has shown that the colonization by barbeiros of the blankets of participants in outdoor communal syncretic religious meetings in infested rocky areas is one of the causes – here cultural – of reinfestation (Noireau, personal communication, 2009). Incidentally, in 1910, the Brazilian entomologist Neiva noted the importance of the circulation of *barbeiros* through carts and the covers of saddles (Neiva 1910). *Barbeiros* thus travel and are able to adapt quickly to a new environment. They invade as easily as they reinfest. In this context, a plausible hypothesis to explain the spread of the disease, probably from an original Andean home, evokes the infestation of equatorial zones, colonized by man a little later, from endemic Andean foci, along rivers flowing to the Atlantic Ocean. The hypothesis is supported by the results of paleoparasitological research on human remains in the Amazonian Andes and in Brazil. Chagas disease can thus be tracked in its establishment and travels for 9000 years (Aufderheide et al. 1993). It affected the first hunter-fisherman-gatherer occupants on the west coast of South America and subsequently developed in radically different sociocultural contexts in a variety of environments that owe nothing in their long existence to the colonizer.

Paleoparasitology also confirms that the mode of contamination is not necessarily always the ocular route. The disaggregation of the mummies' coprolites in the laboratory and the examination of their content revealed undigested *barbeiro* fragments. This means that the insects were ingested either as food in a population exploiting every food resource or at the same time as a food containing bugs. Finally, the examination of Chinchorro food remains shows the consumption of birds and of small terrestrial rodents known currently to be reservoirs of the parasite: the handling of infested animals or whose coat was contaminated by insect droppings may translate to infestation through cutaneous lesions or by oral or ocular means simply by rubbing with the fingers. These findings give historical importance to contamination through the oral mucosa, shown in 1914, but then held to be marginal in man. This mode of contamination has returned to topicality with the observation of oral contamination causing acute Chagasic infections, the first of which was observed in Florianopolis in southern Brazil in 2001. It was then an infection from fresh sugar cane juice contaminated by *barbeiros*. Such observations have multiplied. This is another environmental universe of infection by *T. cruzi* that is revealed through the study of the Chinchorro civilization and the study of human fossils taken from other places (Peru, etc.) and prior to colonization (Araujo et al. 2009). The construction of this Chagas disease universe does not owe much to that discovered in 1909 and has no connection with the Portuguese or Spanish colonizer.

7.7 Economic Pressure as a Pathogenic Environment

Over the past 10 years, the increase in importance of a newcomer has been observed in the epidemiology of Chagas disease. Under pressure from the rapidly growing demand from consumers for acai juice, the cultivation of acai berry palms has grown dramatically in some equatorial regions of Brazil, especially in Para state (Cialdella and Navigantes Alves 2014). Under biological pressure of which we do not clearly know the nature, but certainly encouraged by the influx of workers from infected areas, a particular ecosystem made up of birds infected with *T. cruzi*, and arboreal Reduviidae with a taste for the blood of these birds, puts the acai collectors in direct contact with them and their droppings during the manual picking of the fruit, at the top of the palm trees. A significant number of acute cases of Chagas disease have thus been observed in agricultural workers during the last 10 years (Barbosa et al. 2015). In this case, not only is the pathogenic ecosystem different from anything that we have known so far, but the driver of the epidemic is economic: it is the strong and growing demand for acai, a product decked with admirable virtues. Commercial pressure is certainly a new environment, but it is an environment that bears a great deal of responsibility in the transmission of the disease. Before concluding, let us recall that there are other modes of transmission of *T. cruzi* through the maternal-foetal route, by blood transfusions and organ transplants, which open new perspectives in the history of the disease.

7.8 Conclusion

The word "cnvironment" has been used intentionally to identify a large number of different situations, each of which contributes to the emergence, transmission and discovery of Chagas disease. First, we can conclude from this survey that there is no simple, a fortiori, unique environmental model to account for Chagas disease nor to trace its history since humans settled in South America. Chagas and most of those who have followed him put forward a model, which they observed at Lassance and which exists incontestably. Perhaps even it is still currently dominant. This model has been constantly promoted since 1909, probably because it merely gave a hold over the disease. Plastering a wall and sealing the cracks, and applying insecticides, can be easily planned at the city or state level and then carried out: the major limitation for this local action is resistance to the pesticide and its toxicity, two commonplace phenomena in the anti-vectorial struggle. The historical study of the disease, the multiple conditions of its occurrence and those of its discovery show, on the contrary, an intermingling of environmental causes or if one prefers here an intermingling of different environments: in building the Chagasic endemics, we find the association of biological ecosystems; lifestyles; religious, scientific and political positions; and economic tensions.

We are obviously very far from a causal relationship that is simple to define and manage. We are instead confronted with a multifaceted landscape deconstructed and later recomposed by the observer through selection of certain parameters.[9] Each of its facets is a sort of epistemic object of its own. Each is important only in relation to others, but this importance is not usually predictable at a level wider than the initial observation. Chagas was lucky enough to arrive at a location, Lassance, not a new location – the region has been farmed for a long time – but a location where, around the railway station that has just been built and which would soon serve as a dispensary, organized in a few months a precarious city of manual workers coming from all horizons of the Minas. Without being aware of it, in front of his very eyes, Chagas witnessed the aggregation of a poor population from the surrounding area, who gathered together in "archetypal" houses immediately colonized by wildlife-infected *barbeiros*. These facets literally amassed before his eyes and justified the notion of a new disease. This is also what explains the epidemiological errors of the years that followed. For, after Lassance, epidemiological studies investigated landscapes whose transformation by man is essentially complete, which are now in a sort of equilibrium where a new disease can no longer appear out of nothing: other tools were needed (Kropf 2009). This may explain some of Chagas' difficulties and may serve as a first conclusion: Chagas benefited from the emergence of an "epistemic landscape", the creation of a pathogenic isolate, which made apparent a disease that had existed previously but had never been seen until then, and besides it was difficult to see outside of these very conditions.

The overview of 10,000 years of the coexistence of human beings, *barbeiros* and the variants of *Trypanosoma cruzi* (there are at least four that trace the migration of the parasite from the Andes to Central America) shows an astonishing plasticity in the various actors vis-à-vis the changes that occur in their respective environments. Some of these changes are related to the biology of the relationship that *barbeiros* form with their natural reservoirs, relationships that vary in time and space; some others are related to how human beings come into contact with infected bugs. All

[9] The very particular place that the landscape occupies in parasitology remains to be examined in depth. This is by no means specific to Chagas disease. The attention paid to pathogenic landscapes is evident in the earliest texts on malaria (cf. Celsius for the Pontine marshes) and especially in the research which began with the discovery that Anopheles are the vectors. In all cases, the biology of the vector became an essential element of the description of the disease, and the place – the landscape – where the vector lives is part of the analysis. The description of the landscapes, fauna, flora and the fluids of the studied places became extremely rigorous in the hands of the Rockefeller Foundation from 1913 onwards and in particular with the impetus of the malaria committee of the League of Nations after 1922. Reports addressed to both organizations are examples of very fine ecosystem analysis. The aim of this work was to determine whether the medium is suitable for the proliferation of the vector and its variants or sometimes suitable for the introduction of predators, such as larvivorous fish. Photography became the key tool for communication between malaria specialists. The description of pathogenic landscapes is also found as early as 1900 in the case of the tsetse fly vector for sleeping sickness. The corollary of these "ecosystemic" approaches is the temptation to transform the landscape in order to abolish its pathogenicity, for example, as implemented in Fascist Italy (Snowden 2006, Gachelin 2013) and even more exemplarily in Palestine from 1920 (Sufian 1992), to cite only two well-studied examples.

this has varied constantly under the effect of multiple influences, which in a certain way can be called sociocultural and that from the Chinchorro civilization. This plasticity of the various faces of the actors of a disease is not specific to Chagas disease. It is also found in malaria, the emergence before our eyes of *Plasmodium knowlesi* malaria in Borneo being almost an experimental demonstration.

The plurality of words, arbitrarily grouped under the name of environment, as encountered in the varied contexts of the study of Chagas disease, finally prevents a general sense of what surrounds and determines the disease. Each of these words, on the other hand, has a precise meaning. The environment only serves as a support for a qualification: scientific, religious, economic, political, social, forestry, domiciliary, etc. In the context of the "health-environment relationship" of vector-borne diseases, the environment cannot be defined as a general character, and the word is applied, interchangeably, to a multitude of particular situations. We are confronted with the vague evidence of a role, in the vague categories of "things that surround", which, on balance, has absolutely no meaning. Thus, brandishing the word "environment" alone, without qualifying it, and then linking it to health, a frequent media and political practice, amounts to invoking the spirit of a kind of fuzzy neo-hippocratism.

What does make sense, however, in the course of this discussion on the various meanings given to "environment" is that we must replace vain research for a simple definition and environmental causality with more precise research, maybe renewed each time, of clusters of causes specific to a place, a time and a situation. In other words, the environment in a form that is "usable" both intellectually and practically is reduced to that of which we have an immediate perception and on which we have a possibility of personal action, in brief to the *oikos* of an ensemble that is certainly complex but well-positioned in space and in time. Hence, this point is of vital importance. The practical consequences of this remark (already well-known through the Rockefeller Foundation in its fight against malaria between the two world wars) are in fact applied in the management of other vector-borne diseases. In the case of malaria, for example, studies conducted by the *Cochrane Collaboration Groups for Infectious Diseases* have shown the repeated failure of large-scale interventions to eradicate insect or parasite populations: except in special situations such as on islands, there is first a decline in endemic malaria and then a recovery. The effective response to these repeated large-scale anti-vector policy failures dictated by the simplification of the parameters taken into account is given by local initiatives, particularly developed in Africa. In these cases, small communities take control of the vector and the disease. The results in rural areas are impressive. The notion of a precise and definable local environment makes sense here on the scale of a human group and its particular ecosystem, which is certainly different in the Southern Sahel from what it may be in Kenya but also from one village to another. In any case it is an environment known and controlled by those who live there that is the place on which the analysis is concentrated. The environment-health relationship can thus acquire density when placed in these tenuous geographical spaces, probably close enough to the original definition of ecology given by Haeckel in 1866.

Archive Sources

Archives of the Casa de Oswaldo Cruz, Rio de Janeiro (RJ, Brazil): funds Carlos Chagas and Evandro Chagas Hospital
Carlos Chagas Museum, Lassance (MG, Brazil)
Archives of the Institut Pasteur (Paris): fund Emile Brumpt

Bibliography

Araujo, A. et al. 2009. Paleoparasitology of Chagas disease: A review. *Memórias do Instituto Oswaldo Cruz* 104 (Suppl 1): 9–16.
Aufderheide et al. 1993. A 9000-year record of Chagas' disease. *Proceedings of the National Academy of Sciences of the United States of America* 101: 2034–2039.
Barbosa, M., et al. 2015. Chagas disease in the State of Amazonas: history, epidemiological evolution, risks of endemicity and future perspectives. *Revista da Sociedade Brasileira de Medicina Tropical* 48 (Suppl 1): 27–33.
Benchimol, J., and A.F. Da Silva. 2008. Railroads, disease, and tropical medicine in Brazil under the first republic. *História, Ciências, Saúde – Manguinhos* 15 (3): 719–762.
Benchimol, J., and L.A. Texeira. 1993. *Cobras, lagartos e outros bichos*. Rio de Janeiro: Editora Fiocruz.
Bergard, Laird W. 2006. *Slavery and the demographic and economic history of Minas Gerais, Brazil*. Cambridge Latin America studies. Cambridge, UK: Cambridge University Press.
Bestetti R. and G. Gachelin. *Early Clinics of Chagas Diseases. What did physicians actually see*. In preparation.
Blanchard, R. 1889. *Traité de zoologie médicale*. Paris: J.B. Baillière et fils.
Brumpt, E. 1910. *Précis de parasitologie*. Paris: Masson et Cie.
———. 1913. *Précis de parasitologie*. Paris: Masson et Cie. Second edition 1922.
Chagas, C. 1906. *Prophylaxia do impaludismo*. Rio de Janeiro: Typographia Besnard frères.
———. 1909*a*. *Nova* especie morbida do homem, produzida por un tripanossoma. Brasil medico 16: 227a-b (preliminary note); and, same title, 17: 175–176.
———. 1909b. Nova tripanozomiaze humana: estudos sobre a morfolojia e o ciclo evolutivo do Schizotrypanum cruzi n. gen., n. sp., ajente etiolojico de nova entidade morbida do homem. *Memórias do Instituto Oswaldo Cruz* 1 (2): 159–218.
———. 1911. Nova entidade morbida do homem: rezumo geral de estudos etiolojicos e clinicos. *Memórias do Instituto Oswaldo Cruz* 3 (2): 219–275.
———. 1916. Tripanosomiase americana: forma aguda da molestia. *Memórias do Instituto Oswaldo Cruz* 8 (2): 37–60.
Chagas, C. and Eurico Villela. 1922. Forma cardiaca da Trypanosomiase Americana / Cardiac form of American Trypanosomiasis. *Memórias do Instituto Oswaldo Cruz* 14 (1): 5–61.
Cialdella, N and Navigantes, Alves L. 2014. La ruée vers l' "*açaí*" (*Euterpe oleracea Mart.*): trajectoires d'un fruit emblématique d'Amazonie. Revue Tiers Monde 220: 119–135.
Cockburn, T.A., E. Cockburn, and T. Reyman. 1998. *Mummies, Disease and Ancient Cultures*. Cambridge, UK: Cambridge University Press.
Coluzzi, M., G. Gachelin. et al. 2008. Insects and illnesses: Contributions to the history of medical entomology. *Parassitologia*, 50, special issue.
de Lery, J. 1576. *Histoire d'un voyage faict en la terre du Brésil*. Reedition Paris Hachette, 2000, from the text edited by Edition Lemerre, Paris 1880.
Dean, W. 1996. *A ferro e fogo: a historia e a devastaçào da Mata Atlantica Brasileira*. Sao Paulo: Compahia das letras.
Delaporte, F. 1999. *La maladie de Chagas*. Paris: Payot & Rivages.
Fantini, B. 1994. Anophelism without malaria: An ecological and epidemiological puzzle. *Parassitologia* 36 (1–2): 83–106.

Fiocruz. 2011. *Cinematografo Brasileiro em Dresden. Casa de Oswaldo Cruz*. Rio de Janeiro: Brésil. (https://portal.fiocruz.br/video/cinematografo-brasileiro-em-dresden-audiodescricao)
Gachelin, G. 2013. The interaction of scientific evidence and politics in debates about preventing malaria in 1925. *Journal of Royal Society Medicine* 106: 415–420.
Gachelin, G., and R.B. Bestetti 2017. Early clinics of the cardiac forms of Chagas' disease: discovery and study of original medical files (1909–1915). *International Journal of Cardiology*. https://doi.org/10.1016/j.ijcard.2017.06.102
Gachelin, G., and A. Opinel. 2009. The reception by French physicians of Chagas' discovery of *Trypanosoma cruzi* and American trypanosomiasis (1909–1925). *Historia, Ciencias, Saude, Manguinhos* 50: 35–56.
Gachelin, G., et al. 2013. *La naissance des biotechnologies pharmaceutiques en France (1887–1914). Le sérum antidiphtérique de l'Institut Pasteur*. Paris: Presses universitaires de France.
Gachelin G, P. Garner, E. Ferroni, J.P. Verhave, A. Opinel and I. Chalmers. 2017. Evaluating quinine for treating and preventing malaria. Part 1. *Journal of Royal Society Medicine* 110, 31–40 and 73–82.
Guerra, F. 1970. American trypanosomiasis. An historical and a human lesson. *The Journal of Tropical Medicine and Hygiene* 73: 83–118.
Kraus, R., C. Maggio, and F. Rosenbusch. 1915. Bocio, cretinismo y enfermedad de Chagas. *La prensa medica argentina* 2: 2–5.
Kropf, S.P. 2009. *Doença de Chagas, Doença do Brasil. Ciencia, saude et naçao, 1909–1962*. Rio de Janeiro: Editora Fiocruz.
Löwy, I. 2002. *Virus, moustiques et modernité: la fièvre jaune au Brésil, entre science et politique*. Paris: Archives Contemporaines. Histoire Des Sciences, Des Techniques et de la Médecine.
Neiva, A. 1910. Informaçoes sobre a biologia do *Conorhinus megistus* Brum. *Memorias do Instituto Oswaldo Cruz* 2: 206–212.
Opinel, A. 2008. The emergence of French medical entomology: The influence of universities, the Institut Pasteur and military physicians (1890-c.1938). *Medical History* 52 (3): 387–405.
Opinel, A. and G. Gachelin. 2005. *Parasitic diseases in Brazil: The construction of parasitology. XIXth-XXth centuries*. Parassitologia, special issue, 47.
Perleth, M. 1997. *Historical Aspects of American Trypanosomiasis (Chagas'disease)*. Frankfurt am Rhein: Peter Lang.
Rivera, M.A., and F. Forthhammer. 1991. The Chinchorro people of Northern Chile, a review of their culture and relationships. *International Journal of Anthropology* 6: 243–255.
Sá, M.R. 2005. The history of tropical medicine in Brazil: The discovery of Trypanosoma cruzi by Carlos Chagas and the German School of Protozoology. *Parassitologia* 47 (3–4): 309–317.
Sigaud, J.F.-X. 1844. *Du Climat et des maladies du Brésil, ou Statistique médicale de cet empire*. Paris: Fortin & Masson.
Silverman, H., and W.H. Isbel. 2008. *Handbook of South American Archaeology*. New York/ Heidelberg: Springer.
Snowden, F.M. 2006. *The Conquest of Malaria: Italy (1900–1962)*. New Haven: Yale University Press.
Steverding, D. 2014. The history of Chagas disease. *Parasit Vectors* 7:317. https://doi.org/10.1186/1756-3305-7-317.
Sufian, S.M. 1992. *Healing the land and the nation. Malaria and the zionist project in Palestine, 1920–1947*. Chicago: The University of Chicago press.
Thevet, A. 1557. Les singularitez de la France antarctique, autrement nommee Amerique, & de plusieurs terres et isles decouvertes de nostre temps. Paris: Maurice de La porte. Reedited in 1997, Paris: Editions Chandeigne.
Torres, M. 1915. Alguns fatos que interessam á epidemiolojia da molestia de Chagas. *Memórias do Instituto Oswaldo Cruz* 7 (1): 120–138.
WHO. 2007. Reporte sobre la enfermedad de Chagas. Grupo de trabajo científico 17–20 de abril de 2005 – Actualizado en julio de 2007 – Buenos Aires, Argentina.
Wozniak, E.J. et al. 2015. The biology of the triatomine bugs native to South central Texas and assessment of the risk they pose for autochthonous Chagas disease exposure. *Journal of Parasitology* 101, 5: 520–528.

Part III
Healthy or Unhealthy Environments: Sensorial Experiences of Lands, Spaces, and Milieus

Chapter 8
The Worst Environment in Which to Live in China: A Question of Points of View. The Legendary Miasmatic Far South of China Challenged by Local Doctors in Late Imperial China

Florence Bretelle-Establet

Abstract The South of China, Lingnan in particular (Lingnan = south of the Ling Mountains, including the present provinces of Guangxi and Guangdong), has often been described as the most frightening place in China to live with particular diseases, notably *zhang* or *gu*, usually rendered as "miasmas" and "poisons." This chapter focuses on the "geographical imagination" of disease and health in China and examines how, when, and why the image of a particular unhealthy place was constructed and/or challenged in history. This chapter starts with a survey of the main ideas that were promoted in ancient medical treatises as regards the relationships between particular environments, expressed in terms of geographical divisions, and health issues and how these ideas were later reshaped or challenged. It then examines the historical construction of the Far South as being the unhealthiest place to live in China and sheds light on the peculiarities which were usually attached to the Far South. In the eyes of many actors who wrote about Lingnan, the dangers of the place lay in both its natural and cultural features: the ground was low-lying and humid, and there was an abundant Yang Qi pouring on the ground, which mixed with the Yin Qi of the earth caused suffocating vapors, giving rise to a luxurious vegetation, dense forests, and snakes. And in this particular suffocating Qi, there were aboriginal Man or Yi, who were known to make poisons that could threaten the lives of the Han Chinese. Finally, drawing on medical treatises that were produced in the Far South, and are extant today, it shows how this image of the unhealthiest place to live was challenged by local doctors in late imperial China.

F. Bretelle-Establet (✉)
Université de Paris, SPHERE, UMR 7219, CNRS-Université de Paris, Paris, France
e-mail: f.bretelle@univ-paris-diderot.fr

F. Bretelle-Establet et al. (eds.), *Making Sense of Health, Disease, and the Environment in Cross-Cultural History: The Arabic-Islamic World, China, Europe, and North America*, Boston Studies in the Philosophy and History of Science 333, https://doi.org/10.1007/978-3-030-19082-8_8

Keywords Late imperial China · Zhang · Miasma · Yunnan · Guangxi · Guangdong · Lingnan · Healthy environment · Unhealthy environment · Medicine · Local gazetteer · Health · Disease

8.1 Introduction

Earlier in this volume, Catherine Despeux explains how ancient Chinese scholars, poets, philosophers, and physicians understood the relationship between the environment and humans. She also detailed what a good natural environment for health meant in ancient China. The present chapter focuses on what, in Chinese history, constituted the worst environment in which to live. It also examines the terms in which the relationships between unhealthy environments and human health were understood. The objective here is not to attempt to draw an a posteriori epidemiological picture of disease in Chinese history nor to show the genius of some observations and observers, in the light of what we know today of the relationships between natural environments and disease. Rather, the aim of this chapter is to understand how, in the eyes of many Chinese protagonists, notably medical doctors and scholars, the "geographic imagination"[1] of the worst place to live emerged, grew, and at times was challenged. Before shedding light on what constituted the unhealthiest place to live in China, it is necessary to have in mind some elements brought by Chinese medicine, to understand notably how, in medical thinking, the relationships between human beings, their natural surroundings, and localities were perceived or explained. This chapter thus starts with a survey of the main ideas that were promoted in ancient medical treatises as regards the relationships between particular environments and health issues and how these ideas were later reshaped or challenged in the mainstream. It then examines the case of the Far South margins of the empire, which, during the long run of imperial history, were shaped by many scholars—officials, poets, and physicians—as the unhealthiest place to live. These scholars' records allow light to be shed on the peculiarities which defined the most harmful and frightening place for human health. Finally, drawing on the medical treatises that were produced in the Far South, and are extant today, it examines how local medical doctors saw the environment in which they lived and worked and what they had to say about environment, disease, and health. The comparative analysis of these records reveals that the unhealthiest environment in which to live depended on the geographic and social perspectives from where it was seen and smelt.

[1] I borrow this concept from Marta Hanson and her book on epidemics in Chinese history, *Speaking of Epidemics in Chinese Medicine: Disease and the Geographic Imagination in Late Imperial China* (Hanson 2011).

8.2 Human Beings, Health, and Localities in Chinese Medicine

While the historiography of medicine in China is a field of research that has expanded tremendously in recent decades, it is still impossible to offer an exhaustive picture of how Chinese medical thinkers conceived of the relationship between human health and the environment in history. Medical texts from the ancient past that are extant today are rather scarce and have been well studied by historians, but the medical texts produced since the Ming dynasty (1368–1644) and during the Qing (1644–1911) present a real challenge for historians. In fact, a large number of Chinese medical texts have come down to us since the mid-seventeenth century, but, so far, only a tiny fraction have been studied, with particular attention given to the medical texts produced in the cultural center of Jiangnan (*lit.* south of Yangzi River, an area encompassing the current provinces of Anhui, Zhejiang, Jiangxi, and Jiangsu. See Map 8.1).[2] The work carried out by historians on these medical sources

Map 8.1 Jiangnan and the far southern margins (Yunnan, Guangxi, Guangdong)

[2]Of the 12, 124 medical books preserved today in the 113 biggest libraries in China, 11,129 were written from 1600 onward (Xue 1991). The catalogue of books written (but not always preserved) in all administrative units of the Chinese empire also shows a dramatic increase in titles from the seventeenth century (Guo 1987). These texts had more chance of being preserved thanks to the commercial printing boom which took place in the late Ming (1368–1644) (see Brokaw and Chow 2005; Brokaw 1996, 2007; Chia 1996, 2002; MacDermott 2006).

provides a good starting point for examining the salient features of the medical discourse on the relationships between the body, the natural environment, health, and disease, throughout history.[3]

8.2.1 Man, Disease, Health, and Environment in the Inner Classic of the Yellow Thearch

The idea that the natural environment could impact on human health is mentioned in the first medical treatise transmitted by the written tradition. The *Inner Classic of the Yellow Thearch (Huangdi Neijing* 黃帝內經, abbreviated later *Inner Classic)*,[4] which played a prominent role in the history of Chinese medical theory and practices, mentions links between human beings and their natural surroundings. It is in fact the medical text which, adopting the cosmological thinking which arose in the third century BCE and flourished in Han China (−226; 209), shapes human beings as intrinsically woven into a wider web, the cosmos. As in most ancient cultures, this cosmology was correlative, drawing systematic correspondences between different entities of the cosmos, including the human body, the state, or the celestial configurations.[5] This correlative cosmology relied on Yin 陰 and Yang 陽 and the Five Phases *wuxing* 五行 (Wood, Fire, Earth, Metal, Water). Used independently at first but merged into one cosmological paradigm under the Han dynasty, Yin/Yang and the Five Phases allowed any element or phenomenon in the universe (seasons, sounds, colors, flavors, viscera, emotions, orifices, seasonal Qi such as wind, etc.) to be classified into two or five series. Thanks to the constant movement of increase and decrease of Yin and Yang, and the movements of generation and control, notably, between the Five Phases, it allowed the world and its changes to be explained.

While as early as the first century CE, some critics emerged against this systematic correlative thinking,[6] this cosmology was adopted in all the philosophical

---

[3] See notably (Chao 2009; Grant 2003; Furth 1999; Hanson 2001, 2006, 2011; Scheid 2007).

[4] This text, compiled during the Han dynasty (−206, 220), was commented several times before the Imperial Editorial Office published an authoritative version in the eleventh century which is the first version we have today. This text which continued to be commented is made of two parts, the first, the *Suwen* (Basic Questions), considered the theoretical part, and the second, the *Lingshu* (Magic or Divine Pivot), considered the basis for acupuncture and moxibustion. It has been translated many times, but we refer to Paul Unschuld and Hermann Tessenow's translation (2011; 2016) whose objective was not to provide a guide for today's clinicians but to allow the ideas of the authors who composed it some 2000 years ago to be understood. Since this translation includes many successive comments, it also allows us to see how this text was understood at different times.

[5] On the beginning of a cosmological thinking, see notably (Henderson 2010). On the systematic correspondences between the cosmos, the state, and the human being, see (Sivin 1995), for whom, "In China, the ideas of Nature, state, and the body were so interdependent that they are best considered a single complex."

[6] The most famous critic is the Han scholar Wang Chong (27–97) who underlined that his correlative system could give rise to absurd situations: "The horse is connected with *wu* 午 (itself connected to Fire), the rat is connected with *zi* 子 (Water). If Water conquers Fire, rats normally chase horses…" quoted in (Ho 1985, p. 22).

schools of the time to account for the regular changes in the cosmos, in the state, and in man, who is the hybrid medium between heaven (Yang) and earth (Yin). The ways of conceiving of the body, of diseases, and of the treatments in the medicine that developed at this time attest to this. As Catherine Despeux clearly shows earlier, quoting in extenso one of the *Inner Classic*'s descriptions of the human body, each part of the human body has a counterpart in the natural world and vice versa.[7] Other passages show that they were both regulated by the same dynamics. Because of this correlative thinking, the human body, in all its constituent parts, was intrinsically linked to its natural surroundings, in all its constituent parts. The liver, eyes, sinews, and membranes linked to the Wood phase, for instance, were thought of as having a particular affinity with the environmental elements of East, wind or spring, because these were linked to the same phase. The relationships between morphological and functional units of the human body and the environment were thus explained, on a general and abstract level, by correlative thinking and systematic correspondences within the cosmos.[8] However, the medical treatise mentions a less abstract means of interaction and resonance between the body and its natural environment: the human body could, potentially, be penetrated by natural agents which, in the medical thinking that developed at the time, became the essential causes of disease. Wind, cold, heat, and dampness, among others, were the agents in the natural world able to bring disease, replacing the demonic creatures of the previous medical cultures attested by the Mawangdui manuscripts.[9] These natural or climatic agents were known to enter the body, notably through the skin and the body hair.

As in most ancient medical cultures, the *Inner Classic* reveals that the natural environment and human health were closely related. A person's health and susceptibility to disease were dependent on the ability to cope with the cosmos, and notably with the natural agents, or specific Qi, linked to the seasons, and more generally, to time and the calendar.[10] Parts of this founding treatise also evoke geographic features which, more specifically, accounted for human health, disease, and its treatment. As Marta Hanson (2011, p. 30–35) notes, such links are detailed and explained in Chapters 5, 12, and 70 of the *Suwen* and in Chapter 77 of the *Lingshu*, the two parts of the *Inner Classic*. Chapter 12, "Discourse on Different

[7] See in this volume (p. 84–85) Catherine Despeux's quote and the wider frame of resonances and interactions between the human body and its environment.

[8] (Unschuld 2003, p. 112).

[9] While there is not one single pathogenic model in the *Suwen*, the dominant model is that natural agents enter the organism and cause disease (Unschuld 2003, p. 180–183). On the transition between a demonological thinking in medicine attested in the Mawangdui manuscripts and a medical thinking which relies on natural laws, like in the *Inner Classic*, see (Unschuld 2003, esp. p.180–198).

[10] Coping with the cosmos was also important in the medical thinking of the Arabic-Islamic civilization as underlined in this volume by Katouzian-Safadi and Bensaad, and it was also the case in medieval Europe until the sixteenth century when human health and the human body became less dependent on the invisible powers of the cosmos (Vigarello 1993 [1999], p. 82–84).

[Therapeutic] Patterns Suitable [for Use in Different] Cardinal Points," partly quoted in Despeux's contribution, gives a first understanding of this link. It is introduced by a question by Emperor Huangdi: "When physicians treat disease, one identical disease may be treated differently, and is always healed. How is that?" "Physical features of the earth (*dishi* 地勢) let it be this way," answers his minister Qi Bo who develops his explanations by cutting the earth into five regions, according to the cardinal directions East, West, North, South, and Center[11]:

> The fact is, the region of the East, this is where (or the region yu 域) heaven and earth give life first. [This is] the land (di地) of fish and salt; beaches border on the water. Its people eat fish and crave for salty [meals]. All of them consider their place of residence as comfortable and they regard their food as delicious. As for the salty [meals], they defeat the blood. Hence, its people all have a black complexion and open [skin] structures. Their diseases are always yong-abscesses and ulcers. For their treatment, pointed stones are suitable. Hence, it is for sure that [the therapy with] pointed stones originated in the East
>
> The West, this is the region (yu 域) of gold and jade. It is the location (chu 處) of sand and stones. It is where heaven and earth contract and pull [things together]. Its people live in earthen mounds and the winds are frequent. The water and the soil are hard and strong. Its people do not dress with clothes but wear garments of hair and grasses. Its people [enjoy] rich food and they are fat. Hence, [external] evil cannot harm their physical body (xingti形體); their diseases emerge from within. For their treatment, toxic drugs are appropriate. Hence, it is for sure that [the treatment with] toxic drugs originated in the West.
>
> The North, this is the region (yu 域) where heaven and earth secure and store. Its land lies at a high elevation, [its people] live in earthen mounds. Wind, cold, and piercing frost [dominate]. Its people find joy in living in the wilderness and in consuming milk. Their depots are cold and generate diseases of fullness. For their treatment, cauterization is appropriate. Hence, it is for sure that cauterization originated in the North.
>
> The South, this is [the region] where heaven and earth bestow growth and nourishment; it is the place (chu 處) where yang [qi] is present in abundance. Its land lies low; water and soil are weak. It is [a region] where fog and dew assemble. Its people crave for sour [foods], and they eat [food of] a strong odor. Hence, its people all [have] tight [skin] structures and a reddish complexion. Their diseases are cramps and block. For their treatment, fine needles are appropriate. Hence, it is for sure that the nine needles have originated in the South.
>
> The center, its land (qidi 其地) is flat and damp. It is here that heaven and earth generate the myriad beings in large numbers. Its people eat diverse [foods] and are never fatigued. Hence, their diseases are often limpness with receding [qi], and [spells of] cold and heat. For their treatment, guiding-pulling and pressing-lifting are appropriate. Hence, it is for sure that guiding-pulling and pressing-lifting have originated in the center.

In this chapter, it was acknowledged that regions, places, or lands (*yu* 域, *chu* 處 or *di* 地), defined by their cardinal directions, the quality of their terrain, their altitudes, and climates had relationships with certain cultural habits, such as diet, dress codes and housing which all together produce various complexions and diseases in men, which in turn, required specific therapies. In this chapter, at first glance, the influence of the environment on human health was explained in terms of local cultural lifestyles.[12]

---

[11] (Unschuld & Tessenow (trans.) 2011, vol 1., p. 211–218).

[12] That places, local climates, typography, and local habits are intrinsically interdependent finds, I

In parallel to this geographic division into five specific areas, another division was mentioned as accounting equally for differences in the body and in diseases. In Chapters 5 and 70 of the *Suwen*, in order to explain the corporal differences between men, the medical treatise pinpoints a division of the territory into two portions, Northwest *xibei* 西北 and Southeast *dongnan* 東南, a division which, Hanson recalls, occurred in earlier literary texts and had mythological roots.[13] In Chapter 5 of the *Suwen*, the differences between two opposed places that account for the differences in the body are explained in the cosmological terms of Yin and Yang. The Northwest is described as insufficient in heaven: the altitude is higher than in the Southeast. It is therefore a place where the earth (Yin) dominates. The Southeast is described as a place where earth is lacking, the altitude being low. It is therefore a place where heaven (Yang) dominates. The prevalence of Yin or Yang in these two areas implies Yin or Yang depletions in certain parts of the body, in these two areas, which, in turn, favor specific weaknesses: weakness in the high and right (Yang) parts of the body for those who live in the Northwest (Yin) and weakness in the low and left (Yin) parts of the body in people living in the Southeast (Yang):

> Heaven is not sufficiently present in the North-West. Hence the North-West (*fang*方) is Yin, and the ears and the eyes of man on the right are not as clear as on the left. The earth is incomplete in the South-East. Hence the South-East is Yang, and the hands and feet of man on the left are not as strong as on the right.[14]

This geographic division appears again in Chapter 70 of the *Suwen*, characterized by specific Qi—cold Qi in the high-lying Yin Northwest regions and hot Qi in the low-lying Yang Southeast regions—causing specific diseases in people going to these different places:

> Hence the earth has highlying and low-lying [regions]; among the Qi there is warm [qi] and cool [qi]. In the highlying [regions], the Qi is cold; in the low-lying [regions], the Qi is hot. Hence those who go where it is cold and cool, they experience distension. Those going where it is warm and hot will suffer from sores.[15]

To live in the Northwest or in the Southeast, in high- or low-lying regions, where the Qi was either mostly Yin or Yang, not only caused bodily specificities—like the texture of the skin, whose pores were more closed or open and thus less or more able to let the Yang Qi pour outside or act as a barrier toward the intruders—but also imparts a different life span:

---

think, its illustration in the fact that until the beginning of the 1930s, local gazetteers, a type of official historical source, bring together in a same chapter the geographic descriptions of localities and local customs (fengsu 風俗). Only after the 1930s did local gazetteers report local customs in a new and independent section entitled "society" (shehui 社會).

[13] (Hanson 2011, p. 26–27): this division of the environment is mentioned in various texts from the third century BCE and is explained in the *Liezi*: a mythical fight between Gong Gong and Zhuan Xun, in which Gong Gong, in his ire, destroyed one of the eight pillars of heaven hence making the earth lean toward the Southeast.

[14] (Unschuld and Tessenow 2011, vol 1, p. 116).

[15] (Unschuld and Tessenow 2011, vol 1, p. 326).

> Those [living] in highlying [regions], their Qi gives them longevity. Those [living] in low-lying [regions], their Qi lets them die early.[16]

As a consequence of the relationships between location (Northwest, Southeast, high- and low-lying regions) and cosmological Qi, those who treat illnesses ought to know the specific Qi of particular localities and that of their inhabitants in order to adapt their treatments:

> It is essential to adapt to the respective qi; thereby it is possible to achieve a balance [...] Di: "Good! The qi of one single district [brings about] differences in generation and transformation, in longevity and early death. What is the reason for this?" Qi Bo: "The [differences in the] structures of high-lying and low-lying [regions] are caused by [differences in] the physical features of the earth (地勢). The highest [regions] are governed by the yin qi. The lowest [regions] are ruled by the yang qi. When the yang [qi] dominates, it precedes heaven. When the yin [qi] dominates, it is behind heaven. That is the regularity of the structures of the earth and the Way of generation and transformation [...] Hence those treating an illness must comprehend the Way of heaven and the structure of the earth, the alternating domination of yin and yang, the earlier or later [arrival] of the qi, human longevity and early death, and the recurring times of generation and transformation. This way they are able to understand human physical appearance and qi.[17]

Finally, *Lingshu*, in Chapter 77, mentions a division of the environment into eight areas accounting for the production of variations in diseases. Winds from the eight cardinal and ordinal directions produced specific diseases. This division of the territory into eight portions did not seem to have had the same influence on later medical cultures as the division into two or five areas.

The quick survey of one of the most ancient medical treatises written in China reveals two important points for the argument developed in this contribution. Firstly, it bears testimony to the idea of a continuum between man and his natural milieu, a continuum ensured by the dynamic of cosmogonic forces which permeate everything in the same way, including the human body, and by the permeability of the human body. Human health and disease are intimately related to the environment. Secondly, this medical treatise argues that the environment is not uniform but essentially divided into two, five, or eight different territories, which account for variations in cultural habits, in human complexions, in diseases, and in life spans.[18] These statements are the bases on which later medical thinkers would conceive of the links between the environment, man, and health, an issue which draws greater attention in the second millennium. As Marta Hanson and Angela Leung have shown, the

---

[16] (Unschuld and Tessenow 2011, vol 1, p. 329).

[17] (Unschuld and Tessenow 2011, vol 1, p. 329).

[18] As in most ancient medical thinking, the *Inner Classic* reveals that the human body's structure and functioning were not independent of the place where it lived. What makes this discourse specifically Chinese, and specific of the time in which it was composed and/or compiled, is the way these relationships were understood and the theoretical elements mobilized for linking the natural milieu with the human body. Chapters 5 and 70, by dividing the land into two large portions, and making direct references to the Yin and Yang cosmological forces, attest to a conceptualization of land and the environment inseparable from the correlative cosmology of the time. Chapter 12 does not mention any of the terms founding this cosmology. However, by cutting up the space into five portions (linked to the Five Phases), it is also in line with the cosmological thinking of the time.

question of regional variations accounting for variations in disease, body types, and treatments comes to the forefront in the medical literature, from the thirteenth century onward, and occupies much space in the medical literature of the Ming (1368–1644) and Qing (1644–1909) dynasties.

8.2.2 Northwest, Southeast, and the Frightening Far Southern Margins of the Empire

As these scholars have shown, the growing claims in medicine after the thirteenth century that regional variations accounted for cultural and somatic differences were part of a more general criticism of the medical orthodoxy that was implemented under the Northern Song government (960–1127), with the printing and circulation of orthodox classics, notably the *Cold Damage Treatise (Shanghan lun 傷寒論).* This medical treatise was written by Zhang Ji 張機 (styled Zhang Zhongjing 張仲景, active in the present Hunan province) in the second CE. While drawing on previous medical treatises including the *Inner Classic* and on the systematic correspondences of correlative cosmology, it did not address the issue of geographic variation either to explain some differences in body types or to account for the prevalence of certain diseases. Mostly descriptive and directive, and less theoretical than the *Inner Classic*, this text which puts emphasis on cold and wind as causes of disease focused more attention on temporal or calendrical fluctuations: to explain notably the duration of a disease, the time when it lives one channel, or the daily or yearly time period a treatment ought to be given.[19] While very popular and highly respected in the following centuries, the *Cold Damage Treatise* became the target of criticisms, either for its incompleteness, for its unsuitability for curing contemporary diseases, or for its universalistic approach to diseases and treatments, without any consideration of geographic variations. This latter criticism was notably expressed in an eleventh-century medical treatise in relation to the issue of drugs, which, according to the critic, should not be given to all patients without considering where they lived. But the claim that geographic variations should be taken into account in medicine extended well beyond the sole issue of the use of drugs. This idea became more popular in medical texts after the thirteenth century, as the political center had been transferred from the northern Yellow River Basin (Northern Song, capital Kaifeng) to the southern Yangzi River Delta, after the Song court lost control of the northern part of the empire (Southern Song 1127–1279, capital Hangzhou). Skeptical of the orthodoxy established by the fallen Northern Song court, prominent medical think-

[19] This treatise provides a description of symptoms for a wide range of syndromes according to the channels affected, of therapeutic principles (heating, purging, cooling, etc.) to follow, and of the ways to make the suitable therapeutic formulas. If the original attack by wind or cold was treated too late or improperly, the disease would transmit to the other channels, from the most superficial to the deepest. This text was restored and published in 1065 by the Bureau for Revising Medical Texts, founded in 1057. See (Goldschmitd 2009).

ers of the late Song, Jin, and Yuan dynasties (1277–1367) developed a deeper reflection on disease and on its causes, with, as Leung stresses, a heightened awareness of environmental factors, including geographic particularities, social conditions, and lifestyles.[20]

Angela Leung and Marta Hanson have provided a very detailed analysis of the wide range of ideas and claims relative to this concern among the medical thinkers of this time, who, for the majority, wrote from the new political, economic, and cultural center of the empire, the Jiangnan area.[21] For them, regional differences were expressed less in terms of the five divisions of the *Inner Classic* than in those of its two portions. However, while in the *Inner Classic* this division had opposed the Southeast to the Northwest, in the medical literature of the Ming and Qing dynasties, it was often a South/North division which was put to the fore to account for variations in climates, body types, cultural habits, diseases, and treatments. While known since antiquity as low-lying, humid, and causing early deaths, the South was described more and more as hosting a local Qi, namely, a Qi linked to the ground, and no longer to heaven and its cosmological pattern. The combination of humidity and heat, causing suffocating vapors, formed a local Qi, characteristic of the South, and accounted for the general unhealthy character of the place: "Among the six Qi, hot humidity is responsible for eight or nine out of ten diseases."[22]

That places were governed by different local Qi implied, for many medical thinkers, differences in the prevalence of diseases. While some ancient medical texts, like the *Inner Classic*, had already noted that some diseases were specific to particular environments, this observation was more commonly shared later. Some thinkers argued, for instance, that neither true Cold damage (*shanghan* 傷寒) nor true Wind stroke (*zhongfeng* 中風) existed in the South, which had conversely more Humidity stroke (*zhongshi* 中濕), or that the "athlete's foot" (*jiaoqi* 腳氣, lit. Foot qi), firstly known as a southern disease, was, after its countrywide dispersion, different in the

[20] For (Leung 2003), the increase of this sensitivity was linked to the "Five Periods and Six Qi" doctrine *wuyun liuqi* 五運六氣 that developed during the Song and rose to dominance in the Jin and Southern Song. This doctrine, already present in the *Inner Classic*, but probably added only in the eight century, draws on notions of cyclical recurrence of calendrical terms to explain the occurrence of rain, cold, and other climatic agents, together with the occurrence of other natural facts, in the year (Unschuld 2003, p. 393–488).

[21] To get a precise account of these debates, see (Hanson 2001, 2006, 2011, p. 39–45; Leung 2002 and 2003).

[22] Zhu Zhenheng 朱震亨 (1282–1358) quoted in (Leung 2002, p. 183). Humidity was not the only emanation known to come from the earth or ground. As Leung stresses, other Qi such as *zaqi* 雜氣 or impure qi were also described as definitely emanating from the earth and not linked to the cosmic order. "What we call *zaqi*, while it is said that it is a cosmic Qi, is in fact a Qi that comes from the particular ground 方土之氣 because this Qi does come and grow from the earth" Wu Youxing 吳有性 (1642) quoted in (Leung 2002, p. 185). The belief of the South's unhealthiness was entirely shared by the Manchu emperor Kangxi (r.1662–1722), who expressed many times his fear of traveling south of the Yangzi River. Some of his edicts in which he expresses his observations on the part played by the environment on his own health are translated in (Hanson 2011, p. 105–106).

north and in the south because of different causes.[23] Finally, different local Qi implied that people were also different in their bodily constitution. While opinions could slightly differ among medical thinkers, it was commonly believed that people living in the north were stronger, lived more frugally, and had a longer life, their skin pores were closed tightly, they were thus impenetrable and replete in Yang, and they were rarely affected by external attacks of evil Qi (Yang etiology) but more affected by diseases induced by an internal cause, such as food (Yin etiology). Conversely, people in the South were delicate, used to living richly but died early, their skin pores were often open because of excessive sudation, their Yang was often depleted, and, consequently, they were more affected by external attacks of evil Qi than by internal diseases. In turn, northern and southern people needed different therapies: northern people were very quickly cured with bitter and cooling purgative drugs, while southern people should receive sweet restorative drugs.[24] According to this growing claim of regional variations in the prevalence of diseases and in bodily constitutions, to be a good doctor implies that one takes into account these regional variations. In line with advice already formulated in the *Inner Classic*, the reasons why regions and diseases differ were explained more in terms of geographic and cultural particularities than in terms of the general principles of correlative cosmology.

In the broad brush strokes of this geographic division, with the South or Southeast constituting a less healthy environment than the North or Northwest, it is not very clear what this meant on the ground. For instance, Zhang Heteng 張鶴騰, a seventeenth-century physician in a town in the present Anhui province, Jiangnan area, describes in the same way a very large area comprising the territories around the Yangzi River Delta to the present northern Vietnam in opposition to the northern provinces along the Yellow River: "The climate and customs have all shortcomings and good points. In Qin and Jin (ancient names for the present provinces of Shaanxi and Shanxi, crossed by the Yellow River, considered north), the local Qi is cold, this is why there is a lot of Cold diseases and very few Summer-heat diseases. In Wu (a name for the area comprising the present southern Jiangsu and northern Zhejiang, south of the Yangzi River Delta), Yue (Vietnam), Dian (Yunnan province), Qian (Guizhou province) and in Yue (Guangdong and Guangxi provinces), the local Qi is warm, so there are Cold diseases, but Summer-heat diseases are particularly intense here."[25] In Zhang's words, the area of the Yangzi River Delta was not different from the far southern areas of the empire. However, the southernmost places of the empire, namely, the southeast coast and the province of Yunnan, were usually seen as not fitting this North-South divide. As Hanson writes: "When referring to diseases within China proper [...] Ming physicians embraced the dualities of north-south and northwest-southeast [...]; When confronted with the boundaries between sickness and health beyond their normal clinical world, however, they expressed a

[23] (Leung 2002). On the history of this concept of disease, see (Smith 2017). While some today equate this term with modern beriberi, *jiaoqi* covers a wider range of ailments.

[24] On the development of these ideas in the medical literature of the Ming dynasty, see (Hanson 2006) and (Hanson 2011, p. 49–68).

[25] Quoted in (Leung 2002, p. 196).

different geographic sensibility."[26] As she notes, Xu Chunfu 徐春甫, a physician living in the sixteenth century, native of the present Anhui province, part of the Yangzi River Delta, would give advice to northern people (*beiren* 北人) and to southeastern people alike (*dongnan zhi ren* 東南之人) if they were about to travel or live in Lingnan, this far southern area encompassing the present provinces of Guangdong and Guangxi.[27] For him, as for many observers of his time writing from the Yangzi River Delta, and for the majority of observers from earlier times, Lingnan and the far southwest of the empire, including the present province of Yunnan, were the worst environment in which to live. Not only had this environment been suspected, since the first century CE, of producing abnormal diseases, but it was also considered by many Jiangnan doctors in late imperial China as a place where particularly terrifying diseases like leprosy could be contracted and was the breeding ground of new diseases like the Guangdong rotting sore (*Guangchuan* 廣瘡), usually considered as syphilis.[28]

The Far South was particular not only for doctors writing from the outside but also for nonmedical scholars who very early in history associated the place with *zhang*, a polysemic term encapsulating the images of the worst place to live. Other frightening and abnormal diseases were indeed associated with the Far South, such as *gu* 蠱 "poisons."[29] However, they do not have the same paradigmatic degree of what the worst environment for health meant: *zhang*, best rendered by another polysemic word "miasma," was used in the imperial administration as an official marker of the level of danger of the environments. Places known as and classified as having *zhang* were lands of exile for corrupt or unwanted officials from the Tang dynasty onward.[30] Having/being *zhang* was an additional category to the eighteenth century's administrative division into four groups (*chong* 冲, *fan* 繁, *pi* 疲, *nan* 難) which was used to classify the administrative units of the empire according to the specificity of difficulties they raised and that new officials may encounter.[31] The qualification of administrative units as having *zhang* or not was still used in the provincial gazetteer of Yunnan in 1901.[32] The Far South, stigmatized by the danger *zhang*, was the worst environment for health for most observers. Therefore, it constitutes a good window through which to understand what the unhealthiest environment meant to most Chinese observers.

[26] (Hanson 2011, p. 69).

[27] (Hanson 2011, p. 78).

[28] (Leung 2009, p. 32–33 and p. 179–182).

[29] For a wider range of abnormal diseases considered as characteristic of the Far South, see (Hanson 2011, p. 69–90). On the notion of *gu*, see (Diamond 1988; Obringer 1997).

[30] See (Schafer 1967, p. 38; Constant 2016).

[31] This system had been established during Yongzheng's reign (1723–1736) and classified provinces, prefectures, and subprefectures according to importance for transportation, importance in business, difficulty in gathering taxes, or levels of criminality.

[32] See the first column of tables in the "geography" section of (*Xu Yunnan tongzhi gao* 1901).

8.3 Places with *Zhang* or the Worst Environments for Health

The issue of *zhang* has been the object of numerous studies in the last decades. The identification of occurrences of this term, used alone or combined with adjectives, in a very wide range of written documents, has allowed historians to shed light on the different meanings conveyed by this term and to make historical maps of the environments qualified as *zhang*.[33] In his contribution on the history of *zhang*, the historian Liu Xiangxue noted however that works about *zhang* have often been completed without paying much attention to the historical, geographic, and social perspectives from where *zhang* was discussed and without taking into account testimonies from the actors who were living in areas qualified since antiquity as being *zhang*.[34] Here, I suggest examining the terms in which *zhang* was described in the Far South, in late imperial times, with particular attention to the perspectives from where it was discussed. To do so, I examine two kinds of contemporary sources produced by different actors: local gazetteers of the Far South and medical treatises written by practitioners working in the Far South.

8.3.1 Local Gazetteers and Local Medical Treatises: The Viewpoints of Outsiders and of Insiders

Local gazetteers or *difangzhi* 地方志 were official reports on each administrative unit of the empire, from the province to the smallest canton, and they were expected, since at least the Song dynasty, to provide information on a wide range of subjects, including historical events, the heavens, institutions, people, customs, local cultures, and what today we would call physical geography.[35] Local gazetteers are of paramount importance for learning about China's local history, and as many historians have already noted, they are probably the richest sources as regards accounts of *zhang*, especially in late imperial China. However, when one takes into account the issue of the narrative's viewpoint, one realizes that accounts about *zhang* in local gazetteers are of a different nature. In a small number of gazetteers, these narratives seem to be the fruit of descriptions coming from direct observation, and this is especially the case in the gazetteers linked to the administrative units of Yunnan province.[36]However, in most gazetteers describing administrative units in Guangdong

[33] See (Schafer 1967; Leung 2002, 2003, 2009; Bello 2005, Hanson 2011). Several articles have been published in Chinese only, notably, (Mou and Wang 2003; Zhou 2007; Liu 2013).

[34] (Liu 2013, p. 13).

[35] During the Sui and Tang dynasties, officials at prefecture level were required to send maps and explanations on their administrative unit to the court. However, this type of source grew under the Song dynasty and flourished under the Ming and Qing dynasties. On this type of source, see (Will 1992).

[36] One reason perhaps is that late imperial scholars compiling gazetteers for Yunnan administrative units had fewer historical documents from which to draw. Lingnan, integrated into the Chinese empire since the Han dynasty, was the object of many historical and geographic descriptions since

and Guangxi, accounts about *zhang* are made up of quotes or paraphrased quotes, extracted from earlier gazetteers and, above all, from other ancient central sources—historic, geographic, medical, or poetic. In addition to referring to the official dynastic histories, notably the *Book of Han* (*Hanshu* 漢書, written by Ban gu, 111) and the *Book of Sui* (*Suishu* 隨書, compiled by Wei Zheng, 636), these gazetteers refer to geographic records such as the *Well-Balanced Records of Guihai* (*Guihai yuheng zhi* 桂海虞衡志) written by the Song scholar Fan Chengda 范成大 (1126–1193), the *Critical Commentary on the Lingwai Daida* (*Lingwai daida jiaozhu* 嶺外代答校注) written by Zhou Qufei 周去非 in the twelfth century, or the later *Description of Yue (Yueshu* 粵述) written by the Qing scholar Min Xu 閔敘 (a metropolitan graduate in 1655). The common point to these official or private records is that their authors or compilers were officials who had been appointed in the present provinces of Guangxi or Guangdong for a few years.[37] These gazetteers also often refer to poetry written by scholars who had been sent to the south—like Su Shi 蘇軾 (styled Su Dongpo 蘇東坡 1037–1101) banished to near Guangzhou and later to Hainan—who used poetry to complain about their geographic and cultural isolation. Finally, they frequently refer to the medical guides which had been written during the Song dynasty for people obliged to go to the Far South, like the *Recipes to Protect Life in the South of Guang (Guangnan shesheng fang* 廣南攝生方 *)* or the *Recipes to Protect Life in Lingnan (Lingnan weisheng fang* 廣南衛生方). These medical guides were written during the twelfth and thirteenth centuries by scholars, native to central China and appointed to Lingnan for a few years.[38] In a nutshell, late imperial gazetteers of Guangxi and Guangdong provinces mostly rely on ancient descriptions of *zhang* written by outsiders.[39] As a matter of fact, one often finds in these gazetteers the recurrent motives about *zhang* that one finds in more central sources. One way to access some insiders' viewpoints is to listen to local doctors' voices.

Very few of the medical texts written in this area in late imperial times have survived today. Out of the 276 medical titles identified by Guo Aichun in the bibliographies of more than five hundred local gazetteers of Yunnan, Guangxi, and Guangdong, thirty-five (or 13%) are extant.[40] Moreover, this sample is not

---

the Song dynasty, while Yunnan became a truly Chinese area in the thirteenth century only, after the Mongol conquest of the province that the Ming court (1368–1644) finally consolidated. On the history of Yunnan, see (Bin 2009). (Bin 2009, p. 17) notes, for instance, that the first gazetteers on Yunnan were only produced in the Ming dynasty and more largely in the Qing dynasty, namely, coincidentally with the process of Yunnan's sinicization.

[37] The Song scholar Fan Chengda 范成大 wrote his book while he held a civil service position in the prefecture of Jingjiang 靜江府 (the present town of Guilin 桂林, Guangxi) between 1171 and 1175; Min Xu, native of Anhui, likely graduated between 1644 and 1661, was sent to the province of Guangxi as "Director of scholars."

[38] The former title is difficult to identify. The *Lingnan weisheng fang* was compiled by Li Qiu 李璆 and Zhang Zhiyuan 張致遠 and edited by a Buddhist monk Shi Jihong 釋繼洪 (Yuan dynasty).

[39] Out of the 37 gazetteers on administrative units of Guangdong and Guangxi I have explored, only two gazetteers speak of *zhang* without drawing on earlier sources. (Diamond 1988, p. 9) also notes how ancient accounts on *gu* (poisons) were repeated, expanded, and rarely questioned by officials in later gazetteers.

[40] (Guo 1987). The set of medical titles Guo Aichun identified in gazetteers certainly does not reflect the accurate picture of the situation. It is likely that compilers of local gazetteers were

representative of all that was written. It is mostly biased toward the urban elite of Guangdong province, and only one medical text from Yunnan has survived in libraries that, unfortunately, I have not been able to find so far. There are several reasons why so few of the medical treatises written there have been preserved. One reason is remoteness. It was difficult for authors writing from this area, especially from rural and hinterland Guangxi and Yunnan, to have their manuscript printed and thus preserved in many copies. Massive destruction caused by the wars which have occurred in the Far South of China since the second half of the nineteenth century is another reason.[41] As a consequence, the sample of local medical texts extant today is thus rather small in comparison with what was written. Still, it includes various kinds of texts, written from 1739 to 1909, by authors belonging to various social circles but all hailing from either Guangxi or Guangdong. This literature thus echoes local voices.

The objective of the comparative analysis of these different narratives is not to identify *zhang* in today's categories, whether medical, climatic, or atmospheric,[42] but to give some elements as to what made an environment repellent and terrifying for these different scholars.

8.3.2 *The Localization of Zhang in History*

The first occurrences of the term *zhang* are in the *Book of Later Han* (compiled in the fifth century CE) found in the biography of General Ma Yuan (d.49 CE) who had been sent to Yue (present northern Vietnam) to secure the far southern borders just conquered by the Han emperors and whose soldiers died because of a *zhang* epidemic. From this time on, there were an increasing number of accounts about *zhang* in a wide range of documents, in official and private sources, in historical and geographic records, and in poetry, and from the sixth century onward, the issue of *zhang* was also raised in medical treatises. In all these records, *zhang* was associated with particular places. The first occurrence placed *zhang* in the present North Vietnam. Chao Yuanfang, an imperial physician in the Sui court (581–618), in his *Treatise on the Origin and Symptoms of Diseases* (*Zhubing yuanhou lun* 諸病源候論, 610) located *zhang* in Lingnan (south of Ling (Mountains)) that covers the southern parts of the present provinces of Guangdong and Guangxi. However, in the Tang dynasty

---

unable to record all the medical titles produced in each corner of each administrative unit of China. The 廣州大典 (*Guangzhou dadian*, *The Great Encyclopedia from Guangzhou*), which scanned and reprinted in 2015 writings produced by Guangdong authors extant in libraries, reveals the existence of medical writings that were not identified in the gazetteers Guo Aichun has examined for the province of Guangdong.

[41] For a more detailed story of these medical texts, see (Bretelle-Establet 2010a, b).

[42] Some historians equate *zhang* with malaria. While, certainly, *zhang* may have at times designated this disease, it is difficult firstly to make *a posteriori* a full adequacy between an ancient term and a modern medical diagnosis; secondly, as will be shown later, *zhang* covers a wider range of significations.

(618–907), some northerners thought that all the south of the Yangzi River was *zhang*. Some documents from the eleventh century mention *zhang* in the present Qinghai and Tibet.[43] While the Far South was thus not the exclusive place known as *zhang*, in history, there was a general agreement since the sixth century that the most southern part of the empire, the "south of the Ling (Mountains)" (Lingnan 嶺南), "beyond the Ling (Mountains)" (Lingwai 嶺外), "Ling Mountains" (Lingbiao 嶺表), as well as the south western frontiers of the Chinese empire were notorious hotbeds of *zhang*. The map drawn by the historians Mou Zhonghan and Wang Caiping, on the basis of a systematic investigation of Ming and Qing local gazetteers, makes it clear that, from the fourteenth century onward, the places known to be *zhang* were the southernmost places including Lingnan, Yunnan, the southern parts of Guizhou and Hunan, and the southern part of today's Taiwan (Map 8.2).

Let us examine some of the terms used by scholars to describe *zhang* and the specific environments that were known to favor its prevalence.

8.3.3 The Complicated Identity of Zhang in Late Imperial Gazetteers: From Disease, Death Cause, to Repellent Emanations

As already highlighted by Carol Benedict in her work on bubonic plague in China, where narratives about *zhang* are reported in local gazetteers gives a first hint as to its meaning. *Zhang* was almost always discussed in the "climate *qihou* 氣候" subsection of the general heading "geography *dili* 地理." *zhang* was not seen as something extraordinary like epidemics, for instance, which by contrast were always mentioned in another section of a local gazetteer, usually the "extraordinary facts" or the "list of events" sections. In the minds of the scholars who used to report on it, *zhang* was definitively associated with place, either as an environmental feature, like the fact that snow falls there in winter or not, or as a disease intrinsically linked with a locality, and even more precisely with local ground, as we will soon see.[44] While it is true that the issue of *zhang* was almost always raised in the geographic section of the gazetteers, one could also find it in the biographic section (*renwu* 人物) of some gazetteers about Yunnan. These particular occurrences attest that *zhang* was still considered a cause of death in the eighteenth- and nineteenth-century Yunnan:

The lists of names of soldiers who lost their lives "because of *zhang*" (*zhang gu* 瘴故) during the military campaigns launched by the Qianlong emperor (1735–1796) to secure the Yunnan frontiers against the Burmese, or later military conflicts, are

[43] For Liu Xiangxue, before the Sui (581–618), there are only a few records; during the Sui and the Tang, the number of such records develops; and, from the Song to the beginning of the Qing, the number of accounts about *zhang* explodes. For a systematic investigation of records of *zhang* in Chinese sources, see (Liu 2013; Mou and Wang 2003).

[44] (Benedict 1996, p. 18).

Map 8.2 Prevalence of *zhang* during the Ming and the Qing dynasties. (Map copied and modified from Mou and Wang 2003)

more evidence that *zhang* was considered a cause of death, until the mid-nineteenth century:

> Above is the list of soldiers who perished because of *zhang* during the Qianlong era (以上並乾隆年間軍丁陳亡瘴故)

> Above is the list of foot soldiers who died from *zhang* during the Jiaqing (1796-1820) era (以上並嘉慶年間土練陳亡瘴故).[45]

That *zhang* could kill people is also mentioned in the gazetteer of Tengyue, western Yunnan, in the last decades of the nineteenth century:

> In Tengyue, (in western Yunnan) […], a lot of people affected by the *zhang* poison develop the disease *nue*, the most acute form is called the *zhang* by which you lose your voice, not even 2 or 3 out of ten recover 甚則名為啞瘴, 愈者十無二三.[46]

While not necessarily linked to death, *zhang* was at times defined as a disease and at others as a cause of disease. In Tengyue, as mentioned earlier, people affected by poison *zhang* became sick. In Lin'gao's gazetteer, in Hainan Island, it is noted that "when local people enter (the *zhang* area), they always become sick 土人入之亦輒至疾."[47] When conceived of as a disease, opinions regarding its symptoms differed

[45] (*Guangnan fuzhi* 1906, p. 110–111). We find this type of list in (*Malong xianzhi* 1917, p. 389) or in (*Dali xianzhi* 1916, p. 702) or in (*Xuxiu Shunning xianzhi* 1905, p. 716–717).

[46] (*Tengyue tingzhi* 1888, p. 26).

[47] (*Lin'gao xianzhi* 1892, p. 30).

slightly. A scholar writing from Hainan Island, and quoting an earlier gazetteer, underlined the "hot and cold" symptom: "As soon as villagers enter (a *zhang* area) they are affected by the disease of cold and hot."[48] Some associated *zhang* with *nue*, the name of a disease already existing in the *Inner Classic* (it translates today the biomedical entity of malaria), and reported the symptoms usually associated to *nue*, with notably an alternation of heat and cold. "Most of those who are hit by the *zhang* poison develop the disease *nueji* 中瘴毒多成瘧疾."[49] However the outlines of diseases corresponding to *zhang* are blurred for today's observers as well as for ancient ones who, since the eleventh century, usually complained that this word was interchangeably used in the south to designate disease in general: "People in the south call all diseases *zhang* without any consideration for symptoms" wrote the Song scholar Mei Zhi 梅摯 (994–1059).[50] A statement in the Zhaoping gazetteer of 1934, drawing on a gazetteer edited in 1736, confirms the ambiguity of the term: "at Zhaozhou [...] in one year, five *zhang* (or the five forms of *zhang*) occur alternately. When people are hit, the sudden and violent form is the Sudden Disorder of hot *zhang*; the slow and soft forms are intermittent fevers, lumps in the abdomen, (abdominal) distension, athlete's foot, splenic diarrheas 昭州[...]一歲中五瘴迭作人中之暴則生瘴熱霍亂緩亦生痎瘧痞脹腳氣脾泄諸症."[51]

When described as a disease or as a cause of disease, *zhang* was known to hit (*zhong* 中), to hurt (*hai* 害), to strike (*chu* 觸), to affect (*gan* 感) people, or to enter (*ru* 入, *qin* 侵) the body. One could however resist *zhang*: the most common advice was the adoption of a good lifestyle including moderation as regards food, passions, and sex in order to strengthen one's original Qi. "You must be careful of your life style and limit sex so that your correct Qi triumphs over it. If not, you are sure to become sick."[52] This advice was also mentioned in the gazetteer of Sanjiang, northern Guangxi, with little change—"The Qi *zhang* enters easily, [...], if you do not take care of your eating and living conditions, you are sure to be hit"—[53] or in the gazetteer of Guangnan, in Yunnan province, "Guangnan is part of the smoky *zhang* areas [...] If you encounter this poison, generally, if the body has a deficiency caused by the seven passions, the six desires, hard work, distress, then, this evil enters easily."[54]

Harmful and terrifying, able to enter one's deficient body and kill, *zhang* was not considered as a kind of microorganism. Rather, often used with adjectives convey-

[48] (*Gan'en xianzhi* 1931, p. 41).

[49] (*Tengyue tingzhi* 1888, p. 27).

[50] Quoted in (*Zhaoping xianzhi* 1934, p. 219).

[51] (*Zhaoping xianzhi* 1934, p. 220). The absence of punctuation in the original text makes this sentence difficult to translate. A new edition with punctuation and comments in brackets gives: "一岁中五瘴 (春三月青草瘴、四五月黄梅瘴、六七月新禾瘴、八九月黄茅瘴、又日桂花瘴、菊花瘴) 迭作, 人中之, 暴则生瘴热霍乱, 缓亦生 痎、疟、痞、胀、脚气、脾泄诸症"

[52] (*Tengyue tingzhi* 1888, p. 27) or (*Longling xianzhi* 1917, p. 38).

[53] (*Sanjiang xianzhi* 1946, p. 97).

[54] (*Guangnan fuzhi* 1906, p. 11).

ing the notion of mist *lanzhang* 嵐瘴 *(*mountainous vapors) and *yanzhan* 煙瘴 (vaporous, smoky *zhang*), *zhang* was definitely considered an emanation that one would find in particular natural environments.

8.3.4 Zhang as a Repellent Environmental Emanation Linked to the Sweltering Far South, Its Topography, or Its Dense Vegetation

Scholars writing on the most southern parts of the empire in gazetteers commonly incriminated the specific climate found there, combining burning heat, humidity, and mists. In Yunnan, at Anping "the climate is very hot and there is a lot *zhangli*."[55] Drawing on Song poets' descriptions of Lingnan, late imperial scholars used to link *zhang* to Lingnan belonging to a "burning area" *yanfang* 炎方and to the intrinsic imbalance between its Yin and Yang with the consequences that the climate was hot for more than half a year, that flowers blossomed during the entire year, that plants did not freeze, and that the climate was unstable, changing several times a day.[56]

Zhang was known to develop only in certain seasons or during the whole year, depending on the place and the temperature. For instance, in Zhenyuan (Yunnan) where the winters were not cold, *zhang* was present all the year: "In Zhenyuan, when spring begins it is very hot and during winter it is not cold, there is vaporous *zhang* all the year."[57] In some of the southernmost places of Guangdong and Guangxi, Hainan Island notably, several forms of *zhang* were described as being rampant the entire year too. However in other parts, *zhang* was mostly noted as occurring in certain seasons, summer, or summer and autumn, and was usually known to disappear during winter, like in Baoning, in Yunnan province, "This is a place with rare snow and frost, with a lot of mists in the mountains, there is *zhangli* during the three seasons and it disappears in winter 地少霜雪山多嵐三時瘴癘冬始消,"[58] or as in Guangnan where several forms of *zhang* were known to prevail in spring, summer, and autumn and to disappear with frost.[59] While the combination of heat and humidity, particularly in spring, summer, and autumn, was commonly held responsible for the existence of *zhang* in the entire Far South, Yunnan as well as Lingnan, other factors entered into play which distinguished Yunnan from Lingnan.

For scholars writing on the southwest province of Yunnan, rivers with steep banks and mountainous areas in the province made the environment worse, since mountains and rivers were considered an aggravating factor for suffocating hot and

---

55 (*Yunnan tongzhi* 1901, p. 32).

56 As Catherine Despeux explained earlier, for the *Inner Classic*, a normal place for human beings was under the dynamics of the four seasons.

57 (*Yunnnan tongzhi* 1901, p. 23).

58 (*Yunnnan tongzhi* 1901, p. 32).

59 (*Guangnan fuzhi* 1906, p. 11).

humid mists: "Tengyue is located at the westernmost area of Yunnan, in the remote southern frontiers. Everywhere we find mountains, rivers and basins […] If you cross the mountains toward the north, there is a lot of *zhangqi* […] In the south it turns burning […], there is a lot of *zhangli* […] The vaporous *zhang* of misty mountains is everywhere. It's the Qi of damp mountains that does not move. 騰越處雲南極西炎徼之地山川盤互終[…]過分水嶺而北多瘴氣[…]而南漸炎熱[…] 多瘴癘[…] 山嵐煙瘴無處無之, 山澤之氣不通."[60] Areas near rivers were also particularly feared as the following excerpts show: "in Baoshan […] in the areas close to Pupiao, Luoming, and in the places that border the Lujiang river, it is very hot and on the banks of the Lujiang, it is worse. Upstream-downstream of the river, there is a vigorous toxic *zhang* up to the village of Shidian […] in the places where there is *zhang*, the worst is in summer and autumn, because during these seasons, plants abound. Travelers do not dare stay there. 保山縣[…]蒲縹羅明與潞江相近處多熱至潞江覇則尤為熱甚且沿江上下皆有瘴毒至於施甸[…]凡有瘴處皆甚於夏秋二季時則草皆交頭往來行旅即不敢住宿其處."[61] The same argument was given for the districts of Huize, "near the river, it is extremely hot, there is *zhang* 沿江大熱有瘴," of Xinping "in the area bordering the river there is often a vaporous *zhang* 沿江一帶常有煙瘴,"[62] or of Chuxiong "in the area bordering Malong (and Yong?) rivers, in winter and spring there is also a vaporous *zhang* 馬龍河 (永 ?江) 邊一帶冬春[…] 有煙瘴."[63]

While, in the eyes of scholars reporting on *zhang* in Yunnan province, *zhang* was thus mostly considered as natural emanations potentially leading to disease and death linked to a particular climate and aggravated by the specific topography (rivers and deep mountains), scholars reporting on *zhang* in Lingnan incriminated above all the thick vegetation and dense forests, with the confinement of the fragrance of certain flowers. The association of *zhang* with plant emanations was so closely linked in Lingnan that subcategories of *zhang* associating the name of a plant were progressively coined. Firstly associated with *Heteropogon contortus* (*huangmao* 黄茅), from at least the fourth century, and then with spring "green grass" (*qingcao* 青草), during the Tang dynasty, there was general agreement among Song scholars on four subcategories of *zhang* associated with the name of one plant itself related to one season. In his geographic description of Lingnan, Fan Chengda, the scholar mentioned previously, wrote in the twelfth century: "In Yongzhou, along the two rivers, the environment is particularly bad. In one year, there is no time without *zhang*. In spring, it is the *zhang* of 'green grass', in summer it is the *zhang* of '*chimonanthus praecox*', in the sixth and seventh month, it is the *zhang* of the 'new grain', in the eighth and ninth month, it is the *zhang* of '*Heteropogon contortus*', local people say that the *zhang* of *Heteropogon contortus* is the most toxic 邕州两江水土尤恶一岁无时无瘴春曰青草瘴夏曰黄梅瘴六七月曰新禾瘴八九月曰黄

[60] (*Tengyue tingzhi* 1888, p. 27).

[61] (*Yongchang fuzhi* 1886, p. 29).

[62] (*Yunnan tongzhi* 1901, p. 6 and p. 43).

[63] (*Chuxiong xianzhi* 1877, p. 13).

茅瘴土人以黄茅瘴为尤毒."[64] This sentence was repeated in other Song books, with, at times, little change notably regarding the kind of plant incriminated. For the Song scholar Mei Zhi, whose explanations on *zhang* were quoted in the gazetteer of Zhaoping, the same four categories were mentioned, with more details on the fourth category: "In the eighth and ninth month, it is the *zhang* of *Heteropogon contortus.* Some also call it 'osmanthus flower *zhang'* or 'chrysanthemum *zhang*'; according to locals the *zhang* of *Heteropogon contortus* is the most toxic 八九月黃茅瘴又曰桂花瘴菊花瘴土人以黃茅瘴有毒."[65] The association of *zhang* with plant emanations, in Song dynasty records, was regularly quoted or simply paraphrased in late imperial gazetteers of Lingnan. However, there was not a full agreement on the plants themselves. In the gazetteer of Guangnan, a district in Yunnan bordering the province of Guangxi, for instance, it was a fragrance like that of glutinous rice (*nuofan* 糯飯) that was incriminated.[66] A report from Hainan Island, again taking up the ancient statement of the four types of *zhang*, advocated a rectification based on what seems to be direct observation: "The Gazetteer by Hao says that when osmanthus (*muxi* 木樨) blossoms, the *zhang* of its perfume is particularly violent and its poison easily hits men. That's true. However, osmanthus blossoms after autumn when the *zhang* has already dissipated. Moreover, this kind of plant is very rare here. It is not like *areca catechu* (*binglang* 檳榔) which abounds in mountains and which blossoms during the burning summer. Its poison is far more violent.郝志以木樨開時為香花瘴謂其氣尤酷毒易中人信然然木樨秋後方開瘴已稍舒且此木本地絕少不如檳榔瀰山偏嶺花開又當炎夏之時其毒為更劇."[67]

Some accounts in gazetteers provide explanations on the process by which *zhang* affected people. The Zhaoping gazetteer drawing on the Song scholar Mei Zhi's authority explains: "In the wild mountains, there is no other plant, but only *Heteropogon contortus*. When they are yellow and scorched, and when the summer-heat Qi is suffocating, people who walk among *Heteropogon contortus* have their Qi inflamed, are dizzy, get thirsty, their body becomes hot and they have headaches 荒野岡阜無他草木惟茅生之葉焦黃時暑氣鬱蒸人行茅間為氣所焮昏眩煩渴身熱頭痛."[68] Quoting an earlier gazetteer of the province (1731), the gazetteer of Danxian, written in 1936, gives another explanation of how *zhang* affected people: "In Hainan, there are four types of *zhang* […] They develop from the stuffy Qi of these plants, together with the poisonous insects which hibernate (within them). In the bamboo forests of remote mountains the *zhang* of perfumed flowers is very violent. When osmanthus blossoms, (its perfume) is spread widely by the windy Qi of mountains. Those who smell a particular perfume must leave the forest. A perfume

[64] Fan Chengda in *Guihai Yuzheng zhi* (http://ctext.org/wiki.pl?if=en&chapter=296411&remap=gb, 171, last access, 15/08/2017). Yongzhou is the name of an ancient prefecture corresponding to the present town of Nanning, in Guangxi.

[65] (*Zhaoping xianzhi* 1934, p. 219).

[66] (*Guangnan fuzhi* 1906, p. 11). This is cited from an ancient record which incriminated rainbows as well.

[67] (*Qiongshan xianzhi* 1857, p. 72–73).

[68] (*Zhaoping xianzhi* 1934, p. 219).

like that of chrysanthemum has a poison which hits easily men 海南瘴疠有四 [...] 皆草木之氣挾毒蟲惡蟄蒸鬱而成。深山箐林之中, 香花瘴尤酷, 樨開時山風氣隨之而發。行者聞異香出林, 味如佳菊毒易中人."[69] A narrative in the Yangshan gazetteer, without quoting any earlier source, explains the process of the contamination of plant perfumes in the same terms: "In the town there is a lot of *zhang*. [...] when osmanthus blossoms, the vaporous Qi comes along with it and enters men, we call it the 'perfumed flower *zhang*' 邑多瘴氣[...]木樨花時嵐氣隨之入人謂之香花瘴'."[70]

The oppressive concentration of fragrances emanating from flowers or plants and, more generally, dense and thick forests were, in the eyes (and in the noses) of scholars, the main causes of *zhang* in Lingnan: "in places where forests are dense, where heat and damp are suffocating, it (*zhang*) easily hits men 林木叢薄之處溽暑鬱蒸最易中人."[71] "It is only near the Li where forests are luxuriant that there is a lot of Qi *zhang*. 惟黎地樹木蓊蔚多嵐瘴之氣."[72] "(All *zhang*) are only among damp mountains, where luxuriant (vegetation) confines them, like a fire that could not disperse 唯山澤間蓬蓬勃勃鬱結如火不散."[73] "The south of Five Mounts is famous for being *zhang*. In towns located in mountains, herbs and plants make obstructions and the misty Qi becomes hot and suffocating this is why it can hurt 五嶺以南向號瘴鄉然郡邑之依山者草茅障蔽嵐氣鬱蒸故能為害."[74]

While the features of the natural environment causing *zhang* might differ slightly according to place, the fact that a place was or was not affected by *zhang* was understood as something having no relationship with the regular cosmologic pattern of heaven but more locally with the ground. An explanation in the Hexian gazetteer gives the full account of what comes from the natural pattern of heaven and what does not: "Wind, mist, rain, snow, cold, heat, warm, cool are the constants of heaven (*tiandao* 天道); according to cosmological time, they occur without great change. Then, there are differences, they are due to either the ground (*di* 地) being high or low, it is not a matter of heaven, but a matter of earth (*tu* 土)." The explanation given to account for the prevalence of *zhang* in the district of Tengyue also ascertains: "This is absolutely not a difference brought by heaven, this surely comes from the Qi of the ground (*diqi* 地氣)."[75]

Interestingly, in gazetteers, not all the Far South was affected though. Some places were explicitly recorded as being free from *zhang* even if they were very close to unhealthy places as the gazetteers of Yongchang and Tengyue districts, in Yunnan province, underline: "at a few *li* of distance, one place is peaceful (*heping* 和平), and the other has misty *zhang* (煙瘴)." The area of Tengyue was described as

[69] (*Dan xianzhi* 1936, p. 82).

[70] (*Yangshan xianzhi* 1938, p. 149).

[71] (*Kaiping xianzhi* 1933, p. 33).

[72] (*Lin'gao xianzhi* 1892, p. 30).

[73] (*Dan xianzhi* 1936, p. 82).

[74] (*Dongguan xianzhi* 1911, p. 206).

[75] See (*Hexian zhi* 1934, p. 61) and (*Yongchang fuzhi* 1886, p. 29).

rampant with *zhang* "except in Husa 戶撒, Lasa 臘撒, (where) the climate is peaceful (*heping* 和平)."[76] Places free of *zhang* were peaceful (*heping* 和平), safe (*an* 安), or good/virtuous (*shan* 善). Moreover, some scholars underlined improvements, from the eighteenth century onward, in places which were previously known as *zhang*. The arguments given by these scholars to explain that a few far southern localities were free of *zhang* and the reasons to explain the disappearance of it allow us to go deeper in the ways good or bad environments were contrasted by these scholars.

8.3.5 Exceptional Good Environments in the Far South

In Lingnan, places bordering the sea were usually considered as having a better Qi notably because of wind. As in the seventeenth- and eighteenth-century Europe,[77] wind and air movement was a positive element because it was known to clear confined, repellent vapors, thick mists, or strong fragrances: "In the district boarding the sea the Qi is comfortable, the wind creates movement and no *zhang* develops 近海鄉都, 地氣舒坦, 風濤變動, 諸瘴不作."[78] While wind, in medicine, was one of the most dangerous natural agents, in nonmedical sources it was seen as a good element for human health in that it could disperse all types of confined, harmful atmospheres. The Guangdong scholar Qu Dajun 屈大均, in the seventeenth century, noted in his *New Words on Guangdong* (*Guangdong xinyu* 廣東新語, 1678) that wind was also able to control insects that were, together with plants, at the origin of any sort of *zhang*. "Wind controls insects, and insects are the cause of *zhang*. When wind is not interrupted by mountain forests […] insects have nowhere to copulate, and the poison from insects diminishes daily 蓋風主蟲, 蟲為瘴之本。風不阻隔于山林 […]則百蟲無所孳其族, 而蟲毒日以消矣."[79] Open and large landscapes too were seen as healthier environments because nothing could obstruct the circulation of Qi. "Dongguan is close to the sea, the configuration is unobstructed, there is no *zhangli*, we call it a good/virtuous place 東莞近海山勢平遠無瘴稱為善地."[80] In the hinterland, and despite its name meaning "dense forest," Yulin 鬱林 was con-

[76](*Tengyue tingzhi* 1888, p. 27).

[77]In the eighteenth-century Europe, the circulation of air was constantly praised by medical and nonmedical scholars, with advice to regularly open windows and to change one's air. Lavoisier's discoveries on the chemical elements in air at the end of the century reinforced the importance of air in understanding human health. The circulation of air against any sort of stagnation was so important that it led scientists to elaborate mechanical tools to ventilate and make air circulate (Vigarello 1993 [1999], p. 115, 177, 182; Corbin 1982, p. 141–149; see also Nathalie Blanc's contribution in this volume).

[78](*Yazhou zhi* 1914, p. 15). One finds the same argument, "In prefectures and villages which border the sea, the local Qi is better, the wind makes movement and therefore all *zhang* are rare" in (*Dan xianzhi* 1936, p. 82).

[79](Qu 1678), 目第一卷天語. ctext.org/wiki.pl?if=gb&res=38451 (Last access 15/08/2017).

[80](*Dongguan xianzhi* 1911, p. 206).

sidered a healthy environment too: “The Yulin area is mostly composed of large plains, mountains are not deep, and its forests are not dense. While it belongs to the frontiers there is no *zhangli* Qi 鬱林地多平曠有山不深有林不密雖屬邊疆絕無瘴癘之氣.”[81]

That the climate resembles that of central China was another reason to explain that no *zhang* prevailed there: “The south of the Ling Mountains are famous for being a burning area, where *zhang* sticks to its high summits [...] except in Quanzhou which borders Hunan, and where in the depth of winter, there is a lot of snow, the climate resembles that of central China 五嶺以南, 號為炎方, 其高崗 ?瘴[...]惟全州地近湖湘, 窮冬多雪, 氣候似中州.”[82] More generally, the reasons usually given to explain the improvement of the natural environment in the Far South in history confirm that the more the Far South resembled the features of central China, the better it was. Actually, some scholars recorded that places known from ancient times as having *zhang* were, from the eighteenth century onward, no longer affected by it. Two reasons, often linked, were given to explain these improvements. Increasing population density was one reason to account for the disappearance of *zhang*. In the gazetteer of Hexian, in Guangxi, which drew on an undated provincial gazetteer, we read: “At Hexian [...] in the past, in spring and autumn, there was a lot of rain and water, and therefore there was ‘Green grass *zhang*’ and ‘*Heteropogon contortus zhang*’. Today, the population is dense and the *zhangqi* has entirely disappeared. The environment is good too 賀縣 [...]昔時春秋雨水多則有青草黃茅之瘴今居人稠密瘴氣全消水土亦善.”[83] In Simao, South Yunnan, the same reason was invoked together with economic development thanks to tea production and a concentration of merchants: “Since the Daoguang era (1821-1851) the population density has increased and the vaporous *zhang* has entirely disappeared. In the area ruled by minority headmen, in Yiwu and Yibang, tea is produced, with the large number of merchants, seasons progressively became peaceful (*heping*和平). Elsewhere, like in the area close to the Nanla and Nan’e rivers, the vaporous *zhang* is particularly acute 近城自道光後民居稠密嵐瘴全消土司地惟易武倚邦產茶商賈麇至時序和平餘如南臘南莪兩河一帶煙瘴最烈.”[84] Not far from Simao, the improvement of the Ning’er locality is due to the same cause: “Ning’er is a burning and uncultivated place where we never see frost or snow. After the Daoguang era (1821-1851), people coming to live in the neighborhood have increased, and seasons progressively became harmonious 道光以後近城民居稠密時序漸覺和平.”[85]

The opening up of remote areas and enclosed landscapes by roads was another reason of the improvement of the Far South’s natural environment. The gazetteer of Dan District, in Hainan Island, quoting the earlier provincial gazetteer published in 1731, itself paraphrasing the *New Words on Guangdong*, praises the benefits of opening up the landscape. Recalling that under the Tang or the Song dynasties, having a post there was considered perilous, the gazetteer explains why things have

[81] (*Yulin xianzhi* 1900, p. 63).

[82] (*Sanjiang xianzhi* 1946, p. 97).

[83] (*Hexian xianzhi* 1934, p. 61).

[84] (*Yunnan Tongzhi gao* 1901, p. 35).

[85] (*Yunnan Tongzhi gao* 1901, p. 31).

now changed: thanks to the opening up of the landscape, the pure Qi of central China can reach this remote southern area: "Today, the smoky *zhang* has entirely disappeared. Narrow passes have been made smooth. The landscape is more open. The pure Qi of central China circulates through many routes 今瘴煙大豁, 險隘盡平, 山川流暢, 中州清淑之氣數道相通."[86] The gazetteer of Jiaying published in 1897, quoting an earlier gazetteer of the province, also underlines the benefits of increasing the population density and opening up the landscape: "In the past, people were rare and where forests were dense there was a lot of *zhang* mist. Today everywhere is open (*kaipi* 開闢) and *zhang* mists have all disappeared. People from the north of the Ling (Mountains) find this area pleasant (*letu* 樂土)."[87] Interestingly, the Chinese expression often used in these records for "the population density" is literally "human smoke *renyan* 人煙." While the origin and history of this expression needs further research, "human smoke" is, in some reports, clearly associated with the destruction of trees and forests. Zhao Yi 趙翼 (1727–1814), a scholar native to Jiangsu and appointed in several places of Guangxi and Guangdong, wrote in one of his book: "In the past, the cities' neighborhood was full of mountains covered by trees, which used to produce thick and Yin mists; today human population density (human smoke) increases every day, with the cutting of firewood up to 30 li around the *zhangqi* has entirely disappeared 昔時城外滿山皆樹, 故濃煙陰霧, 凝聚不散, 今人煙日多, 伐薪已至三十裏外, 是以瘴氣盡散."[88]

Increasing population density and human management of the natural landscape, through the construction of new roads and deforestation, were seen, by these observers, as having allowed the improvement of Lingnan's and Yunnan's environment for health. The same arguments were held for the improvement of Taiwan's environment since the beginning of the eighteenth century. In one gazetteer published in 1720, quoted in the work by the historian Liu Ts'ui-Jung, the benefits of the increase in population density were also noted: "The habitations are dense. There are thousands of households. Heavy dew becomes scarce and miasma cannot enter […] This is different from Jilongshan where the threatening mountain vapors are most dreadful." And in another gazetteer, the improvement of the natural landscape for human health through human management was set up as a general rule: "The vapors of mountains and rivers accumulated to become miasma; once human beings have brought them under their management, these vapors have a means of being dissipated […]; this is the common rule of nature."[89]

It is difficult today to ascertain whether the increase in population density, the deforestation, the construction of roads, and the opening of remote areas actually

[86] (*Dan xianzhi* 1936, p. 82) quoting the provincial gazetteer by Hao (1731) paraphrasing the *New Words of Guangdong* published in 1687 (Chap. 1, Item 23) written by the Guangdong scholar Qu Dajun 屈大均 (1630–96), a native of Panyu.

[87] (*Jiaying zhouzhi* 1897, p. 43). The term "kaipi" is perhaps synonymous of other terms using "kai" (to open) to designate the conversion of land for agricultural purposes. See (Marks 2004, p. 86).

[88] Quoted by (Mou and Wang 2003, p. 25).

[89] (Liu 1998, p. 179 and p. 197).

contributed to the disappearance of *zhang*, whatever its meaning, and to make the natural environment of the area less terrifying for outsiders. These testimonies reflect above all a confidence in the environmental changes which were brought by Han immigration into the Far South and its subsequent sinicization. To consolidate Ming rule in Yunnan, conquered by the Mongols in the thirteenth century, the Ming court sent huge numbers of soldiers who were later followed by massive waves of civilian Chinese immigrants. As the historian Bin Yang has shown, the Ming colonization of Yunnan went on with agricultural colonies to support these new incomers, with the confiscation and expropriation of arable land, the opening up of new lands, expansion of farming, and the construction of roads. The massive influx of Han Chinese, according to this historian, exerted an unprecedented influence on local societies and "changed Yunnan's social and cultural landscape."[90] The same happened in Taiwan after massive immigration by Han people in the first part of the seventeenth century. Here too, forest clearance was undertaken on a massive scale for agricultural needs and settlements for the Han migrants.[91] The expansion of Han settlements through land clearance also took place in the province of Guangxi, during the Ming dynasty, at the end of which Han Chinese occupied lands in Guangxi and in the southwestern littoral previously inhabited by non-Han people. More generally, the Ming pattern of economic development, which took at its basis the slogan of the founder of the Ming dynasty Zhu Yuanzhang "Taking agriculture as the basis of the country 農為國本," led to a great deforestation of the Far South. In the eighteenth-century Guangdong, under the pressure of rapid population growth, the amount of cultivated land increased by 75%, and increased even more importantly in Guangxi, encouraged by the state policy of "land reclamation," which allowed the clearance of great amounts of land for the needs of agriculture.[92]

In the eyes and noses of many scholars writing on the Far South in gazetteers, what made the Far South the unhealthiest environment in which to live were its natural and cultural differences from the Chinese central plain: its subtropical climate (mist, heat and humidity, permanent heat), its topography (high mountains, deep, narrow valleys and rivers), its vegetation (dense tropical forests), and its unfamiliar tribes. In these records, the emphasis on the "difference *yi* 異" with regard to central China is in fact often explicit, whether this concerned the cosmological pattern "In Tengyue […] the Yin and Yang pattern of change is also very different 騰越[…]陰陽變態亦異," the climate "Gan'en's climate is very different from central China 感恩氣候迥異中州," the landscape "Guangnan belongs to smoky *zhang* areas. It is because cold and heat are not regular and because the nature of its landscape is different 廣南列於煙瘴之地則以寒暑不常山水異性故也," or the ethnicity and culture "what is the most different, it is the area between the two rivers. In these areas governed by tribal headmen there is *zhang* 其尤異者兩江之間凡土司轄之處皆有瘴癘."[93] To these scholars, often outsiders, the improvement of the

[90] (Bin 2009, p. 141–191).

[91] (Liu 1998).

[92] (Marks 2004, p. 84, p. 282).

[93] See in this order (*Tengyue tinghi* 1888, p. 27; *Gan'en* 1931, p. 41; *Xuwen* 1912, p. 227; *Guangnan fuzhi* 1906, p. 11, *Yongchang* 1886, p. 29). It was also stated in one ancient gazetteer of the Hainan

environment of the Far South for health was in fact understood as the eradication of the strangeness of the place. The terms used to qualify the Far South's improvement, in these descriptions, "peaceful," "safe," and "good-virtuous," are indicative of how they saw the change: once the strangeness of the Far South was reduced by massive Han immigration with its long history of the conquest and management of the landscape, the imbalanced Yin and Yang of the place was balanced and the environment peaceful, virtuous, and healthy anew.

8.3.6 Jin Gui 靳貴 *(ca.1464–1520?)*: The Discordant Voice of an Outside Scholar

Interestingly, the recurring images linked to the harmful environment of Lingnan were noted by a Ming scholar, who, while himself an outsider, did not share the viewpoint of all his counterparts. In a dedication written in homage of his student Wu Jingzhong 吳警眾, after the latter's success in the provincial examinations and his appointment as a magistrate in the district of Guixian in Guangxi, Jin Gui 靳貴 (ca.1464–1520?) opposes the views generally held on the Far South and explains why. For him, this reputation which had pervaded gazetteers since the Song dynasty was justified because the place was particular in comparison with central China from where most scholars came. To the repetitive prejudgment making the Far South the crucible of abnormalities that he enumerates—*zhang*, snakes, low-lying and thin soil, and deficient bodies—he opposes a discourse of truth based on his own experience of the place, repeating that he had never seen any of the supposed dangers:

> People use to say that beyond the Ling Mountains, there is *zhang* everywhere. It is today the provinces of Guangdong and Guangxi. It is because this is the southernmost place where it is very hot. Climate/customs (*fengqi* 風氣) are particular if one compares these with central China (*zhongyuan* 中原). All officials who have lived there, use these words, for a long time […] The *Guihai Yuheng* reports that in the Two Guang, only Guilin 桂林 is free of *zhang*, but beyond, toward the south, all districts are *zhang*. *Zhang* includes (that of) green herbs, *Heteropogon contortus*, new cereal, etc. It is what gazetteers have said for 600 or 700 years, through the Song, the Yuan, the Ming, until today. As regards *zhang*, it is not only Guilin which is free of it. All villages and districts above Wu 梧 (Wuzhou) and 潯 Xun river are free of *zhang*. And even, in Sicheng 泗城, in Xilong 西隆, in the steep mountains where roads are rare there is no *zhang*. The ancient provincial gazetteer says: where vegetation is dense and covering, there are trigonocephalus snakes, I have never seen any. The *Descriptions of Yue* by Min Xu speak of low-lying ground, of thin soil, of a constant Yang Qi which would lead people to have open skin pores because of constant sudation, leading them to be deficient in yin or yang. I have never seen any of this. We must not wrongly accuse (this place) just because the local climate/customs are particular.

Island that on Hainan Island, one would find *zhang* only among the Li and the Dong tribes. This statement was quoted extensively in late imperial gazetteers. See (*Dan xianzhi* 1934), (*Qiongshan xianzhi* 1857), (*Gan'en xianzhi* 1931), and (*Yazhou xianzhi* 1914).

After having invalidated the thesis according to which Lingnan contained numerous particular dangers, Jin Gui removes the abnormal character of *zhang* framing it as anyone of heaven's Qi (cold, hot, wind, etc.)—"*zhang* is a Qi of heaven"—thereby allowing anyone to adopt the preventive measures against any of the Qi of heaven.

> "If your original Qi is strong, even if you are expose to it [*zhang*] every day, you are not hurt. Otherwise, Yulin 鬱林 or Cangwu 蒼梧 districts would not have tens of hundreds of thousands of families. These people who live as closely as trees in the forests would go and live elsewhere. If your original Qi (元氣) is not full, then external Qi can invade (you). But why would this be true only for Southern tribes (百粵)?"

Jin Gui therefore invites his student to be glad of the fear usually attached to this environment, since fear is the seed of moral cultivation leading one to be vigilant and to adopt a good lifestyle which is precisely what strengthens one's individual Qi and maintains one in good health:

> "People must have known fear to become vigilant and then they become good. [...] He who takes care of his life is vigilant, and even if passions are unlimited, he himself limits them; even if a life-style is not imposed on him, he controls himself. His original Qi strengthens and external evil *xie* does not hurt him [...]."

Finally, Jin Gui contests the idea that *zhang* would be a characteristic of the Far South, and being of the opinion that this term actually conveys all kinds of fears caused by the strangeness of the place for northerners or central scholars, he underlines that central China is far more affected by a *zhang* whose metaphorical content is entirely redefined. Drawing on the Song scholar Mei Zhi's 梅摯 (994–1059) "Discourse on the five *zhang*," written when this scholar was appointed in Zhaozhou, the present town of Pingle, in Guangxi, he recalls what the *zhang* of central China is:

> I have often met officials who have *zhang*, but it is not a matter of frontiers. But because they are unscrupulous. [...] '(They) complain violently about land tax and extort and exploit their inferiors and show respect only to their superiors, this is the *zhang* of taxation. They use an abstruse style to show off (to confuse), cannot distinguish between good or bad, this is the *zhang* of imprisonment and punishment. They are drunk in banquets at dawn and dusk, neglect the public affairs, this is the *zhang* of drinking and eating. They despoil people of their advantages to make them their own, this is the *zhang* of wealth. They choose plenty of concubines, they indulge themselves with sensual pleasures, this is the *zhang* of frivolity'. These five *zhang* are not limited to South peoples but they affect in an even greater importance officials from central areas. Why? When you are not afraid, you don't keep a vigilant heart. Those who are glad to have obtained a post in good areas, [...] they avidly search to obtain all that they desire. They give themselves to debauchery, and then the five *zhang* develop into their heart [...][94]

This short dedication contrasts with the descriptions of the Far South in central sources and in late imperial gazetteers which mostly drew on them. For Jin Gui, the bad reputation of the Far South's environment for human health was a historical construction built by outsiders, confronted with the strangeness of the place. As early as the sixteenth century, Jin Gui formulated the opinion that most historians today agree with: while certainly designating diseases that one would find in tropical

[94] (*Guixian zhi* 1934, p. 65–66).

or semitropical areas, including malaria, *zhang* was also often used as a metaphor to designate the geographic and cultural strangeness of the place.[95] Written evidence left by locals seems to confirm Jin Gui's opinion. According to the historian Liu Xiangxue, narratives or songs by local peoples, including today's "minorities" living in these provinces, are totally silent about *zhang*. The medical texts written in the Far South, especially Lingnan, by local people in the last dynasty that I have explored confirm that the worst environment in which to live was a matter of viewpoint.

8.4 Zhang and the Environment of the Far South, Through the Eyes of Local Doctors

8.4.1 The Mysterious Absence of Zhang in Local Medical Texts

Strikingly, almost none of the prefaces of the medical texts written by Lingnan authors include a discussion on the specific and notorious insalubrity of the area. This absence is surprising. As Genette and other literary critics have shown, a preface always has the function of valorizing a text.[96] Lingnan's reputation for being unhealthy was ancient and largely shared, as we saw by exploring local gazetteers; local medical authors could have used these first pages to valorize their text by presenting them as the survival guides to cope with the specific dangers of the area. This had been done earlier in the *Recipes to Protect Life in Lingnan* compiled by Song scholars who had been appointed to Lingnan.[97] With the exception of one text, no author starts his book with the particular issue of *zhang* 瘴 or with any of the abnormal diseases that since ancient times had been thought to prevail in the Far South, such as "poisons" *gu* 蠱 or "athlete's foot" *jiaoqi* 腳氣, or with diseases that had been identified, more recently, with the Far South such as "Guang(dong)'s rotting sore" *guanchuang* 廣瘡. Moreover, none of these medical books open with the idea that the local environment would be particularly threatening for human health. In fact, only one preface signals the peculiarities that were usually attached to the place: it was written by Miao Fuzhao 繆福照 (signed in 1823) for a text *New Recipes to Cure Poisons* (*Zhigu xinfang* 治蠱新方) written by Lu Shunde 路順德, a scholar from the northern part of Guangxi. But Miao Fuzhao was not a local

[95] For (Diamond 1988), the long-lasting association of the Miao ethnic group with *gu* (poisons) by Han scholars comes from the Han's fear of "being culturally swamped by the barbarian majority" and was used to maintain strong ethnic boundaries in the Far South, between the Han and the non-Sinicized peoples. See also (Hanson 2011, p. 80).

[96] (Genette 1987, 1991). For other references on prefaces, see (Bretelle-Establet 2011).

[97] It is by a description of the specificities of Lingnan, of its people, and of the prevalence of *zhang* that Li's chapter actually begins. See (Li & Zhang, p. 1).

scholar. In his preface, he explains that he found the manuscript during the several trips he made between 1818 and 1834, between his native province, Jiangsu, and the South of China. Miao Fuzhao was from Jiangsu (Jiangnan)—where, as he writes, "the population is safe and resources are abundant"—and explains that he found himself in great danger during these trips: "Over 17 years, I made this trip several times, traveling through the barbarians and the *zhang* rains (往來蠻煙瘴雨中), being in great danger every day."[98] By associating in his description of the place, the *zhang* rains 煙瘴雨 and the southern barbarians *man* 蠻, Miao no doubt draws his description of the Far South from the long-lasting imagination of the central scholar who feared the southern frontiers for their strange diseases, landscapes, climates, and cultures. With the exception of this preface, no local author sets his/her medical text in a particular morbid hell. In fact, all prefer to valorize their book by taking over the topos usually displayed in prefaces of medical texts: the importance of medicine, put at a higher level than the "minor way" it had been defined in the past (*xiaodao* 小道 in comparison with the "great way," namely, the classic Confucian learning needed for the imperial examinations), and the importance of the physician, raised at the level of the Confucian bureaucrat scholar.[99]

With the exception of this preface, the other authors never address the issue in the first pages of their books.[100] Further, *zhang* is also absent from the tables of contents and, with a few exceptions, is not discussed at all in these medical texts. It is certainly not because the term *zhang* belonged to vernacular rather than to medical language. If we consider that *zhang* was a vernacular term for diseases that were often referred to as *nue* in medical language, as some doctors and scholars stated in earlier times, and if we examine the heading *nue* in these medical treatises, we do not find the idea that *nue* could have a popular name. Yet to remind his reader that a disease can have a popular name was a very common task for these medical authors who very often framed themselves as interpreters between a popular language (*minsu suowei* 民俗所謂 "it is what people usually call") and a more specialized language, between ancient and new terms, or even between local names of disease and more classical ones. In fact, in the entire corpus, only two authors mention *zhang*. He Mengyao 何夢瑶 (1693–1764), a native of Guangdong, had succeeded in the imperial examinations at the highest level and had therefore occupied several posts in the imperial administration, notably in the nearby province of Guangxi. A prolific author, he was in charge of the compilation of a local gazetteer of Guangxi

[98] (Lu & Miao 1823 [1935]: 1).

[99] (Bretelle-Establet 2011).

[100] As said earlier, almost none of the medical books by Yunnan authors are extant in libraries. We thus cannot ascertain whether the *zhang* issue and that of the insalubrity of the place were addressed in these books or not. It is important to note that a short text was copied in the gazetteer of Longling, West Yunnan, written by the son of a local doctor, Zhao Zhibeng 趙之菶 (active in the nineteenth century) who, informed by Wu Youxing 吳有性's new doctrines on "hot epidemics," specialized in curing *zhang*. In this text, written by a Yunnan insider, *zhang* is discussed as a terrible local disease, causing in Longling the death of 3 to 4 out of 10 people affected, each year, a disease that should be treated in the same way as "hot epidemics" (*Longling xianzhi* 1917, p. 537).

district, the kind of source discussed earlier, and was likely to have been familiar with the earlier central descriptions of the place. He Mengyao includes the issue of *zhang* in his discussion on "nue," and, in fact, he cites some ideas already formulated in the Song medical guide *Recipes to Protect Life in Lingnan*. However, nowhere does He Mengyao underline that *zhang* was a disease particular to Lingnan: "There is also the case of those who contract the Qi *zhang* from the vapors of the mountains, the suffocating vapors of humidity and heat, and who become affected by the *Zhangnue* 又有感山嵐瘴氣, 濕熱熏蒸而為瘴瘧者."[101] Chen Zhen'ge too, a late nineteenth-century doctor, from Guangdong, speaks of *zhang*, but, here again, not in relation to the Far South's environment: "In the wild mountain forests, a humid Qi accumulates. Near the sea, it is even worse and this Qi causes diseases. […] Or there can be a putrid Qi coming from rotten herbs and plants. We call it the toxic mountain *zhang*. When people contract it, it often turns into Sudden Disorder disease *huoluan* 山林曠野常有濕氣集乎其間。而瀕海尤甚其氣病人。[…]或腐草污木淤泥之臭氣。名為山嵐瘴毒。人感其氣多成霍亂."[102]

Zhang was definitely not at the center of local doctors' questions, and more generally, none of these doctors framed their environment as particularly harmful for human health. Furthermore, as we will now see, local doctors did not frame the Far South as a place that was different from the rest of the empire, which, in the dominant geographic imagination of that time, as we saw earlier, was mainly divided in two: north/northwest and south/southeast.

8.4.2 The Far South's Identity Through the Eyes of Local Doctors and Its Influence on Health and Disease

For half the medical authors writing from Guangdong or Guangxi in the Qing dynasty, the issue of the relationships between human health and the environment or locality was not addressed.[103] While the other half did take place into consideration in their medical discussions, very few of them spoke of place or environment in terms other than those generally addressed by doctors writing from Jiangnan, summarized in the first pages of this contribution. Actually, the very names of Lingnan or Yue, ancient appellations for the southern provinces of Guangdong and Guangxi, only appear in four texts (the bold emphasis on place names is mine). He Mengyao, in his *Complete Book of Medicinal Formulas* (*yifang quanshu* 醫方全書 1751?), under the heading "The '*Athlete's foot of the zhang poison*' of Lingnan, West of the River 江東嶺南瘴毒脚氣" writes:

[101] (He 1751 [1994]: vol 1, p. 157).

[102] (Chen 1892, j.1, p. 28–29).

[103] None of the following books mention place or environment as a possible variables to account for the prevalence of disease or bodily variations (He 1751, Guo 1753, Huang 1800, 1876; Liang 1880; Chen 1872, Mai 1876, Huang 1909a, Huang 1933 (a book partly destroyed).

> Doctrines say: The *Inner Classic* says that in the south, the ground is low-lying, the natural environment is weak, and it is where fogs and mists accumulate. **Lingnan, West of the (Yangzi) river**, is like that. In spring and summer, the climate is suffocating and the humid Qi very acute. If the foot contracts it, it develops into the athlete's foot of the *zhang* poison. […] How is it that this disease is so common to the West of the River, in the Ling Mountains 論曰內急謂南方者其地下水土弱霧露之所也**江東嶺南**大率如此春夏之交山水蒸鬱風濕毒氣為甚族或感之逐成瘴毒腳氣 [...]豈非江左嶺表此疾得之為多歟?[104]

While appearing to be a direct observation, this statement is in fact a quotation from an earlier medical encyclopedia, the *Medical Encyclopedia: A sagely Benefaction of the Zhenghe Reign Period* (*shengji zonglu* 聖濟總錄), composed in the eleventh century and commissioned by a Song emperor.[105] It is therefore a statement made by an ancient outsider, copied by He Mengyao. Another author, Huang Yuanji 黃元基, a scholar and a native of Henan, graduated in 1733 and appointed to several administrative posts in Guangdong and Guangxi, in his *Formulas Tested and Collected from Jintai's Cabinet* (*Jingyun zhai ji yan fang* 静耘斋集驗方, 1799, j4), mentions Lingnan in relation to a particular disease:

> In **Lingnan**, there is the disaster of the *Taosheng* Poison 嶺南有挑生之害.[106]

Here again, while looking like a direct observation, the statement is in fact a quotation, already present in the *Great Compendium of Materia Medica* (*Bencao gangmu* 本草綱目by Li Shizhen (1518–1593)), attributed to an earlier text written by the Song scholar Fan Chengda, mentioned above.[107]

Another author mentions the Far South as linked with a particular problem. This is Qiu Xi 邱熺, known as the first person to have introduced the techniques of vaccine inoculation in China, at the beginning of the nineteenth century. In his book *Summary for Smallpox Vaccination* (*yindou lue* 引痘略), written in 1817, he explains the unwillingness of local people to be inoculated because of their fear that vaccine material came from lepers. According to him, lepers were particularly numerous in the province because of the particular environment of the eastern part of "Yue": "The eastern part **of Yue** is low-lying and damp, there is leprosy everywhere 粤東卑濕, 不無瘋疾" (Qiu Xi 1817 [1864], p. 32).

Finally, Wang Xueyuan 王學淵, in his *Guide for Summer-Heat diseases* (*Shuzheng zhinan* 暑症指南 1843), mentions Lingnan and the particular scorching heat that prevails in summer, as the cause of a set of syndromes called Summer-Heat diseases: "It is only **south of the Great River (Yangzi)** that the climate is often hot,

---

[104] (He 1751, [1918], j4, p. 7).

[105] It is in Chapter (juan) 84. (http://ctext.org/wiki.pl?if=gb&chapter=431646. Last access 10/08/2017). On this encyclopedia, see (Goldschmidt 2009).

[106] (Huang 1799, j4, unpaginated). *Taosheng (du)* is a synonym of *gu* 蠱 poisons.

[107] Li Shizhen, Bencao gangmu, p. 508. It is a quote from the Song scholar Fan Chengda 范石湖文集. https://books.google.fr/books?id=deHvCwAAQBAJ&pg=PA508&dq=%E5%B6%BA%E5%8D%97%E6%9C%89%E6%8C%91%E7%94%9F%E4%B9%8B%E5%AE%B3&hl=fr&sa=X&ved=0ahUKEwjyxubA77TRAhWHqxoKHU7dAA8Q6AEIIjAB#v=onepage&q=%E5%B6%BA%E5%8D%97%E6%9C%89%E6%8C%91%E7%94%9F%E4%B9%8B%E5%AE%B3&f=false (access 10/01/2017).

and this is particularly true of **Lingnan** 惟**大江以南**氣候多溫嶺南尤甚." However, a little further on, he removes "Lingnan" and only speaks of a bigger area: south of the Great River. "**South of the Great River**, the ground is low-lying and thin, humidity dominates and there are hot suffocating vapors. Confronted by such a climate, we must beforehand be careful not to be affected. **大江以南**, 地卑氣薄, 濕勝熱蒸, 當此時候更須防患於先." In Wang Xueyuan's words, Lingnan was no different from the "South of Yangzi." This is in fact how local doctors, who took place and environment into consideration in their discussions on health and disease (sharing the common idea at that time that regional differences accounted for the prevalence of diseases and bodily variations), often defined the place where they lived. They usually do not mention Lingnan or Yue. Rather, they used the widespread terrestrial divisions (North versus South or Northwest versus Southeast) which, since the first medical treatise, but even more after the Song dynasty, were seen as the most important opposed territories, responsible for variations in the causes of diseases, in the prevalence of diseases, or in bodily complexions. Through the eyes of medical authors from Lingnan, as we will now see, there was not a special area, Lingnan, but only two different environments: the one that began south of the Yangzi River, south or southeast, and the other, above the Yangzi River, the north or northwest.[108]

Huang Yuanji 黄元基, in his *Formulas Tested Collected from Jintai Cabinet* 静耘斋集驗方 1763, citing Luo Qianfu (a thirteenth-century author), mentions that a single disease "athlete's foot" *jiaoqi* 腳氣 could be induced by two different causes according to regional variations: "The ways we contract 'athlete's foot *jiaoqi* in the **North** or in the **South** differ. In the **South**, which is low-lying and damp […] it is 'athlete's foot caused by an external attack. In the **North**, the ground is high and not damp, but people eat cheese […] this is 'athlete's foot caused by internal prejudice. **南北**腳氣受病不同。**南方**卑濕[…]此腳氣從外受之。**北方**地高無濕其俗飲[…]此腳氣從內受."

Other medical thinkers underline differences in the prevalence of diseases or in bodily complexion owing to this axis of opposition. Yu Tingju 俞廷舉 in his *Medical Anecdotes from Jin Cabinet* (*Jintai yihua* 金臺醫話 1784) underlines the absurdity of giving southern people medicinal formulas that suit northern people for the reasons that "cold damage only occurs in the **North** where it is cold, it is true cold damage. In the **southern frontiers**, it's warm, therefore, cold damage is rare but wind stroke is frequent. People today do not know how to take places into account, they use ephedra and cinnamomum, this is absurd! 傷寒惟北方寒冷, 乃有正傷寒。**南邊**和緩則傷寒少, 而傷風多。**今人**不知地宜, 好用麻黄桂枝, 謬也." Further, he notes also that "all the people from the **North** I see are robust and full, while people from the **South** are always delicate and weak **北方**之人每見實強, **南方** 之人恒多軟弱 ."[109] It is also by opposing the South to the North that Pan Mingxiong 潘明熊 in his *Abstract of Medicine from the Pingqin Library* (*pingqin*

108 (Wang 1843, p. 17 and p. 54).

109 (Yu 1784, p. 287 and p. 294).

shuwu yilue 評琴書屋醫略 1865) explains the differences of the prevalence of disease: "**In the South**, people are usually harmed by wind in their defensive (Qi) and very rarely by cold in their camp (Qi) **南方**風傷衞者多, 寒傷營者少."[110]

Other doctors underline the difference in the prevalence of diseases by referring to the most ancient division, Northwest versus Southeast. Guo Zhi 郭治, for instance, in his *Cold Damage Treatise* (*Shanghan lun* 傷寒論 1827) writes: "Cold damage is an aggravated form of cold (ganmao), and cold is a wilder form of cold damage. In the **Northwest**, cold damage is frequent, in the **Southeast** cold is frequent 傷寒乃感冒之重者, 感冒乃傷寒之輕者。在**西北**則多傷寒, 在**東南**則多感冒." Huang Chihua 黄炽華 in his *Simple Medical Sayings* 醫學芻言 (1909b) calls for a regional medical approach and in his short book, made up of a series of principles, devotes its sixth principle to the issue of "considering the place *zhuodi* 酌地." Quoting the *Suwen*, as well as other later famous Jiangnan doctors, (Li Changyi (1873), Ye Gui (1667–1747), Wu Tang 吳瑭 (1798), Wang Shixiong 王士雄 (1852)), he concludes on the use of some medical formulas particularly suited to the **Southeast**: "Local Qi changes, bodily constitution also changes, the reason is the physical feature of the earth 地氣變遷, 體質移易, 理勢使然矣. [...] Methods and formulas by Wu 吳 and Ye 葉, or formulas created by other contemporary wise men like Wang Xunchen 王勳臣 (Wang Qinren), Yang Lishan 楊栗山 (b.ca.1705), or Wang Mengyin 王夢隱 are very efficient for particular syndromes. [...] For warm and hot diseases of the **Southeast 東南**, they are excellent and very simple."[111]

Not all medical thinkers shared the idea that regional variations accounted for disease or bodily variations. A few medical thinkers like Liu Yuan 劉淵 (ca.1739) or Chen Huantang 陳焕堂 (ca. 1848) truly fought against this view and argued that no regional variation accounted for body, disease, or treatment differences. In both cases, these authors still refer to the common South/North or Southeast/Northwest axis of opposition. Liu Yuan, in his *Compilation of Medicine (Yixue zuanyao* 醫學纂要 1739), opposes the idea that disease would be different in different places and explains:

> Wind is something which exists in the **North** and in the **South**. To say that cold is rare in the **Southeast** will not kill people, it is possible (to say it). However that wind is rare, this is impossible. It is exactly the same for phlegmatic diseases, we find them in the **North** and in the **South**. In the **Southeast**, most are caused by the external humidity of the environment, and in the **Northwest**, most are caused by the inner humidity from (eating) cheese 是風則**南北**俱有, 若云**東南**寒少, 未必殺人, 則可。若云風少則不可也。痰亦**南北**俱有。若水土之外濕, **東南** 雖多, 而乳酪之內濕, 則**西北**尤多也[112]

The most violent critics of the dominant fashion of putting to the fore the issue of regional variations producing variations in diseases, body types, and treatments come from Chen Huantang 陳焕堂. His book entitled *Back to the True Zhongjing* (*Zhongjing guizhen* 仲景歸真 1849) was a clear invitation to return to Zhang

[110] (Pan 1865, p. 3).

[111] See (Guo 1827, p. 1) and (Huang Chihua 1909b, p. 14–15).

[112] (Liu 1739, j.2, p. 5).

Zhongjing's *Cold Damage Treatise* whose explanations and treatments of diseases had no concern with variations in place. In his book written and prefaced in 1849, Chen qualifies as quack doctors those who claimed that regional variations produced variations in human health. What deserves to be underlined is that his author still relied on the general divisions of North and South to explain his viewpoint:

> Today, some say that in the **South** there is no true Cold damage […] How could it be that people from the **South** would all have their essence and spirit solid? And, why, only they would escape from Cold damage? If the topography of the **North** and **South** is respectively thick and thin, dividing men as robust or weak, with the robust in the **North** and weak in the **South**, then, the weak should have more diseases, and thus **Southern people** should have more Cold damage 世有謂**南方**無真正傷寒[…]**南方**之人何術而皆精神完固, 獨舞傷寒哉若以**南北**土地厚薄而分人之強弱則**北**厚而強。**南方**薄而弱。弱者病應多。**南方**人傷寒應多也。[113]

Finally, to give the full picture of how Lingnan doctors in the Qing dynasty saw the relationships between the environment and disease, in this time period when Western doctors and medicine were taking an increasing part in the medical landscape, it is important to mention a last author, Chen Zhen'ge 陳珍閣. In his *Pivot of Medical Threads*, (*Yigang congshu* 醫綱總樞 1892, j1, p. 39), Chen does not evoke the environment's influences on diseases, bodies, or treatments, in terms of regional divisions. His explanations, at times, seem to draw on the descriptions and explanations given in gazetteers by nonmedical scholars, while, at others, they include new elements which seem to be derived from Chen's contacts with Western doctors and medicine. Actually, Chen Zhen'ge had had close contacts with British doctors in a British hospital in Singapore, where he stayed for 3 years from 1886, to learn Western anatomy and attend dissection classes. While Chen Zhen'ge gives a wider understanding of the relationships between environment and diseases, using a wider set of arguments—economic, social, and chemical—it is important to note that, like his counterparts, he did not mention the Far South as special or as particularly dangerous.

For Chen Zhen'ge, humid Qi, whether accumulated in uncultivated mountain forests or even worse, near the sea, was the cause of *zhang* which would make people victims of another disease, sudden disorder, *huoluan* 霍亂, an ancient term which designated many sorts of gastroenteritis as well as true cholera from the beginning of the nineteenth century on:

> **IN the uncultivated mountain forests**, a humid Qi accumulates. **Near the sea**, it is even worse and this Qi causes diseases. […] Also there can be a putrid Qi coming from rotten herbs and plants. We call it the toxic mountain *zhang*. When people contract it, it often turns into Sudden Disorder **山林曠野**常有濕氣集乎其間。而**瀕海**尤甚其氣病人。[…]或腐草污木淤泥之臭氣。名為山嵐瘴毒。人感其氣多成霍亂

Chen Zhen'ge also considered the proximity to water surfaces particularly harmful for people who were in long-lasting contact with its humid Qi. The sea which produced an unhealthy "salty Qi" or "salty fog" was even worse:

---

[113](Chen 1849, p. 15–16)

> **Near rivers, lakes, canals and the sea**, the humid Qi from the water's surface is particularly acute. People who live on boats and make long voyages become poisoned, because of a long exposure to it. Blood is blocked and cannot move, the accumulated water causes edema. A lot of them are incurable. The cause lies in the lack of movements (of people on boats), thus the Qi infuses and cannot unfold, if from outside you receive/contract a salty Qi or a salty fog, all together that makes disease develop. To recover, it is necessary to walk on a green grass lawn, to eat a lot of vegetables to digest **江湖河海**, 水面濕氣最多, 凡人居舟遠渡, 每易為其所襲日久成毒, 阻血不行, 聚水成為腫脹, 多不可治。此由少於行動, 氣鬱不舒, 外受鹹氣鹹霧合集而成。治宜於青草地上行動, 多服菜蔬以化之.

Environments with few humans but with abundant plants, trees, and animals were also suspected of causing particular diseases. To recover, the best thing was to change "one's environment":

> **In new commercial ports** (*xinkai gangbu* 新開港埠), land is large and people are few, plants, trees, and animals abound, a Yin fog develops, a cold humidity prevails […] added to this, the environment (*shuitu* 水土) does not allow the circulation of Yang Qi which could purify the cold poison. When people contract this Qi, their own Qi and blood are blocked, a lot of them have the syndrome of abdominal distension and their feet are swollen because of edema. To cure them, it is necessary to go **to another place** and change **one's environment** 治之者宜遷往別**地方**以換**水土**

Lastly, the nature of the ground, in terms close to chemical analysis, was to be taken into account especially for giving appropriate treatments: "We must cure by taking into account the coolness or hotness of the environment. When the ground/earth produces sulfur and coal, it is hot, and we need to cool down. When the ground/earth produces silver or metals, it is cold, we must warm up 又當分其水土之寒熱以治之, 地產硫磺煤炭為熱宜用清涼, 地產金銀錫為寒宜溫熱."[114]

Chen Zhenge's vision of the relationships between the natural environment and human health while sharing common ideas with nonmedical scholars—the unhealthy mountain fog, the dreadful proximity of rivers, and the unhealthy abundance of plants and trees combined with a low human population—introduced new elements. While more work is needed, it seems that his advice "to walk/move on green grass lawns," to eat lots of vegetables, to take into account underground chemical properties, and finally, to "change one's natural environment to recover from disease" (often mentioned in his book, notably in cases of "consumption disease" 癆瘵 for whom he advises changing one's environment 治宜遷換水土) are new themes which may bear testimony to his 3 years in a British hospital in Singapore.[115]

This brief survey of the medical texts written in Lingnan from the beginning of the eighteenth century to the beginning of the twentieth century reveals two important points. Firstly, this set of medical thinkers and authors did not share a common vision on the issue of the relationships between the natural environment and human health. For many of them, the natural environment was not a relevant issue to discuss in medicine. For some others, the issue of place and human health

---

[114] (Chen 1892, j.2, p. 28–29).

[115] To "change one's environment" recalls the similar idea "to change one's air" which was often promoted in Western medical and nonmedical literature from the seventeenth century on (Vigarello, 1993 [1999], p. 115, 177). More work is however needed to state that these elements were new in Chinese medicine.

was relevant, since to them, regional variations did account for variations in diseases, in the causes of disease, or in bodily complexions. Finally, for some authors, the idea that the natural environment, or regional variations, account for variations in human health was contested and, sometimes, in very harsh terms.[116] The diversity of opinions in these texts testifies that debates occurring elsewhere in China, especially in Jiangnan, were also taking place in this part of the empire. This was likely favored by the massive immigration of Jiangnan scholars into the Pearl River Delta in the eighteenth century as well as by a boom in a market for books in Guangzhou which, by the end of the eighteenth century, made Jiangnan medical literature rapidly available.[117] Secondly, and perhaps even more interestingly, Lingnan medical authors, writing from what was considered the most harmful environment by most outside scholars, whether promoting or opposing the idea that the environment mattered in human health, never, or almost never, referred to their environment as a place different from the broad and abstract "South" or "Southeast." While, in the eyes of many medical doctors from Jiangnan, the South or Southeast did not include Lingnan, which to them was the seat of ancient abnormal diseases and the breeding ground of new diseases, in the words of the doctors who lived and worked there, Lingnan was nothing other than a part of this large South or Southeast area which started south of the Yangzi River Delta. And nearly none of the particular dangers Lingnan was accused of by outsiders were discussed in their medical books. The comparative analysis of records written by outsiders or insiders underlines how the perception of the worst environment in which to live was a matter of points of view and raises a problem that any historian of disease may meet. Actually, in the absence of modern epidemiological studies and measures, it is difficult to ascertain whether the Far South's environment was far more harmful to human's health than that of the rest of the empire. In fact, the comparative analysis of these records, which predate the beginning of epidemiology, only allows us to claim that perceptions of the Far South's environment differed according to the geographic and social perspectives from where it was seen. What these differences of viewpoints account for is also difficult to ascertain. Was the subtropical area particularly harmful for people from outside and less harmful to locals who did not even see the particularity of their diseases, a process that Grmek, studying the history of medicine and disease in ancient Greece, had called "l'invisibilité du quotidien" (the invisibility of daily life)?[118] Was the unhealthy reputation built by outsiders a kind of orientalistic repre-

[116] These authors probably shared the current of the Hanxue movement, a scholarship which invited returning to the original sources and original truths of the Han dynasty that later commentators were accused of having perverted. On this intellectual movement, see (Elman 1984).

[117] (Miles 2006, p. 4–9). The *Catalogue of the Morrison Collection of Chinese Books* attests to the availability in Guangzhou of the most up-to-date Jiangnan medical literature. Further, the network of citations and quotations in Far South medical literature attests to the fact that medical literature from Jiangnan circulated quickly and had a great influence in the whole empire. See (Bretelle-Establet 2017, 2018).

[118] (Grmek 1985). In his attempt to write a history of diseases in ancient Greece, Grmek underlines the different biases of written evidence, notably that of not speaking of the most common ailments which, pertaining to daily life, become invisible to the doctor.

sentation of the Far South margins by central scholars, as suggested by Jin Gui, as early as the sixteenth century? Conversely, was the silence of local medical thinkers on the abnormalities the Far South was accused of being the seat since the most ancient times, the wish to link the cultural and geographically isolated Far South to the less exotic universe of Jiangnan (South of the Yangzi River)? These questions remain open.

8.5 Conclusion

The idea that the human body was permeable to its natural environment was rarely contested in Chinese medical history. This permeability was ensured by several means that were not mutually exclusive: a correlative cosmology that links together celestial, terrestrial, and human structures and a medical conception of the body whose pores were more closed or open and thus less or more able to let the outside enter and the inside pour out. The idea that the structure of the skin allows this permeability was never contested, even when, from the end of the seventeenth century onward, the physician Wu Youxing, a witness of many large waves of epidemics, introduced another explanation to account for the continuum between the natural environment and the body: the intrusion of natural pestilential Qi through the nose and mouth.

A human was thus permeable to his natural environment—"regions," "places," "lands," "environment," "ground," and "landscape" (*yu* 域, *chu* 處, *di* 地, *shuitu* 水土, *tu* 土, *shanshui* 山水)—which was known to vary from place to place and to be less or more healthy. As Leung has shown, at the end of the empire, the perception of the links between the environment and health in China was not very different from the perception that prevailed in Europe from the seventeenth century to the end of the nineteenth century. In Europe as in China, emanations from the natural environment were held responsible for diseases, and local emanations were responsible for the type of diseases one could encounter in a particular place.[119] According to her, what makes the situation different in China and in Europe is the axis of opposition between a good and a bad environment for health. While since the seventeenth-century Europe, the major distinction was between the town and the countryside, with the countryside considered healthier than towns, in the Chinese medical vision, which was also shared by many scholars, the main opposition was based on a geographic division, North/Northwest versus South/Southeast. The North or Northwest, the oldest political and cultural center of the Chinese empire, remained in the eyes of physicians, and even in the eyes of the physicians of the

[119] In fact, many of the elements that were known in China to define an unhealthy environment were also pinpointed in Europe, from the seventeenth century to the last decades of the nineteenth century. Mist, humidity, confined areas, marshy airs, low-lying regions, putrid standing waters used to define unhealthy environments. See notably (Corbin 1982; Wear 1992; Vigarello 1993 [1999]; see also Nathalie Blanc's contribution in this volume).

South, the best environment for health, producing the strongest people who were the least affected by disease.[120] However, the issue of *zhang*, synonymous with a wide range of dangers for human health, reveals, I think, another axis of opposition with, on one side, the wild environment where nature is in excess and culture in deficit and, on the other, the cultivated, populated, and human-controlled environment. This certainly sounds strange to those who have in mind Max Weber's image of a Chinese religious ideal regarding nature and notably the idea that "Confucianism [...] was a rationale ethic which reduced tension with the world to an absolute minimum" or that "Confucian rationalism meant adjustment to the world; puritan rationalism meant rational mastery of the world."[121] The arguments given by Chinese scholars to explain the improvement of the natural environment of Lingnan, Yunnan, and Taiwan—the increase of population density, the clearing of forests, the conversion of uncultivated lands into agricultural lands, and the opening of roads and of landscapes—challenge Weber's vision and corroborate, rather, the work carried out recently in the field of Chinese environmental history.[122] Above all, they provide another illustration that for most late imperial Chinese scholars, wild forests and untamed nature were unpleasant and terrifying.[123]

References

Primary sources

a. Medical sources from Guangdong and Guangxi. The original publication date, when known, follows the title. The edition consulted, in a reprint or in a collection, is in square brackets.

Chen, Huantang 陳焕堂. 1849. *Zhongjing guizhen* 仲景歸真 (Back to the true *Zhongjing*). Guangzhou: Wuyunlou keben.

Chen, Yi 陳義. 1877. *Yifang bu qiu ren* 醫方不求人 (A medical vade-mecum). Hong Kong: Lingnan shuju.

Cheng, Kangnan 程康南. 1893. *Erke miyao* 兒科秘要 (Secrets in pediatrics). Guangzhou: Shoujingtang.

[120] (Leung 2002). On this divide in Europe, see notably (Wear 1992, p. 119–147).

[121] (Weber 1915 [1959], p. 227 and p. 248).

[122] On Chinese environmental history, see (Elvin & Liu 1998), (Mark 2004), and (Elvin 2004). For a more sensitive approach to landscape and nature in China, see (Vogel and Dux 2010). On the basis of recent works in the field of environmental history in China, Elvin (2004, p. 62) corrected Weber's statements and wrote instead: "more than any premodern northwestern Europeans, the Chinese were driven by a desire for the *rational mastery of the world.*"

[123] For (Elvin 2004, p. 78) "the original core of classical Chinese culture was hostile to forests, and saw their removal as the precondition for the creation of a civilized world."

Chen, Zhen'ge 陳珍閣. 1892. *Yigang congshu* 醫綱總樞 (Pivot of medical threads). Guangzhou: Zuijinglou keben.

Guo, Zhi 郭治. 1753 [1981]. *Mairu* 脉如 (About the pulse). Shanghai: Shanghai guji chubanshe.

———. 1827. *Shanghan lun* 伤寒論 (Treatise on cold damage). Place and name of publishing house unknown.

He, Mengyao 何梦瑶. 1751 [1994]. *Yibian* 醫碥 (Stepping-stone for medicine). Beijing: Renmin weisheng chubanshe.

———. 1757 [1895]. *Sanke jiyao* 三科輯要 (Essentials of the three disciplines). Guangzhou: Shijieyua.

———. *Yifang quanshu* 醫方全書 (Complete book of formulas), 1751(?)[1918]. *Guangzhou: Liangguang tushuju.*

———and the Monk Huchan, 僧互禅. 1872 [1885]. *Lezhitang renzi xu zhi* 樂只堂人子須知 (What men need to know from the Lezhi Hall). Foshan: Huawenju.

Huang, Huishi 黄暉史. 1909a. *Yixue xunyuan* 醫學寻源 (Origins of medicine). Place and name of publishing house unknown.

Huang, Chihua 黄炽華. 1909b. *Yixue chuyan* 醫學芻言 (Simple Medical Sayings). Place of publishing: unknown, Jinjian shiyin ben.

Huang, Yan 黃巖. 1800 [1918]. *Yixue jingyao* 醫學精要 (Essentials of medicine). Shanghai: Cui ying shuju.

———. 1879. *Yanke zuanyao* 眼科纂要 (Compilation on ophtalmalogy). Place of publishing house unknown: Jiujingtang.

Huang, Yuanji 黃元基. 1763. *Jingyunzhai ji yanfang* 静耘齋集驗方 (Tested formulas from the Jingyun Studio). Benzhai zangban chaoben.

Liang, Lianfu 梁廉夫. 1880 [1936]. *Bu zhi yi biyao* 不知醫必要 (What a person ignorant in medicine should know). In *Zhenben yishu jicheng*, vol. 53. Shanghai: *Shijie shuju.*

Liu, Yuan 劉淵. 1739 [1873]. *Yixue zuanyao* 醫學纂要 (Compilation of medicine). Foshan: Jinyulou.

Lu, Shunde 路顺德 and Miao Fuzhao 缪福照. 1823 [1935]. *Zhigu xinfang* 治蠱新方 (New formulas for curing poisons). *Congshu jicheng chubian*, vol. 27. Shanghai: Shangwu yinshuguan.

Mai, Naiqiu 麦乃求. 1876. *Shanghan fayan* 傷寒法眼 (True knowledge of cold damage). Place and name of publishing house unknown.

Pan, Mingxiong, 潘明熊. 1865 [1868]. *Pingqin shuwu yilue*, 評琴書屋醫略 (Abstract of medicine from the Pingqin Library). Guangzhou: name of publishing house unknown.

———, 潘明熊. 1873 [1935]. *Pingqin shuwu Ye'an kuoyao* 評琴書屋葉案括要 (Abstract of Ye'an's [Ye Gui's] medical cases from the Pingqin Library (Pingqin shuwu Ye'an kuoyao 評琴書屋葉案括要). Guangzhou: Linjishu zhuangshi.

Qiu, Xi 邱熺. 1817 [1864]. *Yin dou lue* 引痘略 (Summary of smallpox inoculation). Hangzhou: Jingluntang.

Wang, Xueyuan 王學淵. 1843. *Shuzheng zhinan* 暑症指南 (Guide to summer-heat diseases). Place and name of publishing house unknown.

Yu, Tingju 俞廷舉. 1784 [1991]. *Jintai yihua* 金臺醫話 (Medical anecdotes from Jintai). *Zhongguo lidai mingyi yihua daguan*, vol. 1, pp. 281–309. Taiyuan: Shanxi kexue jishu.

b. Gazetteers from Yunnan, Guangxi and Guangdong Cited

Chuxiong xianzhi 楚雄縣志. 1877 [1967]. Chong Qian 崇謙 (comp.). Taibei: Chengwen chubanshe.

Dali xianzhi gao 大理縣志稿. 1916 [1974]. Zhang Peijue 張培爵 (comp.). Taibei: Chengwen chubanshe.

Dan zhouzhi 儋州志. 1936 [1986]. Wang Guoxian 王國憲 (comp.). Taibei: Xinwenfeng chuban gongsi.

Dongguan xianzhi 東莞縣志. 1911 [1957]. Chen Botao 陳伯陶 (comp.). Taibei: Taiwan xuesheng shuju.
Gan'en xianzhi 感恩縣志. 1931 [1967]. Zhou Wenhai 周文海 (comp.). Taibei: Chengwen chubanshe.
Gongcheng xianzhi 恭城縣志. 1890 [1968]. Tao Zun 陶壿 (comp.). Taibei: Chengwen chubanshe,
Guangnan fuzhi 廣南府志. 1906 [1967]. Anon. (comp.). Taibei: Chengwen chubanshe.
Guixian jinshi zhi 貴縣金石誌. 1934 [1986]. Liang Chongding 梁崇鼎 (comp.). Taibei: Xinwenfeng chuban gongsi.
He xianzhi 賀縣志. 1934 [1966]. Long Xianyu (comp.). Taibei: Chengwen chubanshe.
Jiaying zhouzhi 嘉應州志. 1897 [1968]. Wu Zongzhuo 吳宗焯 (comp.). Taibei: Chengwen chubanshe.
Kaiping xianzhi 開平縣志. 1933 [1966]. Yu Qimo 余棨謀 (comp.), Taibei: Chengwen chubanshe.
Lin'gao xianzhi 臨高縣志. 1892. Nie Jiqing 聶緝慶 (comp.). Lingao: Linjian shuyuan.
Longling xianzhi 龍陵縣志. 1917 [1974]. Jiang You 江右 (comp.). Taibei: Chengwen chubanshe.
Malong xianzhi 馬龍縣志. 1917[1974]. Wang Maozhao 王懋昭 (comp.) Taibei: Chengwen chubanshe.
Qiongshan xianzhi 瓊山縣志. 1857 [1974]. Zheng Wencai 鄭文彩 (comp.). Taibei: Chengwen chubanshe.
Sanjiang xianzhi 三江縣志. 1946 [1974]. Wei Renzhong 魏任重 (comp.). Taibei: Chengwen chubanshe.
Tengyue tingzhi 騰越廳志. 1888 [1967]. Chen Songhai 陳宗海 (comp.). Taibei: Chengwen chubanshe.
Xu Yunnan tongzhi gao 續雲南通志稿. 1900. Wang Wenshao 王文韶 (comp.). 岳池: 縣衙.
Xuwen xianzhi 徐聞縣志. 1912 [1974]. Wang Fuzhi 王輔之 (comp.). Taibei: Chengwen chubanshe.
Xuxiu Shunning fuzhi 續修]順寧府志. 1905 [1968]. Dang Meng 黨蒙 (comp.). Taibei: Taiwan xuesheng shuju.
Yangshan xianzhi 陽山縣志. 1938 [1974]. 黄瓚 Huang Zan (comp.). Taibei: Chengwen chubanshe.
Yazhou zhi 崖州志. 1914. Zhang Xi 張嶲 (comp.). Guangdong: Guangdong renmin chubanshe.
Yongchang fuzhi 永昌府志. 1886 [1967]. Liu Yuke 劉毓珂 (comp.). Taibei: Chengwen chubanshe.
Yulin xianzhi 鬱林州志. 1900 [1966]. Feng Decai 馮德材 (comp.). Taibei: Chengwen chubanshe.
Zhaoping xianzhi 昭平縣志. 1934 [1966]. Anon (comp.). Taibei: Chengwen chubanshe.

Others Primary Sources

Qu, Dajun 屈大均. 1678. *Guangdong xinyu* 廣東新語 (New Words on Guangdong). ctext.org/wiki.pl?if=gb&res=38451. Last access 15 Aug 2017.
Li Qiu 李璆, Zhang Zhiyuan 張致遠, Shi Jihong 釋繼洪 13th century 1983. *Lingnan weisheng fang* 嶺南衛生方. Beijing: Zhongyi guji chubanshe.

Secondary Sources

Bello, David A. 2005. To go where no Han could go for long. Malaria and the Qing construction of ethnic. *Modern China* 31 (3): 283–317.
Benedict, Carol. 1996. *Bubonic Plague in Nineteenth Century China*. Stanford: Stanford University Press.
Bin, Yang. 2009. *Between Winds and clouds. The making of Yunnan (second century BCE to Twentieth Century CE)*. New York: Columbia University Press.

Bretelle-Establet, Florence. 2017. Human mobility and books: modes of circulation of medical ideas and doctrines, in the Far South (18th and 19th centuries). In *Human Mobility and the Spatial Dynamics of Knowledge. Mapping Science, Technology and Medicine in Late Imperial China*, ed. Jami, 21–59. Paris: Institut des Hautes Etudes Chinoises.

———. 2018. Writing with and from parts of the discourses of others in Chinese medical texts: Where syntax and layout matter. In *Pieces and Parts in Scientific Texts*, ed. Bretelle-Establet F. and Schmitt S., 223–267, New-York: Springer: coll. Why the Sciences of the Ancient World Matter.

———. 2011. The Construction of the Medical Writer's Authority and Legitimacy in Late Imperial China. *NTM Zeitschrift für Geschichte der Wissenschaften, Technik und Medizin*: Band 19, Heft 4 (2011), pp. 349–390.

———. 2010a. How do documents become sources? Perspectives from Asia and Science. In *Looking at it from Asia: The Processes that Shaped the Sources of the History of Science*, Boston Studies in the Philosophy of Science 265, ed. Florence Bretelle-Establet et al., xxi–xlvi. Dordrecht: Springer.

———. 2010b. Is the Lower Yangzi River Region the only seat of medical knowledge in Late Imperial China? A glance at the Far South Region and at its medical documents. In *Looking at it from Asia: The Processes that Shaped the Sources of the History of Science*, Boston Studies in the Philosophy of Science 265, ed. Florence Bretelle-Establet et al., 331–369. Dordrecht: Springer.

Brokaw, Cynthia J. 1996. Commercial Publishing in Late Imperial China: The Zou and the Ma Family Business of Sibao, Fujian. *Late Imperial China* 17 (1): 49–92.

———. 2007. *Commerce in Culture: The Sibao Book Trade in the Qing and Republican Periods*. Cambridge, MA: Harvard University Press.

Brokaw, Cynthia J., and Kai-Wing Chow, eds. 2005. *Printing and Book Culture in Late Imperial China*. Berkeley: University of California Press.

Chao, Yuan-ling. 2009. *Medicine and Society in Late Imperial China: A Study of Physicians in Suzhou, 1600–1850*. Asian Thought and Culture 61. New York: Peter Lang.

Constant, Frédéric. 2016. Punir par l'espace: la peine d'exil dans la Chine impériale. *Extrême-Orient Extrême-Occident* 40: 14–38.

Corbin, Alain. 1982 [2008]. *Le miasma et la jonquille. L'odorat et l'imaginaire social XVIII-XIXème siècles*. Paris: Flammarion.

Elman, Benjamin. 1984. *From Philosophy to Philology*. Los Angeles: University of California Press.

Elvin, Mark & Liu, Ts'ui-jung. 1998. *Sediments of Time. Environment and Society in Chinese History*, part 1 and 2, Cambridge: Cambridge University Press.

Elvin, Mark. 2004. *The Retreat of the Elephant*. London: Yale University Press.

Furth, Charlotte. 1999. *A Flourishing Yin, Gender in China's Medical History, 969–1665*. Berkeley: University of California Press.

Genette, G. 1987. *Seuils*. Paris: PUF.

———. 1991. Introduction to the Paratext. *New Literary History* 22: 261–272.

Goldschmidt, Asaf. 2009. *The Evolution of Chinese Medicine: Song Dynasty, 960–1200*, Needham Research Institute Series. London/New York: Routledge.

Gong, Chun 龚春. 1983. *Zhongguo lidai weisheng zuzhi ji yixue jiaoyu* 中國歷代衛生組織及醫學教育 (Medical teaching and public health organisations in Chinese history). Beijing: Weisheng bu kejiao.

Grant, Joanna. 2003. *A Chinese Physician: Wang Ji and the 'Stone Mountain Medical Cases History'*. London: RoutledgeCurzon.

Guo, Aichun 郭靄春. ed. 1987. *Zhongguo fensheng yiji kao* 中國分省醫籍考, (Catalogue of Medical Books by Province). Tianjin, kexue jishu chubanshe.

Hanson, Marta. 2001. Robust Northerners and Delicate Southerners: The Nineteenth-Century Invention of a Southern *Wenbing* Tradition. In *Innovation in Chinese Medicine*, ed. Elisabeth Hsu, 262–291. Cambridge: Cambridge University Press.

———. 2006. Northern purgatives, Southern restoratives: Ming medical regionalism. *Asian Medicine* 2 (2): 115–170.

——— (2011). *Speaking of Epidemics in Chinese Medicine. Disease and the Geographic Imagination in late imperial China.* London: Routledge.

Henderson, John. 2010. Cosmology and concepts of nature. In *Concepts of Nature, a Chinese-European Cross-Cultural Perspective*, ed. Vogel and Dux, 181–197. Leiden: Brill.

Ho, Peng Yoke. 1985. *Li, Qi and Shu: An Introduction to Science and Civilization in China.* Hongkong: Hongkong University Press.

Grmek, Mirko. 1985. *Les maladies à l'aube de la civilisation occidentale*. Paris: Payot.

Leung, Ki Che Angela. 2009. *Leprosy in China, a History*. New York: Columbia University Press.

———. 2003. Medical Learning from the Song to the Ming. In *The Song-Yuan-Ming Transition in Chinese History, ed.* Paul Jakov Smith and Richard Von Glahn. Harvard University Asia Center, Harvard East Asian Monographs.

——— (2002). "疾病與方土之關係元至清間醫界的看法" (Observations on the relationships between disease and localities in the medical world from the Yuan to the Qing). 性别與醫療, 165–212.

Liu, Ts'ui-Jung. 1970. Han migration and the Settlement of Taiwan. In *Sediments of Time. Environment and Society in Chinese History*, ed. Elvin and Liu, vol. 1, 165–199. Cambridge: Cambridge University Press.

Liu, Xiangxue 劉祥學. 2013. (Analysis of the issue of 'zhang' in the environmental history of what is today the frontier areas) *Dang jin bianjiang diqu huanjing shi shiye xia de « zhang » yanjiu bianxi* 當今邊疆地區環境史視野下的"瘴"研究辨析, 江漢論壇, vol. 6.

Marks, Robert B. 2004. *Tigers, Rices, Silk, and Silt*. Cambridge: Cambridge University Press.

McDermott, Joseph. 2006. *A Social History of the Chinese Book. Books And Literati Culture in Late Imperial China*. Hong Kong: Hong Kong University Press.

Miles, Steven. 2006. *The Sea of Learning: Mobility and Identity in Nineteenth-Century Guangzhou. Harvard East Asian Monographs 269*. Cambridge, MA: Harvard University Asia Center.

Mou, Zhonghang & Wang, Caiping 牟重行, 王彩萍. 2003. « 中國歷史上的「瘴氣」考釋 » (Textual Researches and Explanations of Miasma in Chinese History) 師大地理研究報告 第 38 期 民國92年5月 Geographical Research No. 38, May.

Obringer, Frédéric. 1997. *L'aconit et l'orpiment, Drogues et poisons en Chine ancienne et médiévale*. Paris: Fayard.

Schafer, Edward H. 1967. *The Vermilion Bird. T'ang images of the South*. Berkeley: University of California Press.

Scheid, Volker. 2007. *Currents of Tradition in Chinese Medicine, 1626–2006*. Seattle: Eastland Press.

Schmitt, Hilary. 2017. *Forgotten Disease. Illnesses Transformed in Chinese Medicine*. Palo Alto: Stanford University Press.

Sivin, Nathan. 1995. State, Cosmos, and Body in The Last Three Centuries B. C. *Harvard Journal of Asiatic Studies*, 55(1) (Jun., 1995):5–37.

Unschuld, Paul U. 2003. *Huang Di Nei Jing. Nature, Knowledge, Imagery in an Ancient Chinese Medical Text*. Angeles: University of California Los.

Unschuld, Paul U., and Hermann Tessenow. 2011. *An Annotated Translation of Huang Di's Inner Classic – Basic Questions*. Vol. 2. University of California Los Angeles.

Unschuld, Paul U.2016., and Huang Di Nei Jing Ling Shu. *The Ancient Classic on Needle Therapy. The complete Chinese text with an annotated English translation*. University of California Los Angeles.

Vigarello, Georges. 1993 [1999]. *Histoire des pratiques de santé. Le sain et le malsain depuis le Moyen Age*. Paris: Seuil, collection Points Histoire.

Vogel, Hans Ulrich, and Gunter Dux. 2010. *Concepts of Nature. A Chinese European Cross-Cultural Perspective*. Leiden: Brill.

Wear, Andrew. 1992. Health and the Environment in Modern England. In *Medicine and society. Historical essays*, ed. Wear. Cambridge: Cambridge University Press.

Weber, Max. 1915. *The Religion of China, Confucianism and Taoism.* (trans. By Hans Gerth). 1951. 2nd edn 1959. Glencoe: The Free Press.

Will, Pierre-Etienne. 1992. *Chinese local Gazetteers, An Historical and Practical Introduction*, Notes de recherche du Centre Chine. Paris: Centre de Recherches et de documentation sur la Chine contemporaine (EHESS).

Xue, Qinglu 薛清录. 1991. *Quanguo Zhongyi tushu lianhe mulu* 全國中醫圖書聯合目錄 (Catalogue of medical books in Chinese libraries). Beijing: Zongyi guji chubanshe.

Zhou, Qiong 周琼. 2007. *Qingdai Yunnan zhangqi huangjing chulun* 清代雲南瘴氣環境初倫 (Discussion on environmental Zhang in Qing dynasty Yunnan). *Journal of Southwest University (Social Sciences Edition)* 33 (3): 44–52.

Chapter 9
Inhabited Lands and Temperaments: Observations and Therapeutic Solutions, the Views of Scientists and Medieval Physicians – Ǧāḥiẓ (9th), Rāzī (9th–10th), Ibn Riḍwān (11th)

Mehrnaz Katouzian-Safadi and Meyssa Ben Saad

Abstract This article focuses on the different ideas that circulated in the medieval Arabic-Islamic world on the natural environment's impact on living beings, plants, animals, and humans and on the interactions between them. We examine the relationship between the geographical location and human life according to the point of view of geographers. Then, we analyze texts reflecting the effects of the environment on plants and animals according to botanists, zoologists, and pharmacists. Finally, we analyze the various medical corpora that focus on the effects of the environment on the mental and physical health of human beings. Our goal here is to highlight the efforts made by scientists (physicians and naturalists) to associate theoretical knowledge with personal observations and experiments in order notably to prevent disease, to determine how diseases may appear, and to know how to treat them.

Keywords Observation · Environment · Adaptation · Therapeutic solutions · Winds · Lands · Rāzī · Ǧāḥiẓ · Ibn Riḍwān

9.1 Geographical Space, Languages, and Cultural Variety

This contribution examines how medieval scholars, in the Arabic-Islamic world, considered the relationships between humans and the geographical, climatic, and more generally, living environment that surrounded them. It also sheds light on how the links that were known to exist between the environment and living beings were

M. Katouzian-Safadi (✉) · M. Ben Saad
Université de Paris, SPHERE, UMR 7219, CNRS-Université de Paris, Paris, France
e-mail: mehrnaz.katouzian-safadi@univ-paris-diderot.fr

F. Bretelle-Establet et al. (eds.), *Making Sense of Health, Disease, and the Environment in Cross-Cultural History: The Arabic-Islamic World, China, Europe, and North America*, Boston Studies in the Philosophy and History of Science 333, https://doi.org/10.1007/978-3-030-19082-8_9

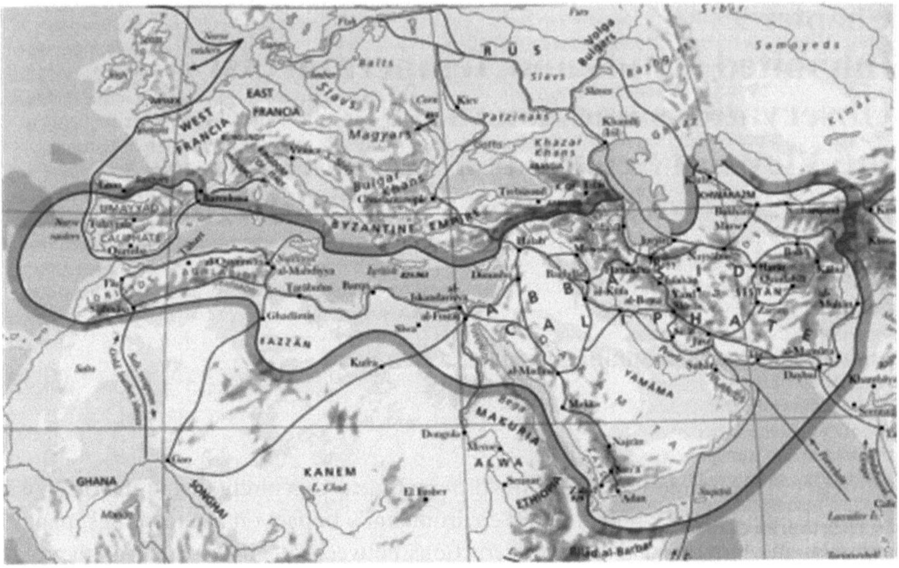

Map 9.1 Map of the Arabic-Islamic world: Abbasid Caliphate (750 CE) https://www.badw.de/fileadmin/pub/akademieAktuell/2013/46/0313_09_dalen.pdf

thought to bring disease or good health. Before giving voice to these scholars, we need to bring to mind some historical, geographical, and language features of the medieval Arabic-Islamic world.

The focus here is on texts written in the Arabic and Persian languages within the cultural space of the Arabic-Islamic civilization in medieval times.[1] In the seventh century, the wars that had lasted for over a hundred years between the Persian Empire and the Eastern Roman Empire (later called the Byzantine Empire) led to the decline of these two powers. These new geopolitical circumstances in the Middle East promoted and prepared the emergence of a new center of power and a new religion, "Islam." With the rise of Islam, the Persian Empire disappeared and the Byzantine Empire's territory began to shrink. However, the cultural influences of these two empires still prevailed in this cultural zone, which we shall designate Arabic-Islamic, even if different languages, ethnic groups, and religions contributed to its emergence and development. This nascent civilization would rapidly extend from the Indus to Andalusia. The geographic area that interests us is located on a band between the 45° latitudes, north and south of the equator (see Map 9.1). At the beginning of the conquests and for more than three centuries, the linguistic tool of these cultural and territorial expansions was the Arabic language. In a short period, Arabic allowed the syncretism of the sciences of that time, by opening new avenues for research and by promoting the diffusion of knowledge and exchanges of ideas

[1]Thanks for helpful discussions to Dr. Anne-Lise Haenni (Jacques Monod Institute, CNRS/Univ. Paris 7) and Dr. Alain Chaffotte (Pasteur institute of Paris).

between peoples of different ethnicities, religions, and languages. Arabic grammarians played a major role in stabilizing the language. Very early in the ninth century,[2] significant Greek, Syriac, and Sanskrit studies were translated into Arabic (see Table 9.1 transliterations of Arabic and Persian words into the Latin alphabet at the end of the article).

Greek scientific texts continued to circulate in this geographical region thanks to scholars who were familiar with Greek, Syriac, and Arabic. Very soon, new original texts appeared in Arabic, following requests from patrons who came from very different social backgrounds (princes, administrators, merchants, scientists, etc.). In the ninth century, research and translation made it possible to have access to many texts by Galen, thanks to Ḥunayn Ibn Isḥāq's translations.[3] From then on, and through these translations into Arabic, a specific medical scientific language stimulated exchange between scientists (see Table 9.2 for the summary of the names of books and authors).

Medieval sciences, written in Arabic from the ninth century, were the result of the meeting of the Greek, Persian, and Indian civilizations as well as of many other civilizations from the Near East, Africa, and Andalusia. These sciences continued to exist through other languages such as Persian, Hebrew, Turkish, and Greek[4] and later through Latin after they were transmitted to Europe. The scholars who participated in this Arabic-Islamic culture were Muslims, Christians, Jews, agnostics, and others. To maintain health and provide care, it was necessary to take into account all the geographical and cultural diversity of this vast region. The flourishing of this Arabic-Islamic civilization played an important role in the syncretism and a renewal of the approach to scientific knowledge.[5]

In this chapter, we present the scientific framework of medical thought in the Arabic-Islamic medieval world. Various factors contributed to construct this framework. Heritage from the Egyptian-Greek worlds took part in this culture. Botanists, pharmacists, and zoologists were aware on the effects of the environment on plants and animals. Geographers contributed greatly to enhance the sharing of knowledge from distant areas and civilizations, and they discussed at length the relationships between geographic area, climate, and living beings. Finally, physicians paid great attention to local climatic and geographical conditions in specific areas. Through their theoretical discourse but also thanks to their therapeutic actions, we have important clues to how physicians considered the question of the relationship between the environment, the body, health, and disease. (See Table 9.2 for a summary of the authors' name and titles of books).

---

[2] Gutas 1998.

[3] For two important translations made by Ḥunayn Ibn Isḥāq, see Ullmann 2002.

[4] In fact, some Arabic books were translated into Greek. For example, *The Book on Smallpox and Measles, Kitāb al-ǧudarī wa al-ḥasba*, (Rāzī 1848) was translated from Arabic into Greek in Byzantine Empire (Katouzian-Safadi 2012).

[5] Gutas 1998.

9.1.1 Theoretical Frameworks and Relations Between the Temperament and the "Milieu"[6]

The bridge between the individuality of the body and its environment was established on the concept of "temperament" and based on humors and elements. This concept gradually developed over a long history that began before the fifth century BCE and continued beyond the fifteenth century CE. The centers of this thinking probably migrated as it developed, from Egypt[7] to Greece and then to the Near East and the entire vast Islamic empire. Later Latin Europe continued along this path by adapting these inherited concepts to its own practical traditions.[8]

Throughout the present study we focus on Hippocrates (ca. 460 BCE–ca. 370 BCE) or rather on the Hippocratic corpus, i.e., on the corpus of texts attributed to him. In the Hippocratic treatise entitled *On Airs, Waters, and Places*,[9] the author tries to find in the climatic conditions such as the seasons, winds, rains, heat of the sun, etc. the causes of the diversity of humans, of their characters and customs on Earth and also the causes behind the appearance of disease. There are three principal themes in this treatise[10]: (a) evidence that climate affects humans; (b) customs, traditions, and political and social conditions leave their mark on humans, and these acquired and modified characters can be passed on from one generation to the next[11]; and (c) Greek citizens suffer the same effects as non-Greeks.

Six centuries later, Galen of Pergamon was born in Asia Minor (129 CE–216 CE). He became an accomplished Greek physician and was a prominent respected practitioner in the Roman Empire and especially in Rome. He wrote a commentary on the treatise of *On Airs, Waters, and Places*. An Arabic translation of this commentary has reached us. In this text, Galen retains some Hippocratic trends and departs from others. He does not keep the ideas expressed in (b) and (c). At the ethnographic level, Galen gives the priority for having good characteristics to the Greeks. Galen seems to reject the effects of modified and acquired characteristics. He retains the major idea regarding action of the climate on all living things. According to Morand,[12] in *De optima corporis nostri constitutione,* Galen indicates that the ideal body depends on two principal factors and issues: good temperament (*eukrasia*) and good organic proportions (*symmetria*).[13] A perfect body requires a

[6] The term "environment" or "milieu" is that which surrounds the human body near or far. In this article, we do not make the distinction of these two terms like the scholars and the philosophers of the twentieth century.

[7] Androlini 2016; Stol 2004.

[8] Nicoud 2006.

[9] Jouanna 2013.

[10] Strohmaier 1997. In this short paper, we can find a summary of this conclusion.

[11] Even if we use terms such as "acquired characters," we are far from evolutionistic thinking and from Lamarckism of the eighteenth century.

[12] Morand 2009.

[13] *Ibid.*, p. 203, ref. 5. See also, Galen 1923.

good mixture of homeomeric parts and a harmonious arrangement of the body. An aesthetic notion is overlapped on the notion of mixture in reference to the Polycrete canon.[14]

According to Hippocrates, medicine was a practice, a *techne*. This idea was expressed in the Middle Ages in Arabic by the art of medicine or *ṣinā'a al-ṭibb* (*sinā'a* = art; *al-ṭibb* = medicine). For Hippocrates it was necessary to know the nature of man or his *Physis*. One of the common theories in the fifth century BCE was the existence of two humors, blood (humor of health) and bile/phlegm (humor of disease). Hippocratic medical theories influenced Aristotle and, thus, Aristotelian thinking. Philip van der Eijk has analyzed some issues of its impact. He writes: "… Il est clair qu'Aristote s'était beaucoup intéressé à la médecine, et que son école avait une connaissance approfondie des textes médicaux 'hippocratiques'[15] et non hippocratiques (par ex. Dioclès de Caryste)" (It is clear that Aristotle had been highly interested in medicine and that his school had a deep knowledge of Hippocratic as well as non-Hippocratic texts [such as those by Diocles of Carystus]). The author adds: "Toutefois, il ne faut pas sous-estimer les différences considérables entre la pensée aristotélicienne et celles des auteurs médicaux" (However, we must not underestimate the significant differences between Aristotelian thinking and that of the medical thinkers).[16] Aristotle made some choices determined by the adaptability of such Hippocratic concepts to his own concepts of "form" and "matter" and the prevalence of form.[17] "En somme, nous avons ici [*La génération des animaux*] un bel exemple de l'utilisation, par Aristote, des idées et analogies empruntées à la littérature médicale. Bien sûr, le Stagirite ne les a pas simplement copiées: il les a transformées et reformulées dans les termes de son propre système, comme à son habitude" (In brief, we have here a nice example of the use, by Aristotle, of ideas and analogies borrowed from medical literature. Of course, the Stagirite did not simply copy them: he transformed and reformulated them in the terms of his own system, as usual).[18] The most important point concerns the theory of humors: "L'exemple le plus frappant est le fait qu'Aristote ne reconnaît pas comme facteurs fondamentaux les quatre humeurs du traité hippocratique *Sur la nature de l'homme* (exposé qui à partir de Galien deviendra le système physiologique canonique dans l'antiquité tardive). Pour le Stagirite, la bile jaune, la bile noire et le phlegme sont des résidus, des *perittômata*, qui ne jouent aucun rôle essentiel dans le système

[14] *Ibid.*, P. 204, the author quotes J. Pigeaud and the article on "The aesthetics of Galen," 1991, p. 2; Barras and Bircher 1994, p. 30.

[15] The "Hippocratic" medical texts or Hippocratic Collection (*Corpus Hippocraticum* in Latin) are a collection of 16 Ancient Greek medical works associated with Hippocrates' teachings, on a variety of medical topics, including diagnosis, epidemics, obstetrics, pediatrics, nutrition, and surgery (Hippocrates 2005). They are assumed to be by several authors, however, probably scattered over several centuries, and different treatises often give contradictory advice. https://www.medicinenet.com/script/main/art.asp?articlekey=20908

[16] van der Eijk 2007, p. 410, l. 23–25 – p. 411, l. 1, p. 411, l. 11–12.

[17] van der Eijk 2007, p. 411, l. 30–33.

[18] van der Eijk 2007, p. 424, l. 13–17.

physiologique humain" (The most striking example is that Aristotle does not recognize as basic factors the four humors of the Hippocratic treatise *On the Nature of Man* [which would after Galen become the canonical physiological system of late Antiquity]. According to Aristotle, the yellow bile, the black bile, and the phlegm are some residues, some *perittômata* without any crucial roles in human physiology).[19]

In the second century, Galen adapted some Aristotelian concepts (e.g., in his pharmacology),[20] and he dismissed others (e.g., the role of male and female in generation). Galen was also greatly influenced by the Hippocratic corpus; he transformed some medical theories and philosophical aspects of this corpus, as it is illustrated by the treatise *On the Nature of Man*, a work by Polybius, who was Hippocrates' son-in law. Galen attributes it to Hippocrates, and he suggests that Plato had read this treatise. Galen declares that the nature of humans consists of four humors, blood,[21] phlegm, yellow bile, and black bile (and not of two as Hippocrates suggested), and tried to establish a relation between the nature of man, humors, temperament, and elements. Jouanna calls this "contresens interprétatif créateur" (creative interpretative misinterpretation).[22] The theories of humors, temperaments,[23] and their relation to elements were not to reach their definite expression with Galen. These theories were developed further in the medieval Arabic-Islamic and Latin worlds (where they had significant ramifications).

According to Morand (2009), the theory of humors is linked to the notion of mixtures. Each species has its own particular mixture, but the best mixture is that found in man. For Galen there were three levels of mixing: elements, qualities, and humors. This theory of mixtures and humors enabled Galen to provide disease etiology. It also served as a basis for general therapeutic choices. Food, age, environment, and physical activity could change the humors. In Galen's work, the transition from composition to function occurs to the detriment of a complete theory of humors. The theory of humors contains gaps and contradictions linked to the fact that it is disseminated in various writings by the author.[24] Galen does not present a

---

[19] van der Eijk 2007, p. 411, l. 30–33 – p. 412, l. 1–3.

[20] Debru 1997.

[21] Hippocrates, ed. J. Jouanna, 1th ed. 1975, 2nd 2002, For a summary of conclusions, see: Jouanna 2011.

[22] Jacques Jouanna (2011): *Dès lors aux yeux de Galien, Hippocrate devient, par sa méthode consistant à décomposer tout être en ses composants (éléments) premiers, le maître de Platon et l'initiateur de toute la recherche philosophique sur la nature poursuivie, après par Platon, par Aristote et l'école stoïcienne.*

English translation: *Thus, by this method [attributed to Hippocrates] that consists in decomposing all living beings to their initial components (elements), in the eyes of Galen, Hippocrates becomes Plato's master, the initiator of all philosophical research on nature, continued later by Plato, Aristotle and the Stoic school.*

[23] Jouanna 2013, ed. J.van der Eijk, Chap. 15, Galen's reading of the Hippocratic treatise, *The Nature of Man: the foundations of Hippocratism in Galen.* p. 313–334; and chap. 16, "The legacy of the Hippocratic treatise *The Nature of Man: the theory of the four humours*," p. 335–360.

[24] Morand 2009.

system ranging from the elements to the humors. These elaborations occurred much later.[25] However, as stated by Galen, various surrounding conditions influence health and disease during the course of daily life.[26] Before any therapeutic treatment is undertaken through medicine or surgery, the physician takes into account the external conditions, the temperament and the type of disease the patient is suffering from and tries to propose the most suitable treatment. These **six factors** are notably the ambient air, food and drink, sleep and wakefulness, exercise and rest, retention and evacuation of wastes, and perturbations of the mind and emotions. Indeed, the state of mind is also considered as an important factor that has an effect on health. References to these agents are found in *Ars médica*, in *Pulsibus ad Tirones* and in four other books.[27] Galen gives different names to these factors.

In Arabic books, the terms *asbāb al-ṣiḥḥa* (*asbāb* = causes or ways, methods; al-*ṣiḥḥa* = health) or *asbāb al-sitta* or six causes are usually employed, even if the number often exceeds six factors. In medieval Latin books, the term "non-natural" or "six causes" is conserved for these factors. They are called "non-natural causes" probably because they are external and not related to the human body. It must be noted that everyday activities, such as sleep, movement, food, bathing conditions (the temperature, duration), as well as cosmological-atmospheric activity, greatly influence the patient's health. During the medieval period, the humoral medical pathology, systematized by Galen, remained the main theoretical framework in the Arabic-Islamic world. However, the analysis of surrounding factors was constantly reconfigured by each physician on the basis of new scientific findings in geography, astronomy, medicine, drugs, etc. The search for the balance of humors *aḫlāṭ* and each patient's individual temperament *mizāğ* became a real therapeutic art and underwent constant revision in the Arabic-Islamic world.

One of the methods for discovering the temperament of the patient is physiognomy,[28] which is an art that allows the temperament to be predicted according to visible signs: skin color, the components of the face, its shape, the voice, the speed of discourse, and the propensity to be talkative. This science, *firāsa* in Arabic, was particularly well developed in medicine by Rāzī. External factors, such as where the patient lived, and his habits and occupations, also helped the physician to evaluate his temperament. The medieval authors knew and used the rules and elements of *firāsa* to classify the temperament of each patient. However, these rules were flexible. They provided an orientation that could be modulated by the physician's practice, observations, and experience. These observations may have contradicted certain rules. To illustrate this complexity and to show cases where practical experiment contradicts the general rules of temperament and physiognomy, we give an example of this medieval thinking. The shining sun produces heat and can make

[25] "In historical periods when the theory of humors became more fundamental than in Galen's time, some Pseudo-Galenic treatises were composed to complete this theory," see van der Eijk 2005.

[26] García-Ballester 1993. https://www.ncbi.nlm.nih.gov/pubmed/8367894 - PMID: 8367894.

[27] Prioreschi 2003.

[28] Rāzī 1987, *al-Manṣūrī...*, *op. cit.*, chap. 2, '*fī al-firāsa*', p. 79–107; see Wilgaux 2008, p. 185–195.

the skin dark. Black populations, living in hot regions, have a hot mood. Thus, black women have an extremely hot temperament[29] because of their constitution. After giving birth, their milk was supposed to be excessively hot too and subsequently not healthy. However, they were considered excellent wet nurses. When a mother had no milk, they were recommended as wet nurses more than other women with a more balanced temperament.[30]

Moreover, to determine the patient's temperament, the physician integrated factors provided by his own life, hypotheses, and personal experience and observations. For example, Rāzī considered the Kurdish population to be very sturdy, and Ibn Riḍwān considered the Egyptians to have some physiological and physical weaknesses associated with their constitution. Rāzī's remark refers to the way the Kurds lived (not to the geographical areas nor to their physiognomy), and Ibn Riḍwān's claims were based on the climatic specificities of Egypt[31] and not on the general theory of climates. Thus, careful reading of each text is essential to understand each author's priorities and choices in a theoretical framework often shared between scholars. In the following pages, we deal with texts concerning natural sciences.

9.1.2 Points of View from Botanists, Pharmacists, and Zoologists on the Relationship Between the Environment and Human Beings

The Effects of Different Types of Soil on Plants in pharmacopeia, the properties of a plant depended on where it was cultivated. Thus, Ibn Bayṭār (d. 1248, in Syria), a thirteenth-century botanist and pharmacist, preferred to use aloe (*al-ṣabr*) from the island of Socotra, cumin (*Kammun*) from the Kirmān region (a province in the southeast of the Iranian plateau), and myrobolan (*hlilağ* or *ihlīlağ*) from India.[32] For physicians, there was a real difference between the quality of the plant depending on where it grew and also between the cultivated plant and wild varieties. Thus, the environment influences plants[33] and therefore the qualities of the plant and how they act on animals and humans. Two important examples illustrate this important relationship. The honey produced by bees foraging on the same types of plant will pres-

[29] Ibn Buṭlān 2010, *Risāla fī..., op. cit.* p. 132, 206, 233, 248.

[30] Katouzian-Safadi et al. 2017.

[31] Ibn Riḍwān, 1984, *Risālah fī daf' maḍārr al-abdān...*

[32] Ibn al-Bayṭār 1877, *kitāb al-ğāmi' li-mufradāt*, French transl., L. Leclerc, *Traité des Simples*, aloe (*ṣabr*), vol.2, p. 361–367, n° 1388, Cumin (*Kammun*), vol.3, p. 198 – 200, n°1967, myrobolans (*hlilağ*), vol. 3, p. 393–397, n° 2261.

[33] Ibn Sīnā 1877, *al -Qānūn*, (Avicenna, *Le Canon*), on when to harvest pharmaceutical plants, see vol. 1, p. 228.

ent different properties depending on the area where the plants grow.[34] The medical therapeutic uses of honey depend on the areas and the soil where it was produced. Similarly, the quality of milk depends on the species, on the age of the animal producing it and on the food (the grass) it eats. During the childhood after breastfeeding and during the process of weaning, the entire food chain of the animal producing the milk is taken into consideration when feeding the infant and the young child.[35]

The Effects of Different Types of Soil on Animals the zoologist Ğāhīz (776–869) was one of the naturalists who collected data from books and popular oral traditions.[36] His life spanned his century, and he became a very popular author and an important figure frequented by scholars and men of power. He is one of the founders of zoology written in Arabic.[37] In his *Kitāb al-Ḥayawān* (*The Book of Animals*),[38] Ğāhīz proposed to define the life and the admirable world of the living, to present its order, its organization, and its diversity through a classification of animals. The living world described by Ğāhīz is dynamic. Plants, animals, and human beings are affected by external conditions such as climate, soil, water, and habitat. Moreover, the relationships between plants, animals, humans, and their surroundings influence the life of each of them. The influences exercised by environmental factors may be favorable or hostile to certain species. On the basis of his observations and resemblances between plants, Ğaḥiẓ believed that the shape of a plant could change according to where it grows. For Ğaḥiẓ the coconut tree or *nārğīl*[39] results from the transformation of the doum palm or *muql* when it grows in a particular territory.[40] Like plants, animal species acquire the characters of their geographical area:

> وترَى جرادَ البقولِ والرّياحِين وديدانَها خَضراءَ، وتراها في غيرِ الخُضرة على غير ذلك، وترى القمْلة في رأس الشابِّ الأسودِ الشّعر سوداء، وتراها في رأس الشّيخ الأبيضِ الشّعرِ بيضاءَ، وتراها في رأس الأشمط شمطاءَ، وفي لون الجملِ الأورق، فإذا كانت في رأس الخَضِيب بالحمرة تَراها حمراء، فإنْ نَصَلَ خضابه صار فيها شُكْلةٌ، من بين بيضٍ وحُمْر، وقد نرى حَرّة بني سُليم، وما اشتملت عليه من إنسانٍ، وسبع، وبهيمةٍ، وطائِر، وحشرة فتراها كلَّها سوداء.
>
> "When it lands on vegetables or on fragrant plants, the locust takes a green color and it takes another color on other surfaces; the same applies to worms. A louse in the black hair of a young man is black. It is white in the white hair or the graying hair of an old man, and it is gray in the hair of an ash-gray colored camel. If it is on the head of someone who has

---

[34] Katouzian-Safadi and Bonmatin 2003.

[35] Ibn al-Bayṭār , *kitāb al-ğāmi'. op. Cit.*, Arabic into French translation, see *Laban* or "le lait" (milk), vol. III, n° 2007; Katouzian-Safadi et al. 2017.

[36] Aarab et al. 2003. See also in this volume Aarab, Katouzian-Safadi, Lamouchi-Chebbi.

[37] Aarab and Lherminier 2015.

[38] Ğāḥiẓ 1938–1945, *Kitāb al-ḥayawān*.

[39] Nārğīl is the Arabic form of the Persian term "nārgīl" that means coconut tree.

[40] Ğāḥiẓ 1938–1945, *Kitāb al-ḥayawān*, vol. IV, p. 130. *Phoenix dactylifera*, commonly known as date or date palm species, and the coconut tree (*Cocos nucifera*) are considered members of the Arecaceae family. Although the place of origin of the date palm is unknown because of its long cultivation, it probably originated from lands around Iraq and the doum palm is native to the northern part of Africa. http://www.theplantlist.org/tpl1.1/record/kew-152659- https://en.wikipedia.org/wiki/Date_palm

> applied henna, it is red, and when the dye begins to disappear, the red color gets mixed with white. It should also be noted that in the region of Banu Sulaym, that is known to have black soil, all creatures, whether humans, wild animals, cattle, birds, reptiles or '*ḥašarāt*'[41], have adopted the same color."[42]

Ǧāḥiẓ does not discuss the reasons for this change, and we do not know whether the author is considering a chain of causalities or not. Similarly, when animals leave their "natural" habitat, their behavior changes:

> **عير العانة اذا أقام في غير بلاده احتاج الى الأخذ من حافره، والى أن يختلف به الى البيطار, والطائر الوحشي من هذه المغنيات والنوائح، لو أقام عندنا دهرا طويلا لم يصوت اذا أخذناه وقد كرز. وكذلك المزاوجة والتعشيش والتفريخ.**
> "As for the male of onager [Asiatic wild ass], called *al-'āna* [in Arabic], if it is in a territory other than its own, it its hooves lengthen, to the point that it becomes necessary to cut them at the veterinary (or *bayṭarī* in Arabic). When wild birds that sing and weep live with us (humans), they stop singing and they lose their feathers. They also lose the capacity of mating, nesting and brooding."[43]

For the zoologist Ǧāḥiẓ, the territory inhabited profoundly influences the nature of living beings, be it an animal or a human being. An attempt to explain this is given in some cases. Thus, the environmental conditions can cause "alterations to the air," (*yafsadu al-hawā'* يفسد الهواء), provoke alterations to water and soil, and lead to changes in nature or character (*ṭibā'* - طباع):

> **لا ننكر أنْ يفسُدَ الهواءُ في ناحِيةٍ من النواحي فيفسدَ ماؤهم وتفسُدَ تُربتهم، فيعملَ ذلك في طباعهم على الأيّام، كما عمل ذلك في طباع الزّنج، وطباع الصّقالبة، وطباع بِلادِ يأجوج ومأجوج،وقد رأينا العَرب وكانوا أعراباً حينَ نزلوا خراسانَ، كيف انسلخوا من جميع تلك المعاني، وترى طباعَ بلاد الترك كيفَ تطبَعُ الإبلَ والدّوابَّ وجميعَ ماشيتهم: من سبُعٍ وبهيمةٍ، على طبائعهم.**
> "It is undeniable that when the air becomes corrupted in a given area, it may lead to a corruption of the water and soil in this area, and in the long run, these changes are thought to have lasting effects on the (character) nature (*ṭibā'* طباع) of the individuals living in this region of the world. Such is the case for the nature of blacks (*zanǧ*), of slaves and for the inhabitants of the land of Gog and Magog.[44] We see how the Arabs, who were nomads, at the time of their settlement in Khurāsān, have molted (metamorphosed) themselves (*insalaḫa* انسلخ) of most of the features we know. We see how the (character) nature of the country of the Turks has marked the camels, horses, and all the carnivorous or herbivorous animals that live there."[45]

Here Ǧāḥiẓ uses a very drastic term for the human modifications of nature and behavior when he changes his living environmental and cultural area. Actually, different forms of the verb *salaḫa* (*insalaḫa* انسلخ) mean "to moult" in British English or "molt" in American or "to metamorphose."[46] Here Ǧāḥiẓ uses the term *insalaḫa*

[41] In Ǧāḥiẓ classification, *ḥašarāt* stands for a class of animals over insects and contains other little animals.

[42] Ǧāḥiẓ 1938–1945, *Kitāb al-ḥayawān*, IV, 71 et V, 368–369.

[43] Ǧāḥiẓ 1938–1945, *Kitāb al-ḥayawān*, VII, 100, l. 5–9.

[44] The term *Zanǧ* could refer to a geographical area in Southeast Africa or to its inhabitants. Gog an Magog it's being between reality and fiction, see *E.I.*

[45] Ǧāḥiẓ 1938–1945, *Kitāb al-ḥayawān*, Vol. IV, p. 70–71.

[46] Lamouchi-Chebbi and Katouzian-Safadi 2016.

انسلاخ (that could be translated as "metamorphose") that refers to a very drastic transformation of nature.

It is important to note that, for the author, physiology and psychology are intimately linked and the transition from one to the other occurs naturally. The next paragraph concerns the change of mood in humans when they change their territory. Cities act on man and thus, a person who goes to Tubbat becomes playful and happy as long as he remains in there. Someone who stays in Mosul will feel full of vigor, and he who stays in Ahwāz sees his vigilance diminish significantly:

> الا ترى أنّهم يزعُمون انّ مَن دخَل أرضَ تُبّتَ لم يزَلْ ضاحكاً مسروراً، من غير عَجَبٍ حتّى يخرجَ منها، ومن أقام بالموصِل حولًا ثم تفقّد قوّته وجد فيها فضلًا، ومن أقام بالأهواز حَوْلًا فتفقّد عقلَه ذُو فِراسةٍ وجد النُّقصانَ فيه بيّنًا.
>
> "Don't you see that they say that whoever goes to the land of Tubbat stops being joyful and he becomes joyful again only when he leaves it. And whoever settles in Mosul,[47] will feel strengthened, and whoever settles in Ahwāz[48] sees his vigilance diminish significantly …"[49]

Finally, an example of an eighth-century social pact and peace treaty can be given. It was well known that in Sīstān, a region in the southeast of Iran, venomous snakes[50] were the cause of many fatalities in humans. Ǧāḥiẓ reports that during a war in 81 of the Hejira calendar (716 CE), a peace treaty was signed in Sīstān between the warring parties. One important section of this treaty dealt with wild animals. The inhabitants living in this region were aware of the necessity of hedgehogs for their own survival, because the main source of food for the hedgehogs was the venomous snakes that were plentiful there. With this in mind, a section was added to the peace treaty that recommended a prohibition on the killing of hedgehogs to preserve their numbers (hedgehogs in Arabic *qunfuḏ and* in Persian *khār pošt*).[51] This testimony illustrates the inhabitants' consciousness of the food chain and of the interdependency between humans and wild animals.

If by "environment" we follow Canguilhem, in that it is "the set of actions that are exerted on a living being from outside,"[52] Ǧāḥiz's contribution, through descriptions of several phenomena, attests to a conception of the surrounding environment quite close to Canguilhem's definition: for Ǧāḥiẓ, the natural environment affects life, behavior, and morphology and plays a role in the physiological characteristics of living beings.[53] After examining the observations of the naturalist Ǧaḥiẓ, we now focus on the opinions of geographers.

---

[47] Muṣūl or al-Mawṣil (or Mosul today) is a city in the North of Iraq. It was the capital of the province of Ninive in high Mesopotamia.

[48] Ahwāz in Arabic or Ahvāz in Persian is a city in southeast Iran near the border with Iraq.

[49] Ǧāḥiẓ 1938–1945, *Kitāb al-ḥayawān*, vol. IV, p. 135.

[50] Ǧāḥiẓ 1938–1945, *Kitāb al-ḥayawān*, vol. IV, p. 117.

[51] Ǧāḥiẓ 1938–1945, *Kitāb al-ḥayawān*, vol. IV, p.168, l. 4–5, p. 169, l. 1–3. Thanks to Dr. Younes Karamati for discussion about this paragraph; see Karamti, *Daneshnameh* (Persian Encyclopedia).

[52] Canguilhem 1992, p. 131.

[53] Interestingly, the editor Harūn (edition from 1938 to 1945) added the title *bī'a* to this chapter, which nowadays means "environment." That title does not exist in the manuscript (see index for "environment" terms).

9.1.3 Geographers' Points of View

As soon as the Umayyad caliphate (661–750) seized power, the center of political power moved to *Bilād al-Shām* or Greater Syria with the demands of an imperial court with Damascus as the capital. With the arrival of the Abbasids (750–1258) and the consolidation of the Empire, large urban centers were created; in 762, the political capital was moved to Baghdad in present-day Iraq. Rapidly, the centralized power was replaced by multiple centers of political decision-making that preached and promoted the same cultural unity through literary and scientific exchanges.[54]

The relationship between an individual's human nature and the place where he lives is addressed in geographical and medical texts that influenced each other. Theories and studies in these fields influenced each other. However, we will deal independently and in different paragraphs with these two inseparable approaches. Geographical texts present a land comprising seven latitudes, seven *Iqlīm*[55] or seven climates. The lands occupied by the Arabic-Islamic civilization were often situated at the center of this map, providing a good environment for human life. More detailed studies of medieval geography reveal that these considerations are not static; the geographies corroborate the theoretical presuppositions with the realities on the ground. Geographers and cartographers did their utmost to describe the environment around these new centers of power. This is the case, for example, for the geographer Idrisi[56] in Sicily.

In addition, there are different models such as the pre-Islamic Iranian model of *Kešvar*[57] (territory, country) which divides the earth into seven concentric areas. Geographers' texts present and discuss these very different models and views on the world that coexist, overlap, and influence each other.[58] This mapping is changeable and adaptable and scholars reveal a pragmatic and ethnological approach to the populations. Their points of view depend on where they lived and their needs, their interests, and their cultural practices. Thus, the fourth climate, which is considered as the most favorable zone, has no precise limits and its area fluctuates according to

---

[54] Lapidus 2002.

[55] Definition of the term *Iqlīm* in *Encyclopaedia of Islam (E.I.)*: *Iqlīm* – "clime, climate" or more generally, "region"[…]. Inherited from Greek tradition, the idea of climate refers to a zone extending, in longitude, from one bound to the other of the inhabited world and included, in latitude, between two parallels: the latitudes themselves are determined by the length of the day at the summer solstice or at the equinox; […] the limit between two climates allows a certain margin of uncertainty […]: the general opinion is that the boundary remains in any case a theoretical one and does not correspond to any concrete reality.

(For complete definition of *Iqlīm*, see Miquel, *Iḳlīm*, *Encyclopaedia of Islam.* The Arabic letter *qāf* (ق) is indicated by the Latin letter "q" in universal transliteration used in this paper and by ḳ, in *E.I.*).

During the medieval period, the term climate did not refer to any atmospheric or meteorological description, as it does nowadays.

[56] Nef 2000, p. 61–78.

[57] Baghdadī 1966 and 2011 *Tārīḫ Bagdād.*

[58] Olsson 2014, p. 508, note 118.

the geographers.[59] Furthermore, some characteristics attributed to populations do not stem from their place of residence but rather from their real or legendary history. This vision, which maintains a certain dynamism, was superimposed on the humoral theory in order to attribute a more favorable physiological and psychological temperament to the inhabitants of this central zone compared to other zones situated in the south and in the north.

The influence of geography extended beyond plants and animals to all living beings; according to medieval scholars, humans are also subject to such influences. Before beginning to apply the therapies, the medieval physicians were interested in the temperament of their patient. Thus, the seven *iqlīm* map or the seven "climates" often serve as a support to guide them on the temperament of their patient. While each zone of the map or *iqlīm* is associated with one category of predispositions as defined in the Hippocratic and Galenic texts,[60] this mapping of human geography was not static but rather dynamic and continually had to incorporate the experiments and the observations of the medieval naturalist physicians. The crucial point is that most of the research and deductions necessary to discover these temperaments took place in the field, based on the observations of scholars as presented below.

9.1.4 Points of View of Physicians

A Look at the Sky: The Moon and Atmospheric Conditions

The lunar and solar calendars jointly regulated time in human societies in the Middle East and the Arabian Peninsula before the advent of Islam. The solar calendar predominated in the pre-Islamic Iranian world. However, the Hegira calendar (that started with the stabilization of the Arabic-Islamic empire) is based on the movement of the moon.[61] Thus, there was a popular and scholarly familiarity with these two calendars in the region.[62]

The humoral medical pathology systematized by Galen also includes the moon and its movement. In his writings, Galen develops the notion of "critical days"[63] or "crises" which correlate the movement of the moon and diseases.[64] This book was

59 Al-Azmeh 1992.

60 Strohmaier 1997.

61 Saliba 2001 p. 27–49.

62 Katouzian-Safadi 2012, p. 23–51; p. 38–40.

63 Grmek 1983, *Les maladies à l'aube de la civilisation occidentale*, p. 13: lines 10–17. Grmek suggests two motivations for using critical days: the first is the strong wish of introducing numbers into the explanation of nature. The second, based on observation and experience, refers to intermittent fevers in the infectious or parasitic diseases that induce repeated feverish episodes separated by periods without fever.

64 Galen 2011, ed. Gleen Cooper, p. 129, note 24, and see Galen, *Critical Days*, Khün, 844 11.8–12.

translated by Ḥunayn Ibn Isḥāq as *Kitāb al-Buḥrān*,[65] and the Arabic authors interacted with these theories based on their philosophy and medical practices.

Despite the rejection of medical astrology,[66] the moon is present in Rāzī's (d. 925) observations and drug therapies. We find traces of these considerations in his works in reference to periods of crises, that is to say, during those crises that intervene in the evolution of the disease in an abrupt way. Changes in the moon may improve the condition of the patient, but they can also lead him onto a fatal path.[67] Changes take place in the air during these days, which can provoke diseases and affect the life of animals and plants. Rāzī addresses and develops causes of these changes[68]: observations show that the causes of the occurrence of these changes, which happen during these days, can be found in the universal change that takes place in the moon and that is detected in the air, as well as in the position of the moon in relation to the sun. Rāzī establishes a hierarchy of these days. He distinguishes the days that are likely to be harmful or fatal, and others that rarely cause danger. Some days may also herald future crises. Rāzī, who defends a book-based medicine accompanied by practice at the patient's bedside, who rejects medicine based mainly on philosophy or mathematics, and who does not resort to numerology or astrological predictions, accepts however the existence of cycles in the diseases or changes in diseases linked to lunar cycles (e.g., cycles of fever with malaria).

As to drug therapies, Rāzī draws our attention in his aphorisms to compound drugs whose preparation requires special knowledge and skill. In another book, the author attributes this skill to knowledge on the environment and the stars. In the book on *The Secrets of Medicine* (*Kitāb al-sirr al- ṣinā'a al-ṭibb*), he quotes Aratus (third century BCE). For Rāzī, the movement of the moon induces changes in the air, and the air influences everything that is subject to generation and decay on earth; all this is known to farmers, to astronomers (*munağğimūn*), and to philosophers. Physicians agree that the movements of the stars influence the air, water, habits, natures, on land in all longitudes and latitudes. A physician must take into account the state of the stars (*kawākib*), the moon, and the seasons during the preparation of drugs.[69]

However, these lunar effects should not be confused with a medicine-related astrology neither in Galen nor in Rāzī. Youssef Mourad[70] and other researchers such

---

[65] Cooper, ibid. p. 144. Cooper refers to the book *Crisis* from Galen translated into Arabic by Qustā Ibn Lūqā (d.912) and to The *Canon of Medicine* from Avicenna (d. 1037). Cooper does not mention different books from Rāzī on this topic.

[66] Ventura 2008, 278 p.

[67] Rāzī, *Rhazes' Kitāb al-muršid...*, *op.cit.*, ed. Arabe (1961) and French transl. (1980), *op.cit.*, aphorisme n° 235.

[68] Rāzī, *Ibid.*, aphorisme n° 234.

[69] Rāzī, *K. Sirr sinā'a al-ṭibb*, éd. Kh. Harbī, (http://content.wdl.org/7460/service/7460.pdf); p. 112–113. Katouzian-Safadi 2012, p. 39, l. 25–7, p. 40, l. 1–2.

[70] Mourad 1939, *op. cit.*, chap. 1, "Objet et évolution de la physiognomonie," p. 8–9.

as Glen Cooper[71] recognize that astrological practices are not monolithic. He distinguishes a "natural astrology" and another that he describes as being "judicial" and that we translate here as "premonitory." According to Cooper, the latter form of astrology as well as what is close to magic and divination is not present in Galen.

The same can also be said about Rāzī. Cooper bases his reasoning on the references made to Aratus by Galen.[72] Indeed, these texts by Aratus did not help in defining the position of the planets nor in determining their influence to make medical forecasts possible. Rāzī, like Galen, speaks of a natural astrology. The latter is part of the Hippocratic world where the stars influence atmospheric conditions that can affect a sick person's recovery, without really enabling the physician to predict the future and to act on it.

A Look Toward the Land: The Territories

To discover the importance given to environmental elements, we will first examine a propaedeutic book of medicine. This is the work *al-Manṣūrī fī al-ṭibb* (abbreviated *Manṣūrī* hereafter) written by Rāzī.[73] This book shows the continual research of these medieval physicians who observed the environment to be able to adapt to it as much as possible in order to preserve the good health of the human body. This book is intended for the daily practice of medicine during the medieval period; it was commonly used in the Islamic and Latin worlds. Indeed, this work was translated into Latin in Toledo by Gerard of Cremona (d. 1187) under the name of *Liber ad Almansorem*. It circulated widely in Europe in its entirety, in chapters (the ninth chapter in particular, which deals with therapeutics, was widely circulated on its own without the other chapters) or even in leaflets written in vernacular.[74] Its influence continued to be felt until the Renaissance, and an edition was printed with commentaries by Andreas Vesalius. *Manṣūrī* was one of the most widely read medieval medical manuals in Europe. This book was of paramount importance for medieval medicine in the Arabic-Islamic world and in Europe in medical practice and for Latin university teaching.

In his introduction, Rāzī gives the definition of medicine as being the Science and the Art of preserving health (*hifẓ-al ṣiḥḥa*) and of getting rid of disease. To achieve these two goals, the author constantly considers the links between the body

[71] Cooper 2011, see p. 123.

[72] Cooper 2011, *op.cit.*, p. 132; Aratus (300 BCE) described, in his work *Phenomena*, constellations and other celestial phenomena. His work, very popular in the Greco-Roman world, was not scientifically based on observation or mathematics. *Encyclopedia Britannica*, 2 (11th ed.). See Aratus, Cambridge University Press. p. 321. (https://en.wikisource.org/wiki/1911_Encyclop%C3%A6dia_Britannica/Aratus_of_Soli).

[73] Rāzī dedicated this book to al-Manṣūr, the governor of the city of Ray.

[74] Jacquart 2003, p. 207–220.

and its “environment,” that is, all that surrounds the body. Rāzī’s intention in his *Manṣūrī* is to clarify some general rules[75]:

وقد قلنا أننا لا نقصد في كتابنا هذا إلى مثل فى هذا التفضيل والتحديد إذ كان يحتاج فيه إلى وغول وإغراق في الصناعة بل إلى الأمر الأعمّ الأكثر.

“As we said, our goal when writing this book was not to carry out refined research and study but to examine general conditions.”

The author gives us here common scientific knowledge to indicate the framework and the general factors the physician had to take into account in order to treat people.[76] These chapters are reminiscent of the medical instructions of the Hippocratic *On Airs, Waters, and Places* treatise. However, Rāzī’s words describe the environmental conditions that the ninth-century author experienced. Chapter 5, “Care Health,” provides an overview of the daily care package to better adapt the body to the various conditions of the external environment. Chapter 7 describes the precautions to be taken when traveling. We begin by analyzing this chapter.

Disturbances Linked to Traveling

There is a corpus of medical texts dedicated to the theme of traveling and caring for the body. They are referred to as *zād al-musāfir* or “the meal of the traveler.” In such medical texts, the physicians are concerned with the entire life of the person on the move, with his corporal, physiological, and psychological changes during traveling. The physicians deal with the permanent adaptation of the body and with new geographical and atmospheric circumstances. This theme can be treated in an independent book or in a chapter of a general medical book such as Rāzī’s *Manṣūrī* where the theme of traveling and caring for the body composes chapter six (p. 281–300), *fī tadbīr al-musāfirīn* or “Traveler’s diet.” This chapter provides guidelines to maintaining health during travel in general. However, physicians must adapt these guidelines to the bodily characteristics of each person.

Traveling for various purposes (leisure, trade, war, education or teaching, etc.) is considered by this physician as a real change imposed on the body. Each person must somehow get used to these new conditions by totally adapting all parts of his life (habitat, food, beverages, the place of his temporary accommodation, etc.). We consider below the case of the movements of the army.[77] Depending on the season, the area occupied by the camp will be different.

ينبغى أن تُنزل العساكر في الصيف على التلول و الربوات. وتستقبل وجوه الخيم ريح الشمال، ويتباعد بينها، وتبعد الدواب ما امكن. [...] وأما في الشتاء فليكن التدبير بالضد في ذلك[...] . ومتى كان الهواء ومداً جنوبياً رطباً

[75] Rāzī 1987, *Manṣūrī*, *op. cit.*, p. 281–282 l. 1–2.

[76] In the introduction to *Manṣūrī*, Rāzī says that he made a selection of texts from Buqrāṭ (Hippocrate), Ǧālinūs, Āramsūs, and other scholars such as Bolos and Ahrūn and also refers to his contemporaries such as Ḥunayn Ibn Isḥāq and Ibn Massawayh.

[77] I thank Dr. Mohammad Sadr for drawing my attention to this paragraph.

فليقلل الغذاء ويهجر الشراب البتة و يكثر من الرياضة. [...] ومتى مرض خلق كثير فلينزلوا جانباً بعيداً ولا يكونوا فوق الرّيح بل تحتها. [...] وليتفقّد الغذاء والسبب الذي يمرض منه الكثير في العسكر فتضاد و تجتنب.

"During the summer, the army camps must be on hills and on raised areas, and the tents must be oriented towards the north wind and far from each other. Quadrupeds must be far from the tents. In winter, it is necessary to resort to the opposite positions. […] When the weather is wet, (with wind blowing from the south), it is necessary to reduce food, avoid wine and to increase physical activity. […] When many people fall sick, they must be moved away [from the camp and positioned] so as not find themselves upstream but downstream (and also downwind) […] Food must be checked again and the causes that lead to disease must be investigated among the soldiers so that these causes can be avoided."[78]

The Winds, Rains, and Climates

The wind is an element that can have beneficial or harmful effects on the body. According to Rāzī, the north wind is beneficial; it eliminates epidemics but causes sore throats, lung disease *(al-nazl)*, and colds *(al-zukām)*; it reduces stools and increases urine. The warm southern wind leaves the lungs and throat intact but causes other diseases. It relaxes the body that feels exhausted, causes ophthalmia and headaches, fevers[79] appear, and epileptic attacks increase.[80] Depending on the season, the winds have different effects. Moist air containing a lot of water vapor (*buḫārāt*) can be found in the vicinity of lakes or ponds as well as very hot and stagnant air favor fast-spreading diseases (present-day epidemics) (*amrāḍ al-wabā' īyya*),[81] such as pestilence (*al-wabā'*). Such cases of illness occur in valleys surrounded by high mountain ranges. The Hippocratic treatise, *On Airs, Waters, and Places*, already mentioned winds; however, the winds described by Rāzī, with their own qualities, are likely to correspond to those experienced by Rāzī on his own travels.

In other works, Rāzī is particularly interested in the danger of pestilence or fast-spreading diseases. By studying the atmospheric conditions of a region (seasons, rains, state of the air), a physician can make some predictions about the appearance of diseases. In the first translations of medical books from Greek to Arabic, by Ḥunayn Ibn Isḥāq in the ninth century, the Greek words *λοιμικός - λοιμός* are rendered into Arabic by the term *tā'ūn,* and *tawā'īn* in the plural; the term *λοιμώδης* is translated as *al-wabā'*. Dols[82] usually translates *tā'ūn* by plague[83] and *wabā* by pes-

[78] Rāzī 1987, *Manṣūrī*, p. 295, l. 3–15.

[79] The author uses expression "الحميات العفنية" or *al-ḥumīyyāt al-'afnīyya* , where *al-ḥumīyyāt* means "fevers" and *al-'afnīyya* means "miasmatic or purified." The qualifications of fevers must be translated carefully. Here, it is not possible to identify the fever and to deduce the related disease. See Katouzian-Safadi 2014, p. 339, col. A, l. 3–33.

[80] Rāzī 1987, *Manṣūrī*, p. 162, l. 5–25 - p. 163, l. 1–5.

[81] Rāzī 1987, *Manṣūrī,* p. 163, line 3, "هذه الأهوية كلها وبائية," *hāḏihi al-ahwīyya kulluhā wabā'īyya*, "all these airs are pestilential or epidemic." For the terms *al-wabā'* see Dols 1974.

[82] Dols 1974.

[83] Ibn Ḫātima al-Anṣārī 2010.

tilence, and these translations are often accepted.[84] The only certainty is that these terms refer to diseases that affect a large number of individuals and that can be described as epidemics. It is however safer to keep a definition with fuzzy boundaries for these translations and to reconsider the meaning of these terms by a deeper analysis of the context according to the corpus, author, and time period under examination.[85]

The succession of the seasons and the variation of temperatures and of rainfall partly regulate the diseases and their prediction. In his book *The Secrets of Medicine*,[86] Rāzī establishes links between these acute fast-spreading diseases and the natural conditions of the environment.[87] He refers to measles and smallpox and other forms of disease, the *tawā'īn*, plague, (plural of *tā'ūn*), or the *al-wabā'* pestilence (pestilence). If rainfall is frequent in a region and if flies (*ḏubāb*) quickly develop there, these diseases threaten humans. If such frequent rains fall in winter and spring, the *al-wabā'* gains ground, and the number of human deaths will be quite high. If during autumn the rain is not frequent, then there will be disorders of the black bile that invades bodies. The frequency of fevers due to *wabā* increases when the air is turbide (*kadir*) and unclean. When winter is warm and spring is cold, the symptoms of leprosy appear. In areas close to shores, near water, there is more balance in autumn; the physician must then prepare for a cold winter using emollients and soothing perfumes.

Concerning psychological illnesses, Rāzī states that change of air and a long trip can be beneficial. He illustrates his thinking, in aphorisms 20–35, in the chapter dealing with "air." He provides some tips for therapies:

> إذا حدث فى الهواء العَفَن، و استبات فيه أراييح منكرة و كثر الجدرى والحصبة والطواعين، فينبغى أن يُفزع ألى الأسراب الغائرة، بعد أن تكون يابسة، أو يكون فيها أسِرّة، والبيوت البعيدة من الهواء مثل التى فى وسط الأبنية. وترش الأماكن بالخل، وتبخر بالبان، وورق الآس، والسعد، ويكثر استعمال الخل فى الطعام، و يُشرب الماء.
>
> "If the air is miasmatic, with unpleasant smelly odors, if the number of cases of smallpox (*ğudarī*), measles (*ḥasba*) and plagues (*tawā'īn*) multiply, one should take refuge deep underground. Once the premises have dried out, beds must be laid outside, and rooms must be arranged away from the surrounding miasmatic air. All places must be sprinkled with vinegar; olibanum (*al-bān*), myrtle leaves (*waraq al-ās*) and cypress (*al-sa'd*) must be burned. The frequent use of vinegar in food and drink water is recommended."[88]

[84] Conrad 1982.

[85] Sublet 1971. See p. 142. The author says *tā'ūn* and *wabā* refer to the plague; see also Conrad, *op. cit.* p. 272, note 10.

[86] Rāzī, *Kitāb Sirr ṣinā'a al-ṭibb*, p. 91–92.

[87] Katouzian-Safadi and Bonmatin 2014.

[88] Rāzī 1961, *Rhazes' Kitāb al-muršid*, aphorisme n° 32.

The Water of the Different Lands

In *Manṣūrī*, one of the major problems encountered when traveling seems to be the change of the sources of water. At each stage of the journey, the author proposes mixing the water of the place reached with that of the previous encampments. Before the trip, it would be best to collect water from different stages of the trip. If such precautions are not taken, the consumption of water must be combined with that of vinegar, onion, and garlic. Another solution is also proposed[89]:

> وينفع من ذلک أيضاً أن يتزود المسافر من طين بلده، فاذا نزل على ماء أخذ منه شياً وألقاه فيه وأقرّه حتى يصفو ثم يشرب منه.
>
> "It would be beneficial for the traveler to keep some clay (*ṭīn*) from his country to add to the water, let it settle and then drink it."

In the medieval period, many texts were written on the great diversity of qualities attributed to clay and its therapeutic preparation and uses.[90] It may be noted that even today there are popular expressions to explain sickness due to traveling. In Persian we hear people say*: ab beh ab shodeh* which could be translated as "The person has gone from one water to another" and became sick. In Arabic we know the expression "travel sickness" which corresponds to diarrhea among travelers (*dā' al-musāfir* داء المسافر). It should be noted, that while water and wind could be sources of disease, they were also known to protect human health. When they are associated to masterful architectural knowledge, they can constitute a barrier against extreme heat at the edge of the desert. This is illustrated by the wind towers in Yazd. These ancient constructions induced a temperature decrease of 20 to 30 °C inside the house (Fig. 9.1) during extreme summer heat at the edge of the desert.

9.1.5 *The Nature of the Body and Its Relationship with Its Surroundings*

Aware of regional specificities that account for variations in health and disease, some physicians wrote books that dealt exclusively with the health situation of a population in a given region. This is the case for Ibn Buṭlān[91] (1066) who gave a description of his observations and of his therapies specifically adapted for Baghdad and its surrounding. It is also the case for As'ad al-Maḥallī (1202) who specialized in the temperaments of the inhabitants of the city of Damascus.[92] The work of Ibn Riḍwān (998–1061) focuses mainly on the geographical and medical situation of Cairo and its surroundings.

---

[89] Rāzī 1987, *Manṣūrī*, p. 293, l. 11–13.

[90] Rautureau et al. 2017, p. 36–46.

[91] Ibn Buṭlān 1990.

[92] As'ad al-Maḥallī 2000, ed. Khal.

Fig. 9.1 Architectural construction: cooling system based on wind and water. The wind tower in the city of Yazd, in the center of Iran, at the edge of the desert. (Photo, François Caudron)

This author lived under the Fatimid dynasty in Egypt (969–1171), at the time when the centers of political power multiplied at the expense of the Abbasid caliphate of Baghdad, but the cultural and linguistic unity was preserved. Ibn Riḍwān, famous for one of his observations in the field of astronomy, is especially recognized as a medical practitioner in Egypt whose geography he knew well. In the book that interests us, "The preservation of health in Egypt," he links the health of the inhabitants of Egypt with their geographical location. After years of field experience, he reflects on the results of his observations. In this study, he addresses his criticism to another physician, Ibn al-Ğazzar (895–979) who lived in Kairouan (in present-day Tunisia) and who discussed health and the environment in Egypt without ever having been there.[93]

Cairo and its surroundings enjoyed great economic prosperity, until the middle of the eleventh century when famines and diseases (pestilences) overwhelmed the population. For, Ibn Riḍwān, one of the main and natural reasons for these calamities was the fluctuations of the Nile. There were also social causes such as a period of war and the fact that these areas become routeways where many travelers, traders, and pilgrims from various places meet. The recurrence of epidemics was at the center of the controversy between Ibn Ğazzār and Ibn Riḍwān who contradicted his colleague's proposals. His arguments are based on his observations; on his knowl-

[93] This book of Ibn Ğazzār has been lost. It's title is: *Kitāb fī na't al-asbāb al-muwallida lī al-wabā, Book on the Causes of Epidemics.*

edge of the geography of the region which, he says, is constantly changing, and on his daily practice that guides him in his therapeutic decisions.

In his autobiography, Ibn Riḍwān boasts of possessing and knowing the books of Hippocrates and Galen translated into Arabic. In the text we are examining, the author is careful to sprinkle all the chapters with references to the texts attributed to Hippocrates and to those of Galen; he quotes these authors before presenting his own observations. He thus establishes his profound knowledge of these two authorities in medicine before proposing his own opinions. However, Ibn Riḍwān's descriptions are entirely the result of his own findings and experiences in the place where he practiced as a physician. The text is entirely based on his cultural milieu and time. Ibn Riḍwān's book is composed of 15 chapters. Of these, the first 8 chapters describe the geographical peculiarities of the country. Let us illustrate how Ibn Riḍwān incorporates the theoretical framework inherited by Hippocrates and Galen to build his own theories.

In the first chapter, Ibn Riḍwān delimits the country making references to Ptolemy, Hippocrates, and Galen, but he adds his personal and contemporary knowledge when he indicates the position of towns in Egypt.[94] Ibn Riḍwān gives the latitudes that correspond to the astronomical studies of Battānī[95] (d. 929) who was his contemporary,[96] and he does not use Ptolemy's figures. In chapter four, "On the seasons of the year in Egypt," Ibn Riḍwān first describes Galen's position concerning the balanced nature of spring. Then he deviates from this standard position and presents the specific situation of Egypt using the sixth months of the Coptic solar calendar. He describes the particularities of spring in Egypt, the influence of specific winds and the importance of the Nile. In fact, for him, the fluctuations of the latter are largely determined by the environmental situation.[97] Overall, Galen's theoretical texts draw a framework for the expression of personal and historical observations accumulated over the years.

Ibn Riḍwān also criticizes his contemporaries. The title of chapter five is "On the incorrectness of most of Ibn al- Ğazzār's reasons for the unhealthy air in Egypt." Ibn Riḍwān underlines the sources of all of his respectable predecessor's mistakes in different chapters, for example: "It suffices to say that Ibn al-Ğazzār, despite his stature in the profession, wrote a book about Egypt in which he failed to explain the temperament or condition of Egypt."[98] In fact, as Ibn Riḍwān says, Ibn Ğazzār had an excellent theoretical formation and practical skills in his country (Tunisia today). However, he had never been to Egypt. He collected his information from his compatriots who traveled to Egypt and who became sick. Ibn Riḍwān cited Ibn Ğazzār[99]:

---

94 Ibn Riḍwān 1984, Ar., p.2, l. 19, Eng. p. 80., l. 8.

95 Ibn Riḍwān 1984, Ar., p.2, l. 19, English p 80, l. 8–9.

96 For al-Battānī astronomy, see Ibn Riḍwān, Ibid., p. 80, ref. 9: Al-Battānī, ed. and trans. Nalino, C., A., Rome, 1899–1907, vol. 2, p. 43–45, vol. 3, p. 239.

97 Ibn Riḍwān 1984, Ar. p. 9, l. 15–36, Eng. p. 95, l. 3–32.

98 Ibn Riḍwān 1984, Ar. p. 22, l. 18–20, Eng. p. 121, l. 17–19.

99 Ibn Riḍwān 1984, Ar. p. 12, l. 3–7, Eng. p. 99, l. 3–10.

"The cause of illness in those who traveled from Tunisia to Egypt is the great variation of the air. The journey exhausted them and made them susceptible to illnesses. Then, when the air [around them] was changed, they fell sick and death was swift." Ibn Riḍwān replies that the inhabitants of Egypt were accustomed to variations in the air and were not afflicted by illness. Ibn Ǧazzār blamed the humidity[100] and the dense fog[101] of Egypt as a source of illnesses that spread rapidly (epidemic). Ibn Riḍwān contradicts this claim. He suggests this fog could be unhealthy for travelers (from Tunisia). He asserts that for the inhabitants of Egypt, this fog in winter coincides with a period without disease in Egypt. Moreover, he accuses Ibn Ǧazzār of confusing endemic diseases and epidemic diseases (*wāfid*, *wabā*).

In the remaining chapters, Ibn Riḍwān uses his own observations to assess the specificities of the milieu and its diseases and then goes on to discuss the treatments that should be given to people of Egypt.

Chapter six describes in detail the following capitals of Egypt: al-Fusṭāṭ, al-Qarāfah, Cairo, and al-Gīzah, and implicates the influence of the position of the mountains and the fluctuations of the Nile in the variations in climate and in health. In chapter seven,"on the cause of pestilence and other epidemic diseases("*fī al-wuqūf 'alā asbāb al-wabā' wa sā'ir al-amrāḍ al-wāfida*"" فى الوقوف على أسباب الوباء وسائر الأمراض الوافدة ")", Ibn Riḍwān addresses the issue of epidemics. For Ibn Riḍwān, there were multiple causes of epidemic diseases, and notably a change in the qualities of air, water, food and also psychic events. Ibn Riḍwān gives a description of each after referring to Hippocrates and Galen. He then describes them using his personal practice of medicine in relation to the surrounding variation of the climatic conditions. We should point out here Ibn Riḍwān's efforts in distinguishing different various kinds of epidemic diseases. Ibn Riḍwān defines as:

- *al-amrāḍ al- wāfida*: general term, epidemic[102] illnesses that affect many people in one land at a given time
- *al-wabā'*: a kind of epidemic disease (translated by *pestilencia*)
- *al-mawātān*: a disease "in which the mortality rate is high."[103]

In chapter eight, the author recapitulates the surrounding conditions and concludes that the temperament of Egypt is hot *(ḥārr)* and moist (*raṭib*) and the corruption (*'ufūna*) of the Nile is less in the south than in the north. He introduces the six manners that protect the body (see above "From humors to temperament," ref. 22 and 23 in this article). He describes the "six causes" (*al-asbāb al-sitta*)[104] and gives a summary of some of them: air (*al-hawā'*), food and drink (*al-akala wa al-ašrība*), sleep (*al-nawm*), and wakefulness (*al-yaqẓa*). While emotions can favor the emer-

---

[100] Ibn Riḍwān 1984, Ar. p. 12, l. 22–25, Eng. P. 100, l. -7.

[101] Ibn Riḍwān 1984, Ar. p. 12, l. 32–35; Eng. p. 100, l. 22–23, p. 101, l. 1–3.

[102] Conrad 1982. According to Conrad, *wabā* refers to the pestilential corruption of the natural environment that causes specific diseases, one of which is the *ṭā'ūn*. *ṭā'ūn* is the name of plague today.

[103] Ibn Riḍwān 1984, Ar., p.17, l. 18–34, Eng. p 112, l. 10.

[104] Ibn Riḍwān 1984,Ar., p.17, l. 9, Eng. p 113, l. 23–6, p. 114, l. 1–2.

gence of disease, they are not a cause of disease if they do not exceed a certain limit: "Psychic events, such as anger, sadness, and joy do not create illness if they do not go beyond the proper bounds. It is desirable for people of Egypt to increase their happiness and joy in order to strengthen the natural heat of their bodies for the digestion (*haḍm*) to improve, and the congestion of their bodies to lessen."[105] Here the psychic state belongs to surrounding factors. The psychic state is in direct relation to "natural heat" (*ḥarāra ġarīzīyya*) which is the principal source of life. For Rāzī, Ibn Riḍwān's great predecessor, the patient's enjoyment and the moderate satisfaction of his desires are a crucial element for recovering and an important factor of his surroundings. We find this lesson in numerous books by Rāzī.[106]

Finally, chapters nine to fourteen give a medical response to health preservation, care, and drug therapies by taking into account local geographic conditions. Ibn Riḍwān describes 25 points for health preservation. The first five recommendations are about the depth and careful knowledge of the local situation and health problems. The physician needs to know (1) the temperament of the country (each country has a specific temperament), (2) the indigenous illnesses, (3) the current season, (4) the temperament of this season, and (5) the prevalent epidemic illnesses.[107] All care, diets, and drug treatments must be adapted to these particularities.

Although in his book, references to Hippocrates are numerous, Ibn Riḍwān draws his recommendations and prescriptions from his personal observations and not from his theoretical reading. In chapter ten, "What is necessary for physicians to do in the land of Egypt" (*fī mā yanbaġī li al-ṭabīb an yaf'aluhu bi 'arḍ Miṣr*), he indicates that one must take care of one's food and take simple or compound drugs from the area of Egypt. He advocates:

> وأن (يجعل) علاجه ملائما لما عليه الأبد ان بأرض مصر [...]. فإن نكاية هذه فى الأبدان سريعةٌ لا سيّما ابدان المصريين سريعة الوقوع في النكايات. ويختار ما كان من الأدوية المسهلة وغيرها ألين قوّة حتى لا تكون على طبيعة المصريين فيها كلفه و لا يلحق أبدانهم منها مضرّة. ولا يقدم على الأدوية الموجودة فى كتب الأطباء اليونانيين و الفرس فإن أكثرها عملت الأبدان قويّة البنيّة غليظة الأخلاط.
>
> "The physician's treatment should be adapted to the bodies in Egypt […]. The harmful effect of such drugs can be rapid in the body, especially in Egyptians, whose bodies are quick to suffer from injury. He should select from purgative drugs and others the gentlest in strength, so that there is no discomfort or harm to the bodies of Egyptians. He should not prescribe drugs found in the medical books of Greeks and Persians, for most of them are aimed at bodies with strong constitutions and with strong coarse humors."[108]

In the chapters that follow, he makes a selection of the food and drugs used by physicians who lived in the Islamic cultural empire such as Ibn Māsawayh, Rāzī, Ibn Ǧazzār, Sāhir, and Ubaydallah, among others. The author adds new items to the list of foods and describes the variation of their properties according to their mode of preparation. Ibn Riḍwān explains that the physician must consider all the specificities of the patient's body. The strengths and the weaknesses of each part of the

---

[105] Ibn Riḍwān 1984, Chap. 8, Ar., p. 20, l. 24–26, Eng., p. 117, l. 24–28.

[106] Rāzī 1961, *Rhazes Kitāb al-Fuṣūl*, aphorisms n° 275, 128, 279, 280.

[107] Ibn Riḍwān 1984, Chap. 9, Ar., p. 21, l. 34–37, Eng., p. 120, l. 24–27.

[108] Ibn Riḍwān 1984, Chap. 10, Ar., p. 25, l. 4–11, Eng., p. 128, l. 12–16.

body (stomach, head, intestines) must be evaluated in relation to the season and the particularities of Egypt climate. Ibn Riḍwān's description of the relation between the health situations and the surroundings demonstrates, above all, his commitment to medical practices and his excellent mastery of Egypt's climatic changes and, more generally, of geographic knowledge of Egypt.

9.2 Conclusion

The studies that informed physicians about their distant or close environment (the sky, the sun, the earth, plants, animals, etc.), in the medieval Arabic-Islamic world, belong to very different bodies of texts. The books of natural philosophy, medicine, and pharmacopeia are just one part of the physician's cultural background. Indeed, the texts concerning geography, astronomy, and astrology are part of a practical knowledge that the physicians had to study during their training. This knowledge defines the intellectual framework that allowed physicians to describe and interpret the effects of the environment on the bodies of healthy or sick people. However, this framework was flexible and as our analysis of these texts show, medical thinkers had the freedom to finally integrate their own local experiences and observations. The term *iqlīm* translated to climate could sometimes designate today's concept "environment." The medieval author referred to the concept of the environment by using terms such as territory, habitation, or climate, depending on the situation and the element that were examined. We can summarize our comments.

The term "concept" is from the Latin *concipere* that means to contain and to receive. A concept is a complex cognitive construction that must replace and substitute some direct sensory and perceptible experience. The use, the propagation and the acceptance of the term "environment" instead of "milieu" in Europe and the United States, has its own history,[109] depending on the situation of each country in the eighteenth to nineteenth centuries. Even if this term itself did not exist in the medieval period, we can say, **all perceptions** covering today's terms "environment" or "milieu" invaded the everyday life and occupied the intellectual sphere of medieval scholars such as physicians, geographers, botanists, and zoologists. It was not necessary then to create an abstract name to subsume all that could be perceived by the five senses from the surrounding natural world. To know and to understand the surrounding world (of all living beings in general and human beings in particular) sensitive and sensory experiences were effective and sufficient. Thus the very names of the elements in the environment's that could be perceived by the senses, such as the moon, water, wind, earth, etc., were used to speak of the environment.[110]

[109] Taylan 2014.

[110] Thanks to Pauline Agator (student in Master/University Paris 7 Denis Diderot) for her interest in this subject and for helpful discussions and exchanges during 2016–2017.

In the present study, we have not taken into account literary texts (in prose or poetry) or texts addressed to the educated administrative staff (*adab* in Arabic). However, the latter category of texts forms a whole that reveals the powerful relationship with the staff in the service of the prince and the surrounding environment in a broad sense. These texts should also be considered as shedding light on the relationship between man and his environment through various levels of society and throughout various historical times of the medieval period. With this study we have attempted to highlight the relation between practices and theories, and how theoretical knowledge has shaped practices and produced new shared theories and practices and also practices specific to a region or to an ethnic group. These new practices have themselves interacted with, and rehabilitated, existing theories. This circularity is difficult to restore in a linear script. In addition, references to local habits and practices passed on in the oral tradition are very scarce; we sometimes discover them in the written texts of great scientists such as Rāzī or in certain geographical texts. Another difficulty for the researcher is to understand the value of observation because modern science associates the scrutiny of the observer to future interventions/modifications of the object observed. Even if he shares this dream, the medieval scholar – such as Ǧabir Ibn Ḥayyān – was aware of his technical and material limitations in the immediate time. In such a situation, the observations serve to imagine permanent and man-made adaptations, to adapt his work, lodging, food, and movements to the seasons, winds, rivers, weather, plants, and animals. On these observations and the resulting rules depend the preservation of health. These continuous collections of careful, fine, and thorough observations are essential to rehabilitate theories and reinvent practices to strengthen health in connection with the surrounding world.

Table 9.1 Transliterations of Arabic and Persian words into the Latin alphabet

Usual characters in the Arabic alphabet	Arabic international transliteration
أ	ʾ
ا	ā
ة	a
ث	ṯ
ج	ǧ
ح	ḥ
خ	ḫ
ذ	ḏ
ش	š
ص	ṣ
ض	ḍ
ط	ṭ
ظ	ẓ
ع	ʿ
غ	ġ
ق	q
	ḳ (in *Encyclopaedia of Islam*)
و	ū/w
ي	ī

Table 9.2 Summary of the names of books and authors

	Name	Geographical area	Life	Work	Abbreviation	Work category
1	Ğāḥīẓ or Jâhiz	Basra and Baghdad	776–869	*Kitāb al-Ḥaywān (Book of animals)*	*Ḥaywān*	Zoology
2	Rāzī	Ray (near Tehran) and Baghdad	865–925	*Kitāb al-Manṣūrī Latin translation: Liber ad Almansorem*	*Manṣūrī*	Medical science
3	Ibn Ğazzar	Ifrīqīyya (Tunisia)	895–979	*Kitāb fī na't al-asbāb al-muwallida lī al-waba'* (*Book on the Causes of Epidemics*)	–	Medicine: preservation of health in Egypt
4	Ibn Sīnā or Avicenna	East and west of Iran	980–1037	*Kitāb al-Qānūn (The Canon of Medicine)*	*Qānūn*	Medicine
5	Ibn Riḍwān	Egypt, Cairo	998–1061	*Kitāb daf' maḍār al-abdān, bi arḍ Miṣr (Treatise on the preservation of bodily ills in Egypt)*	–	Medicine: preservation of health in Egypt
6	Ibn Buṭlān	Iraq, Egypt, and Syria	ca. 1001–1066	*Risāla fī širā' al-raqīq wa taqlīb al-'abīd (Treatise on the Purchase and Examination of Slaves)*	–	On the salves
				Taqwīm al-ṣiḥḥa Tacuini Sanitatis (Health care)		Health care and therapy
7	Ibn Bayṭār	Andalusia and Syria	1197–1248	*Kitāb al-ğāmi' al-mufradāt (The book of simple drugs)*	*mufradāt*	Pharmacopeia

Various Terms that Could Design the Humans' "Milieu"

Contemporary terms		
البيئة	*al-bī'a*	Environment
المحيط	*al-muḥīṭ*	Milieu
المجال الحيوي	*al-mağāl al-ḥīyawī*	Living area
المنظومة البيئية	*al-manẓūma al-biy'īyya*	Ecosystem
Medieval and current terms		
الماء	*al-mā'*	Water
الهواء	*al-hawā'*	Air
الطبيعة	*al-ṭabī'a*	Nature
التربة	*al-turba*	Ground

Medical Terms
al-aḫlāṭ الأخلاط pl., sing. الخلط *al-ḫilṭ*: humor
al-asbāb الأسباب pl., sing., السبب *al-sabab*: ways, methods
al-ṣiḥḥa الصحهة: health
al-asbāb al-ṣiḥḥa - الصحهة الأسباب: non-natural causes method of health preservation
al-haḍm الهضم: digestion
al-ḥarāra ġarīzīyya الحرارة غريزية: natural heat
al-mizāğ المزاج: temperament, complexion – *mazağa*: to mix
al- 'ufūna العفونة: corruption or putrefaction according to the context

Biological Terms
salaḫa سلخ: to molt/to shed
insalaḫa انسلخ: to metamorphose
al-nārğīl النارجيل: coconut tree: *Cocos nucifera*
yafsadu al-hawā' يفسد الهواء: It corrupts the air
al-ṭibā'- الطباع pl.: nature, character, sing. *al-ṭab'*, الطبع

Bibliography

Antique and Medieval Primary Sources

As'ad al-Maḥallī, Ya'qūb Ibn Isḥāq. 2000. *Maqāla fī mazāğ al-Dimašq, Treatise on the errors of the physicians in Damascus*, ed. Arabic text & English trans, by Olivier Khal, Oxford/New York: Oxford University.

Baghdadī, al-Kahtīb. 2011. *Tārīḫ Baġdād*, Bayrūt: Dār al-Kitāb al-'Arabī; 1978, Beirut: Dār al-Kutub al-'Ilmīyah.

Battānī (Al-), *Al-Battānī sive Albatenii Opus astronomicum. Ad fidem codicis Escurialensis arabice editum*, ed. Carolo Alphonso Nallino, Milan, 1899–1907 (3 vol.)

Galen. 1923. *De sanitate tuenda*, ed. K. Koch, Corpus Medicorum Graecorum V, 4, Leipzig-Berlin.

———. 2011. *Galen, De diebus decretoriis, from Greek into Arabic: a critical edition, with translation and commentary, of Ḥunayn ibn Isḥāq, Kitāb ayyām al-buḥrān,* ed. Cooper, Gleen, Farnham: Ashgate.

Ğāḥiẓ (1938–1945) *Kitāb al-ḥayawān,* éd. 'Abd al-Salām Muḥammad Ḥārūn, 1964–1979, ed. Le Caire; 1992, ed. Beirout.

Hippocrate. 1975. reed. 2002. *La nature de l'homme,* ed., trans, comment., Jacques, Jouanna, 2nd ed., anastatique augmentée et corrigée. Berlin: Akademie Verlag.

Hippocrates. 2005. van der Eijk Philippe, ed., *Hippocrates in context* (= Studies in ancient medicine 31) Leyde, Boston.

Ibn al-Bayṭār, *kitāb al-ğāmi' li-mufradāt al-adwiya wa al-aġḏiya*, French transl., L. Leclerc *Traité des Simples*, 3 vol, édition du XIXème siècle, Paris, 1877, réimpression Beyrouth 1th ed., 1291/1874, Le caire, 2nd ed., Bagdad, 1384H/1964, 3nd ed., Beyrouth 1422 H/2001Leclerc, 1992.

Ibn Buṭlān. 1990. *Le Taqwim al-sihha (Tacuini Sanitatis) d'Ibn Butlan: Un Traité Medical Du XIe Siecle.*, trad. Hosām, Elkhadem, Louvain, Peeters.

———. 2010. *Risāla fī širā' al-raqīq wa taqlīb al-'abīd,* (Epître sur l'achat et le commerce des esclaves), French trasl., Floréal Sanagustin, Médecine et société en islam médiéval - Ibn Buṭlān ou la connaissance médicale au service de la communauté – Le cas de l'esclavage, Geuthner.

Ibn al-Ḫatima al-Anṣārī and Abū Ǧa'far. 2010. *Taḥṣīl ġaraḍ al-qāsīd fī tafṣīl al-maraḍ al-wāfid, La grande peste en Espagne musulmane au XIVe siècle, éd.* et trad. Suzanne Gigandet, Damas, Ifpo.

Ibn Riḍwān. 1984. *Kitāb daf' maḍār al-'abdān, Medieval Islamic Medicine: Ibn Ridwan's Treatise "On the Prevention of Bodily Ills in Egypt"*, English translation & introduction by Dols, Michael W., Arabic text, ed. by Gamal, Adil, S., Comparative Studies of Health Systems and Medical Care. University of California Press.

Ibn Sinā, *al-Qānūn fī al-ṭibb*, Cairo, ed. Būlāq, 1297/1877, 3 vol. Beirut (reprint), Dār Ṣādir.

Rāzī, Muḥammad Ibn Zakariyyā. 1848. *Kitāb al-Ǧudarī wa al-ḥiṣba* éd. Trad. Anglaise, Greenhill, William Alexander, *A treatise of the smallpox and measles* Rhazès, London: Sydenham Society. https://archive.org/details/39002086344042.med.yale.edu. French translation. 1866. Leclerc, Julien et Lenoir, De la variole et de la rougeole. Persian Trans. (1992) Mahmūd Naǧmābādī, Université de Téhéran, 1040, 3ème ed., 1371 H.S./1992. English trans (1848): http://books.google.fr/books?id=LC0IAAAAIAAJ&dq=A+treatise+of+the+smallpox+and+measles.

Rāzī, Muḥammad Ibn Zakariyyā. 1987. *al-Manṣūrī fī al-ṭibb*, ed. & notes by H. B. al-Siddīqī, Manšurāt ma'had al-maḫṭūtāt al-'arabiyya, al-Kuwait.

———. 1961. *Rhazes' Kitāb al-muršid aw al-fuṣūl*, ed. Arabe, A.Z. Iskandar, Maǧallat ma'had al-maḫṭūtāt al-'arabiyya, vol. 7, fasc. 1; French transl., 1980 and 1996, *Guide du médecin nomade*, by El-Arbi Moubachir, Paris.

Rāzī, Muhammad Ibn Zakariyyā, *K. Sirr ṣin'a al-ṭibb* (ou *Le livre du secret de la médecine*), éd. Khalid Ḥarbī, Alexandrie, date not indicated. http://content.wdl.org/7460/service/7460.pdf and Kuhn Brabant Rosa, ed. Madrid, *al-Qantara*, 1982, p. 358.

Secondary Sources

Aarab, Ahmed, Provençal, Philippe and Idaomar, Moḥamed. 2003. La méthodologie scientifique en matière zoologique de Ǧāḥiẓ dans la rédaction de son œuvre *Kitāb-al-Ḥayawān*. *Anaquel de estudios arabes* 14: 5–19.

Aarab, Ahmed and Lherminier, Philippe. 2015. *Le livre des animaux d'Al Jâhiz*. Paris: Harmattan.

Androlini, Isabelle. 2016. "Crossing the borders between Egyptian and Greek medical practice", chap. 7, in *Popular Medicine in Greco-Roman Antiquity: Exploration*, ed. William V. Harris, Leiden/Bonston: Brill.

Al-Azmeh, Aziz. 1992. Barbarians in Arab Eyes. *Past & Present* 134 (Feb.): 3–18.

Barras, Vincent, and Terpsi Bircher. 1994. La perfection de l'homme chez Galien. *Equinoxe* 11: 27–36.

Ben Saad, Meyssa. 2010. Ph.D., *La connaissance du monde Vivant chez le savant al-Djâhiz (776–868)*, Université Paris 7 – Denis Diderot, Paris.

Canguilhem, Georges. 1992. *La connaissance de la vie*. Paris: Vrin.

Conrad, Lawrence I. 1982. Tā'ūn and Wabā', Conception of Plague and Pestillence in Early Islam. *Journal of the Economic and Social History of the Orient* 25 (/3): 268–307.

Cooper, Gleen M. 2011. Galen and Astrology: A mésalliance? *Early Science and Medicine* 16: 120–146, see p. 123.

Debru, Armelle. 1997. Philosophie et pharmacologie: La dynamique des substances *leptomères* chez Galien. In *Galen on pharmacology. Philosophy, History and Medicines* Proceedings of the Vth International Galen Colloquim, Lille, 16–18 March 1995, ed. A. Debru, 85–102. Leiden, New-York, Köln.

Dols, Michael. 1974. Plague in early Islamic history. *JAOS* 94 (3): 371–383.

Eijk (van der), Philip, J. 2007. "Les mouvements de la matière dans la génération des animaux selon Aristote", in *La science médicale antique, nouveaux regard*, (dir.) Véronique Boudon-Millot, Alessia Guardassole and Croline Magdelaine, (Beauchesne) Paris, 407–424.

García-Ballester, Luis. 1993. On the origin of the "*six non-natural things*". *Sudhoffs Arch Z Wissenschaftsgesch Beih., n°32*: 105–115. https://www.ncbi.nlm.nih.gov/pubmed/8367894 - PMID: 8367894.

Gutas, Dimitri. 1998. Greek thought, arabic culture. The Graeco-Arabic translation movement in Baghdad and early 'Abbāsid society *(2nd-4th/8th-10th centuries)*. ed. Routledge, London an New York,; French transl., A. Cheddadi, 2005, *Pensée Grecque, culture arabe, Le mouvement de traduction gréco-arabe à Bagdad et la société abbasside primitive (11°-IV°/VIII°-X° siècles)*, traduit de l'anglais par Abdesselam Cheddadi, Aubier, Flammarion, Paris.

Grmek, Mirko. 1983. reed. 1994, *Les maladies à l'aube de la civilisation occidentale,* Paris, Payot.

Jouanna, Jacques. 2011. "Aux racines de la nature de l'homme",. discours in Institut de France, http://seance-cinq-academies-2011.institut-de-france.fr/discours/2006/jouanna.pdf.

———. 2013. *Greek Medicin from Hippocrates to Galen*, ed. John Scarborough, Philip J.van der Eijk, Anne Ellis Hanson, Joseph Ziegler, vol 40, selected paper by Jacques Jouanna, translated Neil Allies, chap. 9, "Water, health and disease" in the Hippocratic treatise *Airs, Waters, Places*. p. 155–172; Chap. 15, "Galen's reading of the Hippocratic treatise *The Nature of Man: the foundations of Hippocratism in Galen*. p. 313–334; and chap. 16, "The legacy of the Hippocratic treatise *The Nature of Man: the theory of the four humours"*, p. 335–360.

Jacquart, Danielle. 2003. "Les manuscrits des traductions de Gérard de Crémone: quelques caractéristiques formelles", in J. Hamesse (éd.), *Les traducteurs au travail, leurs manuscrits et leur méthode (moyen Age–Renaissance)*, Tumhout, 207–220.

Karamati, Younes. 2015. *Daneshnameh (Persian Encyclopedia), see "snake in old Persian an Arabic book" (af'ī dar matūn kohan farsī va 'arab)*. Vol. 4, 649, col. B, l. https://www.cgie.org.ir/fa/publication/entryview/40164.

Katouzian-Safadi, Mehrnaz & Bonmatin, Jean-Marc. 2003. L'usage du miel dans les médicaments simples et composés chez Rhazès (m.925). *Revue d'histoire de la Pharmacie, N°* 337: 29–36.

Katouzian-Safadi, Mehrnaz. 2012. "La lèpre, la variole, la mélancolie: soigner les déséquilibres du corps", in *La contamination: lieux symboliques et espaces imaginaires,* V. Adam, L. Revol-Marzouk (dir.), 23–51. Paris: Garnier, "classiques Garnier".

Katouzian-Safadi, Mehrnaz & Bonmatin, Jean-Marc. 2014. Du Diagnostic Differential aux Therapies Prudentes: Le traité de la rougeole et de la variole de Râzî, in *Medicine and Healing in the Ancient Mediterranean World*, D. Michaelides (Dir.) 334–349. Oxford & Philadelphia.

Katouzian-Safadi, Mehrnaz, Meyssa Bensaad, and Jean-Philippe Izard. 2017. Soigner l'enfant dans les textes médicaux arabes et persans du Moyen Âge. *Annales de Bretagne et des Pays de l'Ouest* T. 124 (3): 203–227.

Lamouchi-Chebbi, Kaouthar & Katouzian-Safadi, Mehrnaz. 2016. Métamorphoses et transformations subies par les insectes dans *Kitāb al-Hayawān* de Jāhiz, *Actes du Congrès l'animal et l'homme,*. 141e congère CTHS, http://cths.fr/co/index.php?id=163, in Press.

Lapidus, Ira M. 2002. *A History of Islamic Societies*. Cambridge: Cambridge University Press.

Miquel, André. *"Iqlīm"* or *"Iḳlīm" Encyclopaedia of Islam,* vol. III, ed. Lewis, b., Menage, V. L., Pellat, ch., Schacht, J., Leiden (Brill) London (Luzac & CO.), 1986, p. 1076B–1078A.

Morand, Anne-France. 2009. "Mistura das qualidades e determinação da saúde em Galeno: aspectos químicos e cósmicos, in *A saúde dos antigos. Reflexões Gregas e Romanas*", ed. M.C.D. Peixoto, Loyola, São Paulo, French translation, 203–216. http://bsa.biblio.univ-lille3.fr/upload/couverture-saude-morand.pdf.

Mourad, Youssef. 1939. *La physiognomonie arabe et le kitāb al-firāsa de Fakhr al-dīn al-Rāzī.* Paris: Geuthner.

Nef Annliese. 2000. Les relations entre les pays d'Islam et le monde latin: la place de la Sicile du milieu du Xe siècle au milieu du XIIIe siècle. *Cahiers d'Histoire* 78: 61–78.

Nicoud, Marilyn. 2006. Savoirs et pratiques diététiques au Moyen Âge. *Cahiers de recherches médiévales* 13 (spécial): 239–247. http://crm.revues.org/864.

Olsson, Joshua. 2014. The world in Arab eyes: A reassessment of the climes in medieval Islamic scholarship. *Bulletin of the school of Oriental an Africa Studies* 77 (3): 487–508.

Prioreschi, Plinio. 2003. *A history of medicine*, vol. V, *Medieval medicine*. Omaha: Horatius Press. https://www.ncbi.nlm.nih.gov/pmc/articles/PMC4448*513/*.

Rautureau, Michel, Gomes de Sousa Figueiredo Celso, Nicole Liewig, and Mehrnaz Katouzian-Safadi. 2017. *Clays and Health–Properties and Therapeutic Uses*. Cham: Springer.

Pigeaud, Jackie. 1991. L'esthétique de Galien. *Mètis. Anthropologie des mondes grecs anciens* 6 (1): 7–42. http://www.persee.fr/doc/metis_1105-2201_1991_num_6_1_960.

Saliba, Georges. 2001. Science before Islam, in A. al-Hassan, M. Ahmad, and A. Iskandar (dir.), *The Different Aspects of Islamic Culture, vol. 4: Science and Technology in Islam,* Part 1, *The exact and natural sciences.* Cambridge, MA/London: MIT Press, UNESCO.

Stol, Marten. 2004. An assyriologist reads Hippocrats. In *Magic And Rationality In Ancient Near Eastern And Graeco-roman Medicine*, ed. Herman F.J. Horstmanshoff, Marten Stol, and C.R. Van Tilburg. Leiden/Boston: Brill.

Strohmaier, Gotthard. 1997. La question de l'influence du climat dans la pensée arabe et le nouveau commentaire de Galien sur le traité hippocratique des airs, eaux et lieux. In *Perspectives arabes et médiévales sur la tradition scientifique et philosophique grecque: actes du colloque de la SIHSPAI*, ed. Ahmad Hasnawi, Abdelali Elamrani-Jamal, Mārūn ʿAwwād. Paris: Leuven.

Sublet, Jacqueline. 1971. La peste prise aux rêts de la jurisprudence; le traité d'Ibn Hağar al-ʿAsqalānī sur la peste. *Studia Islamica* 33: 141–149.

Taylan, Ferhat. 2014. Ph.D., *Introduction à la "Rationalité Mésologique. Connaissance et gouvernement des milieux de vie (1750–1900)".* Thèse de doctorat en philosophie, Université Bordeaux Montaigne.

Ullmann, Manfred. 2002. *Wörterbuch zu den griechisch-arabischen Überzetzungen des 9. Jahrhunderts.* Wiesbaden: Harrassowitz Verl.

Ventura, Iolanda. 2008. *Astro-Medicine. Astrology and Medicine, East and West*, éd. Anna Akasoy, Charles Burnett et Ronit Yoeli-Tlalin, Florence, SISMEL/del Galluzzo (Micrologus' Library 25),. 278 p.

Wilgaux, Jérôme. 2008. La physiognomonie antique: bref état des lieux. In *Langages et métaphore du corps dans le monde antique*, ed. Véronique Dansen et Jérôme (dir.). Rennes: Presses Universitaires de Rennes.

Chapter 10
Health and the Environment: Aldo Leopold, Land Health, and the First-Person Ecology Approach

Catherine Larrère

Abstract In the current conditions of environmental degradation, the most crucial question is to know how this deterioration affects our health. There is no doubt that environmental pollution is a serious threat to our health. But can the environment be safe in itself? And what can a safe environment tell us about our health? How can the loss of contact with our natural environment affect our health? These are the kinds of questions that Aldo Leopold had in mind, when, in his preface to his most famous book *A Sand County Almanac* (1949), he wrote: "Our bigger-and-better society is now like a hypochondriac, so obsessed with its own economic health as to have lost the capacity to remain healthy." How can one be healthy in a healthy environment? To what extent can we speak of healthy environments? The answer to these questions does not only require scientific, objective, and quantified analyses. Above all, it needs an examination of first-person ecology – the study of the subject in his/her environment. It is from this perspective that we propose to study Aldo Leopold's texts on the health of the environment.

Keywords Ecological health · Environmental health · Ethics · Esthetics · Aldo Leopold

According to a 1989 document by the World Health Organization (WHO), "environmental health" addresses "those aspects of the human health and disease that are determined by factors in the environment. It also refers to the theory and practice of assessing and controlling factors in the environment that can potentially affect health."[1] For its part, the purpose of ecosystem health and ecological health is "to

[1] https://health.gov/environment/DefinitionsofEnvHealth/ehdef2.htm. Last access 11/01/2018.

C. Larrère (✉)
ISJPS, UMR 8103, Université Paris 1 - Panthéon-Sorbonne, Paris, France
e-mail: larrere@wanadoo.fr

F. Bretelle-Establet et al. (eds.), *Making Sense of Health, Disease, and the Environment in Cross-Cultural History: The Arabic-Islamic World, China, Europe, and North America*, Boston Studies in the Philosophy and History of Science 333, https://doi.org/10.1007/978-3-030-19082-8_10

determine the ecologic parameters within which land may be humanly occupied without making it dysfunctional."[2]

These two fields of research are symmetrical: one studies the effects (and primarily the negative effects) of the environment and its components on human health; the other seeks to assess to what extent the environment cannot be too severely degraded by human activities. But do they overlap? It seems clear that these two fields of research are developing independently of each other. This may be because each one only seeks to understand the impacts or the effects: those of the environmental factors on human health and the human impact on the environment. In both cases, they disregard the relationship between a living being (it can be a human being) and its environment, a reciprocal and interactive relationship. However, is it not the relationship that matters when we are concerned with health, whether environmental or ecological?

To understand this relationship, it is not enough – this is the hypothesis that we will make here – to have an objective ecology, in search of "substantial" standards of health. We need an ecology that makes this relationship its very object and questions the role that human beings – beginning with the speaker – give themselves according to the way they fit into their environment and through which they preserve it. We need, according to Michel Bitbol's[3] expression, a first-person ecology.

It is this "first-person ecology" that Aldo Leopold, an American forester and a guiding figure in the protection of nature and environmental ethics, practiced in his most famous work, *A Sand County Almanac*, and also in his writing on land health or the ecological conscience. This first person intervention is by no means exclusive to scientific work and objective study, but it is critical in situating human beings in their environment and in understanding ecology as a relationship.

We will therefore present this relationship, as Aldo Leopold addresses it, in the first person and concerning the health of the Earth and its inhabitants, in three stages. In its ethical dimension, first of all, it allows us to understand why the protection of nature is not only a technical task but, primarily, a moral responsibility. Then we will study it in its esthetic dimension, where the relationship with health is formed. Finally, we will study it in its political dimension, which will lead us, without leaving the first person, from the "I" to the "we." This is what, as a comparison between Leopold and Rousseau – and also Thoreau – will show us, makes the author of the *Almanac* so original.

10.1 Ethics

The book that made Aldo Leopold famous, *A Sand County Almanac*, was written at the very end of his life and published after his sudden death in 1948. It was the result of a commission. In the early 1940s a famous New York publisher, Alfred A. Knopf,

[2] Callicott 1999, p. 339.

[3] Michel Bitbol organized a study day on this subject on 6 April 2010 at the Institute of Complexity Science. See Michel Bitbol 2014, Chapter 5, for a general view of the problem.

asked him to write “a personal book recounting adventures in the field …., warmly, evocatively, and vividly written … a book for the layman … [with] room for the author’s opinion on ecology and conservation … worked into a framework of actual field experience.”[4] Aldo Leopold accepted and with this in mind assembled and amended a number of articles published since the 1930s; while the final result did not please the publisher who had asked for it (the book was finally published by Oxford University Press in 1949), for us, today’s readers, it corresponds well to the description of the commission (personal, addressed to a broad lay audience, and opinionated), which brings it into the very American tradition of “nature writing” inaugurated by Thoreau with *Walden*.[5] The *Almanac* progresses with a series of short stories, or vignettes following the months of the year, where Aldo Leopold tells of his morning walks on the rundown 80 acres of Wisconsin farmland (the “sand county”) he had bought, the animals he encountered, and all those who used the same area. These lively and endearing accounts led to the presentation of an environmental ethic (which Aldo Leopold calls *Land Ethic*). A famous maxim summarizes it as follows: “A thing is right when it tends to preserve the integrity, stability and beauty of the biotic community. It is wrong when it tends otherwise.”[6]

One of the most celebrated essays in the *Almanac*, “Thinking Like a Mountain,” enables us to understand the maxim. It deals with the extermination of wolves. Should they be made to disappear and hunting relied on to regulate game? The answer is no. The eradication of the wolves would probably lead to the proliferation of ungulates, for the greater benefit of hunters, but in the longer term, overgrazing of the slopes would lead to erosion, which would make the game disappear and would cause lasting ecological damage: neither the integrity, the stability, nor the beauty of the biotic community (the mountain) would be preserved. From the point of view of objective ecology, things are simple. This, Callicott points out, “is all about predator-prey population dynamics and the relationship of those dynamics to vegetation cover and soil stability.”[7] A classic ecology lesson, so to speak.

But this lesson was given in the form of an account in the first person. In two ways, this extermination of the wolves actually took place. The federal forestry agency carried it out in the 1920s, from Arizona to New Mexico, and Aldo Leopold, then Regional Manager of Forest Services for the South West of the United States, participated in it and personally witnessed the consequences. But this is not how he described it in “Thinking Like a Mountain.” He presents himself as a young hunter who, in the company of a few friends, fired on a she-wolf crossing a stream. It was on reaching the wolf “in time to watch a fierce green fire dying in her eyes”[8] that Leopold understood what he did not know, but the mountain did: fewer wolves meant not more deer but too many deer. It was at the level of the community of the whole that it forms and of its temporality that it is necessary to evaluate the action.

[4] Callicott 2013, p. 18. See also Dennis Ribbens, “The Making of a Sand County Almanac” in Callicott 1987, pp. 91–109.

[5] On this tradition, see ecocriticism studies, and in particular Lawrence Buelle 1996.

[6] Leopold 1970, p. 262.

[7] Callicott 2013, p. 30.

[8] Leopold 1970, p. 138.

Told in the first person, the story has much more power than it would have had if presented, in a textbook, as an ecological case study or even the account of his experience as a forest service official ("I could see that the destruction of the wolves had not ultimately had the expected effects …"). By putting himself personally in the staging of his account, Leopold mobilizes the literary resources that allow us to become aware of nature: we hear, see, and feel the hunt in the mountains; we are moved by the agony of the she-wolf. Inside, we adhere to what Leopold says. From this point, the allusion to professional experience can be both discreet and effective: "Since then, I have lived to see state after state extirpate its wolves…."[9]

"Thinking Like a Mountain," however, is more than a living object lesson by a gifted pedagogue. It is with Aldo Leopold, with the person himself, and with his experience that we identify. And that is exactly what he wanted. In 1944, while working on the *Almanac*, he had an exchange of correspondence with a former student,[10] who advised him to personalize his presentation, and especially with regard to the wolves: "You have sometimes followed trails like anyone else that lead up wrong alleys… Your lesson is much stronger, then, if you try to show your own attitude towards your environment has changed… That's why I suggested the wolf business. I hope that you will have at least one piece on wolves alone."[11]

Following this suggestion, Leopold chose to recount how his errors had changed his mind and presented this "wolf business" in the same text by bringing together three periods from his life: the narrative (1909), the experience of the forester (1920), and the general reflection, at the time of writing (1944). The compression of time emphasized personal commitment. We witness a moral experience, with which, thanks to the literature, we can identify. It is a true conversion: Callicott, in his commentary, does not hesitate in comparing Leopold with Saint Paul on the road to Damascus.[12] By participating in the extermination of the wolves as a forestry officer, Leopold not only made a mistake, he made himself an accessory (this is the term he used in the preface of a first handwritten version of the *Almanac*[13] where he refers to this episode of his professional life) to a wrongful act. At the heart of "Thinking Like a Mountain" is an assertion of individual responsibility for a moral error, the understanding of which requires a correction, or a personal conversion. This account thus illustrates what Leopold, in a text written at the same time, in which he also presents some short first-person narratives on his work as a protector of nature, calls "the futility of trying to improve the face of the land without improving ourselves."[14]

---

[9] Leopold 1970, p. 139.

[10] In 1933 Aldo Leopold was recruited as Professor of Game Management within the College of Agriculture's Department of Agriculture Economics, at Madison, The University of Wisconsin, where he was head of the Department of Wildlife Management.

[11] Albert Hochbaum, cited in Callicott 2013, p. 31, Callicott 1987, p. 96.

[12] Callicott 2013, p. 32.

[13] Leopold, 1947 version of the foreword to the *Sand County Almanac*, in Callicott 1987, p. 284.

[14] Leopold 1991, p 340.

Nature conservation requires scientific knowledge (ecology, conservation science), the application of which is both technical and political (setting up private associations and public institutions). But to attain the sought-after "state of harmony between men and land"[15] requires the personal involvement of those who engage in it and the recognition of moral responsibility. And this does not only concern professionals, hence Leopold's emphasis on his own history as a hunter. All those concerned with nature (and especially the farmers who were Leopold's constant points of contact in his activities in favor of nature conservation) are committed to its protection, to the health of the environment, and have a personal responsibility which cannot be delegated: "A land ethic reflects the existence of an ecological consciousness, and this in turn reflects a conviction of individual responsibility for the health of the land."[16]

What separates scientific ecology from first-person ecology is a standpoint. The objective lesson of ecology is that of the general interdependence of living beings and their environment. Taken up by first-person ecology, this general lesson becomes an injunction to state what one's place is in this interdependence, to testify that one is aware of the role that one plays there. And this is indeed how Leopold defines a "conservationist" in the *Almanac*:

> I have read many definitions of what is a conservationist, and written not a few myself, but I suspect that the best one is written not with a pen, but with an axe. It is a matter of what a man thinks about while chopping or while deciding what to chop. A conservationist is one who is humbly aware that with each stroke he is writing his signature on the face of his land. Signatures of course differ, whether written with axe or pen, and this is as it should be.[17]

To sign is to commit one's responsibility. But the metaphor of the signature also involves esthetic judgment.

10.2 Esthetics

An ecological conscience, to which first-person ecology gives us access, is not only moral. It also has an esthetic dimension. It is even this that has the most importance in the first two parts of the *Almanac*, before we arrive at the "upshot" in the setting out of the purpose of *Land Ethic*. The sketches for the different months of the year on his Wisconsin holding (first part) and then the sketches taken from a few episodes in various American states (part 2, where the Arizona of "Thinking Like a Mountain" is found) are invitations to discover the beauty of nature. This is not stated in an objectivizing discourse in the third person but suggested, through descriptions, but just as much through accounts, of which Leopold is the leading actor and which he relates in the first person. For Leopold does not parade the beauty of the sites before

[15] Leopold 1970, p. 243.

[16] Leopold 1970, p. 258.

[17] Leopold 1970, p. 73.

us, as if he were commenting on the visit, he presents himself in action: walking with his dog, fishing, and hunting. To the extent that, as Raphael Larrère has shown, the beauty of the movements is an integral part of the beauty of nature.[18] This is true of hunting, one of Leopold's favorite activities, and which is related to beauty, not only because the hunter is an attentive observer of the places he passes through, sensitive to their beauty, but also because the hunter does not put as much pride in the efficacy of his action as in the formal perfection of the movements which accomplish it.[19] Thus from the first partridge Leopold fired on in flight: "It was a swinging shot of the sort the partridge-hunter dreams about, and the bird tumbled dead in a shower of feathers and golden leaves".[20]

As noted by his commentators, what makes Leopold's esthetics original is that it does not pass – or very little – by the mediation of art.[21] Whereas the Western tradition mostly subordinates the beauty of nature to its pictorial representation (the "picturesque"), Leopold's esthetics put us directly in contact with nature. It calls into question the primacy of vision, this sense of distancing, of "point of view", to the benefit of the senses of proximity, of contact, such as touch and smell, or of immersion, like hearing, which restores the etymological meaning to esthetics (of the Greek verb which means to feel): it is indeed a question of making us sensitive to nature.

Because it does not derive from art but is linked to the experience of the world in which it plunges us, Leopold's esthetics is inseparably sensitive and cognitive. Leopold often laughed at those who stuck to the surface of things, which is the case with the dominant esthetic vision, that which can only see nature through the clichés (in every sense of the term) of its initial representation:

> The taste for country displays the same diversity in aesthetic competence among individuals as the taste for operas, or oils. There are those who are willing to be herded in droves through 'scenic' places; who find mountains grand if they be proper mountains with waterfalls, cliffs and lakes. To such the Kansas plains are tedious.[22]

As Callicott points out, what can give the Kansas Plains an esthetic appeal is not so much what we see but what we know about them: their evolutionary history and their ecological composition.[23] It is necessary that this knowledge is not only book-based, otherwise those who see, for lack of sufficient knowledge, see nothing:

---

[18] Larrère 2013, p. 21.

[19] Leopold, we saw this with the story of the wolf, hunt, which may shock the ecological conscience of today. For Leopold, hunting was not a useful activity; it was a sport, that is to say, a vestige of a way of life that had disappeared, which made it possible to retain its merits, provided that it was practiced with respect for the biotic community. The beauty of the movements is also a consequence of the difficulty, which is the case for shooting birds on the wing.

[20] Leopold 1970, p. 129

[21] Callicott 1987, 1989, 1994; Larrère 2013.

[22] Leopold 1970, pp. 179–180.

[23] Callicott 1987, p. 162.

They see the endless corn, but not the heave and the grunt of ox teams breaking the prairie. History, for them, grows on campuses. They look at the low horizon, but they cannot see it, as de Vaca did, under the bellies of the buffalo.[24]

This superficial and naive approach has practical consequences. Seeing only scenery in nature, we find only recreational objectives in nature conservation: these are beautiful landscapes that we want to see. To do this, roads are built, which have destructive effects: the nature that we want to protect dies from the influx of those who want to contemplate beauty. Thus, Leopold concluded an essay devoted to "Conservation Esthetic": "Recreational development is a job not of building roads into lovely country, but of building receptivity into the still unlovely human mind."[25]

The conclusion is the same at the esthetic level as at the moral level: there is no point in wanting to conserve nature if we do not change ourselves. Objective ecology and first-person ecology are inseparable, in esthetics as in morality. They need each other. "Ecological science has wrought a change in the mental eye" noted Leopold.[26] Here we may rightly see the effect of objective knowledge on esthetic appreciation.[27] But the reciprocal is true: knowledge does not only produce a change in perspective, it demands it. Without a change of outlook, without a "profound revolution in general sensibility,"[28] visitors to the national parks will continue to see them as a spectacle and will be insensitive to the beauty of the prairie, and even more so to that of the marshes, these place repulsive a priori into which one must sink to discover the fascination.[29] To bring them to a different perspective, it is not enough to teach them a course on ecology; they must be attracted, through esthetic pleasure, to another experience of nature, in which they will position themselves differently.

Moving from the usual esthetic understanding, that perspective of a remote view of landscapes, to the immersion in nature where Leopoldian esthetics leads us, is indeed to perform a real shift. By opening ourselves to a new historical dimension, that of the long term of evolution, we perceive in these seemingly ordinary landscapes what the immediate gaze does not see: thus some wild flowers in a kind of wasteland recount the history of the Great Prairie, which goes back well before the time of colonization.[30] The historical perspective changes our view of the beings and the esteem we have of them: "our appreciation of the crane grows with the slow unraveling of earthly history," and the long past of the crane, whose tribe "stems out of the remote Eocene," bestows "a paleontological patent of nobility" to "the endless mediocrity of a common place."[31]

[24] Leopold 1970, p. 179–180.

[25] Leopold 1970, p. 295.

[26] Leopold 1970, p. 291.

[27] Larrère 2013, p. 20.

[28] Callicott 1987, p. 163.

[29] Afeissa in Barbanti and Verner 2016, pp. 205–224.

[30] Leopold 1970, pp. 47–53.

[31] Leopold 1970, pp. 102–103

There is therefore a convergence between ethics and esthetics in the awakening of ecological as well as conscience: we understand that everything does not revolve around us, that we must position ourselves in a larger whole, of which we are only one element among others, in order to understand our world and to act on it. In the first two parts of the *Almanac*, esthetics has a propaedeutic function (awakening in us the desire to protect the places and the beings we discover), but its role does not end there: it is still present in the third part, that of objective ecology (in its descriptions and its prescriptions). Esthetics does not only call on feelings, it is also a matter of judgment. "Examine each question in terms of what is ethically and esthetically right, as well as what is economically expedient."[32] And in the statement of the maxim which follows this advice, to the two criteria of stability and integrity, is added that of beauty. What, then, does esthetic appreciation, which is necessarily subjective, have to do with knowledge of scientific ecology and the objective statement of a just ethic?

Callicott puts forward an interpretation: "by 'beauty', in reference to land, Leopold means pretty much what he means by 'land health' [i.e.] self-sustaining and self-renewing normal[33] ecosystemic functioning."[34] This is a very rapid transition from the subjective to the objective and holds to the propaedeutic role of Leopoldian esthetics: its superiority over the ordinary view of things is that while the ordinary view leaves us with appearances, an esthetic vision like Leopold's allows us to move on to the objective content. It seems to us, however, that esthetic appreciation has a role which is peculiar to itself when it comes to the hierarchy of the various ways in which we evaluate our environment and nature more generally. When Leopold invites us, when deciding what to do, to not only attune ourselves to the criteria of economic utility, but also to those that are right, the esthetic dimension is as important in understanding value, and perhaps even more so, than the ethical dimension.

Moral conscience, which is that of our responsibility toward our actions and their harmful consequences, is linked to regret: "One appreciates what is left only after realizing how much has already disappeared."[35] But this is not enough: the disappearances, Leopold often emphasizes, are irreversible; we must not wait for it to be too late. "Conservation, then, is keeping the resource in working order, as well as preventing over-use. Resources may get out of order before they are exhausted, sometimes while they are still abundant."[36] This is where esthetic conscience goes further than moral conscience. The moral eye is sensitive to degradation, to the wrongs done to the earth. It sees the earth sick. Beauty shows us that nature is in good health, vital, and vigorous. However, we do not need only to fight against degradation but to keep the earth healthy, whole. It is the capacity for self-renewal of the whole that we must preserve.

[32] Leopold 1970, p. 262

[33] The question of "normal" when we speak of "health" will not surprise the reader.

[34] Callicott 2013, p. 96.

[35] Leopold 1999, p. 91.

[36] Leopold 1991, p. 257.

Leopold returns untiringly to the interdependence of the parts and the whole, their interconnection, their unity. To state this unity, he makes use of several models. One of the most important is that of the organism. This is what leads him to speak of land health and, from this point of view, to compare medicine and nature conservation (and also agriculture): "There are two organisms in which the unconscious automatic processes of self-renewal have been supplemented by conscious interference and control. One of these is man himself (medicine and public health) and the other is land (agriculture and conservation)."[37] Another model is that of harmony, a more abstract model, but one whose esthetic, and more precisely musical reference, is classic: the ear perceives the consonance that characterizes musical harmony with pleasure, a sign of a good relationship. Musical references, which bear witness to the perception of a dynamic unity, are frequent in the *Almanac*, especially in the "Marshland Elegy."[38]

Beauty is therefore the perception of health, and both refer to the same objective, that of nature conservation, conceived of as a "state of harmony between men and land."[39] Health and beauty are linked in part, and this is why esthetic appreciation contributes to nature conservation. For beauty is understanding the whole. If our attention is caught by a bird in flight, little by little we appreciate its environment. You cannot have one without the other: "We cannot love cranes and hate marshes," summarizes Callicott.[40]

But it is this unity, of which we have objective knowledge and an intuitive grasp, that the use of nature for economic purposes calls into question, beginning with agriculture. For land to be economically profitable, it must be in good health, but economic use is selective: it does not need all the aspects of the whole that it uses for its own benefit. Consequently, while there may be economic arguments for nature conservation, they are not sufficient:

> Healthy land is the only permanently profitable land, but if the biota must be whole to be healthy, and if most of its parts yield no salable products, then we cannot justify ecological conservation on economic grounds alone.[41]

To pretend otherwise would be to reason sophistically, by confusing the part and the whole. There is a "partnership" relationship between man and the earth that makes both prosper together.[42] However, while this relationship is reciprocal, it is unequal because it is a relationship between the whole and a part: we are included in nature. And beauty, through which we grasp the whole that encompasses us, far exceeds our partial economic evaluation. It has something of a free gift. It is simply a question of pleasure. In "A Landowner's Conservation Almanac," Leopold explains in detail to the owners of a farm what they need to do to attract wild plants

---

[37] Leopold 1991, p. 287.

[38] Leopold 1970, pp. 101–108.

[39] Leopold 1970, p. 243.

[40] Callicott 1987, p. 162.

[41] Callicott 1999, p. 224.

[42] Leopold 1991: 255.

and fauna: on which trees the migratory birds like to land, why they have to keep some dead trees, how to preserve flower-covered refuges, etc.[43] Gradually, he emphasizes the beauty that will then be offered to the farmers, the pleasure they will obtain in all these encounters with the wild. Keeping "some remnants of non-domesticated plant and animal life"[44] within the farm itself in this way, does not undermine its economic profitability, and above all offers the possibility of a richer, more diversified life: "the wholly tamed farm offers not only a slender livelihood but a constricted life. (…) There is pleasure to be had in raising wild crops as well as tame ones."[45] To make way for the wild even in one's domestic life is ultimately to give meaning to it, to open it to the diversity of possibilities.

Is it possible to find the link between environmental health (the effects of the environment on the human organism) and ecological health – or land health – the ability of an environment to absorb the effects of human actions?

10.3 Political Dimension

One of the principal obstacles to the establishment of a *Land Ethic* is, according to Leopold, that we are not aware of our connection to nature ("consciousness of land" says Leopold).[46] Mediations interpose themselves between nature and us; we no longer have a "vital relationship with the earth" ("no vital relation to it").[47] Restoring this link means working for the health of the land and also for the health of those who inhabit it. The two are linked: a purely "recreational" view of nature protection is at the expense of the nature, which becomes exposed to destructive hordes of visitors, but it does not particularly do the visitors much good either, because they do not find the opportunity for a mental switchover which would open up new horizons for them. Rather than travel long distances to seek a wildness of which they understand nothing, it is better to know how to make space for it at home, following the advice of Leopold in "A Landowner's Conservation Almanac." In Leopold, one finds the basic idea of what would later become ecopsychology, that of the necessary link between individual well-being and the health of the planet.[48]

The roots of this idea are not new, as a comparison with Rousseau shows. Leopold invites the owner-managers to use a natural art in order to attract wild plants and animals, which was already practiced in Julie's garden, as presented by Rousseau in *La Nouvelle Héloïse*.[49] In her estate at Clarens, on the shores of Lake Geneva, Julie was able to turn an old rundown orchard into a luxuriant garden with a mix of wild

[43] Leopold 1999, pp. 84–160.
[44] Leopold 1999, p. 78.
[45] Leopold 1970, p. 203.
[46] Leopold 1970, p. 261.
[47] Leopold 1970, p. 261.
[48] See, in particular, Theodore Roszac 2001 [1992].
[49] Rousseau 1761, 1964, pp. 471–488.

flowers and cultivated plants. The birds found themselves there in abundance without being forced to stay by shutting them in an aviary, as they freely visited places which had been rendered attractive to them by the food which had been judiciously grown there. It was this garden, which she called her "Elysium," that Julie showed her former lover, Saint Preux, around and where he returned alone in the evening. He found peace and joy there: "I passed two hours in Elysium, which were not inferior to any I ever spent."[50]

Julie's garden was not created by chance. It can be considered as an experimental device that ensures, to perfection, the balance between the soul and its sensitive environment. This is precisely what Saint Preux admires: "I said to myself: peace reigns in the depth of her heart as it does in this asylum she named."[51] The idea is that our soul is regulated in its moral dispositions not only by itself but just as much, if not more so, by the outside. It is then a question of turning this external influence from objects on our soul, via our bodies, into continuous moral refinement by the environment.

Such is the bond that binds us to our environment and affects our mental, as well as our physical, health. Nature preserved, or, conversely, the denaturing of the environment, is of profound importance to my soul: disorder introduced into nature disorders my own nature as much as natural order shapes and strengthens my own internal order. Recognizing my weakness in the face of sensitive immediacy can thus enable me to use it by arranging things differently, in order to favorably determine my moral dispositions and thus mediately increase my strength over myself. In controlling external nature, at the same time, I control myself.

This is what Rousseau calls "morale sensitive" (sensitive morality) or "matérialisme du Sage" (the wise man's materialism). In *The Confessions* he explains that he intended to make a book from it and that external circumstances prevented him from carrying this out.[52] But besides the fact that the model of Julie's garden shows that he remained attached to it, his plan of "providing an external regimen which — varied according to circumstances — could put or maintain the soul in the condition most favorable to virtue" was not exclusive to him: it is found in the medical discourse of the time. It is a question of making use of the influence of physical sensibility on moral sensitivity for the purpose of improving human beings. One of the principal representatives of these medical conceptions was Antoine Le Camus who in 1753 published *Médecine de l'esprit, où l'on traite des dispositions & des causes physiques qui, en conséquence de l'union de l'âme avec le corps, influent sur les opérations de l'esprit, & des moyens de maintenir ces opérations dans un bon état, ou de les corriger lorsqu'elles sont viciées* (*Medicine of the mind; where we treat the Dispositions and the Physical Causes which, as a consequence of the union of the soul with the body, influence the operations of the mind; & the means of maintaining these operations in good condition, or correcting them when they are flawed*).

[50] Rousseau 1761, 1964, p. 487, transl 1810, p. 303.

[51] Rousseau 1761, 1964, p. 487.

[52] He makes reference to it several times in *Les Confessions* (Rousseau 1959, pp. 409, 516, 608).

Such a project certainly calls for a causal vision of the impacts of the environment on our soul but also depends on the idea that we are not independent islands, that the limits of our body do not really exist, and that our virtues, as well as our happiness, are relationships that incorporate us into the whole of nature. As Rousseau says in a fragment about climate: "l'homme tient à tout ce qui l'environne" (man is attached to everything in his environs),[53] he is attached to it in the different meanings of the term. In this sense, "the wise man's materialism" or "sensitive morality" can be considered as variants of ecopsychology; it is a way of positively reinforcing our attachment to our environment.

Is a similar device found in Leopold? Not exactly. Admittedly, he adhered to the basic principle that "Ecology tells us that no animal-not even man- can be regarded as independent of his environment"[54] and this general lesson can be stated objectively, but is also being experienced in the first person, like attachment. Here, however, it is not only individual; it does not merely encourage one "to please oneself," as Saint Preux does (Rousseau 1964, 487). Its reach is much more collective.

What characterizes the effect of Julie's garden is the harmony that is realized between the interior and the exterior and between nature within me and nature outside me. In the *Reveries of a Solitary Walker*, the harmony between the interior and the exterior goes on until every boundary disappears. "Plus un contemplateur a l'âme sensible" (the more sensitive a soul a contemplator has) – like Rousseau – that is, the more he can be affected by sensations and feelings and by the interaction and concord of the two, "plus il se livre aux extases qu'excite en lui cet accord" (the more he gives himself up to the ecstasies this harmony arouses in him). This takes him back to himself but without confining him. His soul is in fact in equilibrium with the exterior; it is "expansive."[55] This is what permits the "ecstasies," those moments of fusion with the environment, which is no longer truly exterior.

The blurring of the boundary between the interior and the exterior can work in both directions. Thus, Rousseau, traversing the forest of Saint-Germain, discovered there the principles of the state of nature by consulting his "heart" (Rousseau 1959, 388). But more often than not, it is much more a question of human nature than of the nature outside, and this is all the more so as the relationship is understood in the first person. In general, one does not study *The Rêveries* to look for Rousseau's relationship with nature and his conception of it; we see above all an autobiographical text, an exploration of the ego. Were *The Reveries* not published in the same volume as *The Confessions*? The couple of nature outside me and nature within me is therefore made to the benefit of the latter and leads to making the landscape outside only a projection of the interior landscape. What Saint Preux likes in Julie's garden is that the whole garden, beginning with the name she gave it, "peignait en quelque sorte l'intérieur de celle qui l'avait trouvé"(in a certain manner, it painted for me the interior of the woman who had chosen it).[56]

---

[53] Rousseau 1966, p. 530.

[54] Leopold [1934] quoted by Callicott 2013, p. 41

[55] Rousseau 1959, p. 1074

[56] Rousseau 1964, p. 487.

While Thoreau's botanical or even ecological (in objective ecology) reflection has been more remarked and commented on than Rousseau's, it is nevertheless true that we find in him the same connection between the nature within me and the nature outside me, and the same indication is that there is nature outside me only because there is nature inside me, an idea expressed in the often quoted verses:

> It is in vain to dream of a wildness distant from ourselves.
> There is none such. It is the bog in our brains and bowels,
> The primitive vigor of Nature in us; that inspires that dream.
> I shall never find in the wilds of Labrador any greater wildness.
> Than in some recess of Concord; i. e., than I import into it. (Thoreau, *Journal*, 30 August 1856)

There is no such thing with Leopold. First-person ecology is an ecology of the I, but not of the ego: it does not take itself as subject matter. As Raphael Larrère remarked: "Leopoldian esthetics are not romantic. Leopold does not subjectivise nature, he does not make it the space for the projection of his states of mind, his passions, his fantasies."[57] He does not invite us to enter into his intimacy but to discover forms of interiority where we would not expect to find them: in the nature outside me, in exteriority. Leopold does not make us share his feeling for nature, but he makes us sensitive to nature, but it is a nature that thinks like a mountain or that has a soul like that the "fierce green fire" he let us glimpse in the eyes of the dying she-wolf. To say the "soul" of living beings and natural entities, rather than the scientific language of objective ecology which detests animism, suits first-person ecology which can use the literary styles of "ecopoetics,[58]" those of metaphor and of irony. These, by soliciting the imaginary, or by forcing thinking beyond what is said explicitly, make understood what the direct expression of objective language cannot say.

As first-person ecology does not bring us back to ourselves, but opens us up to an exteriority in which we discover interiorities, rather than ecopsychology, we should speak of Leopold's ecosociology. By making us experience the general interdependence between living beings and their environment, first-person ecology allows us to discover ourselves as members of a collective, what Leopold, in a term that borrows from the (objective) ecology of his time, calls the "biotic community." In this way a vital relationship with the earth is restored, the loss of which we suffer from. The damage is not only individual. It is also social. In the foreword to the *Almanac*, Leopold deplores that "our bigger-and-better society is now like a hypochondriac, so obsessed with its own economic health as to have lost the capacity to remain healthy."[59] The economy, in fact, is not self-sufficient; it cannot find in itself what allows its self-renewal, as it depends on the whole natural world that encompasses it. This is why the solution is on the side of an ecosociology. It is a question of understanding how and in which ways the well-being of our societies is linked to that of the earth.

---

[57] Larrère 2013, p. 25.

[58] Following the term introduced by Pierre Schoentjes 2015.

[59] Leopold 1970, p. XIX.

This means that the problem is political. Certainly, Leopold emphasizes the individual responsibility each of us has in the protection of nature, on the fact that it depends on everyone's capacity to engage in it, consciously, from personal conviction. He therefore criticizes the division of tasks which established itself, in the United States, between what is within the sphere of individual interests (economic rationality) and the government, to whom the management of the general interest is delegated. "We have spent several generations teaching the farmer that he is not obligated to do anything on or to his land that is not profitable to him, as an individual." The consequence of this "fallacy" is the relegation of "the conservation of what is merely beautiful to the state."[60]

A society oriented toward economic functionality and the rationality of interested individual action comes to think that what pertains to the state or to beauty is also frivolous or superfluous. It is wrong. Leopold had experienced this: again, he learned from his mistakes, but this time the lesson is not only individual and moral, it is collective and political.

It is the story of the Flambeau River, whose beauty had been admired by early settlers, and that the Wisconsin Conservation Department wanted to restore to its wild state "for the use and enjoyment of young Wisconsin."[61] But the project ran up against the interests of local livestock farmers who wanted to build dams on the river to benefit from cheap hydroelectricity. Impressed by the weight of private interests and their priority over collective interests, the defenders of the river wanted to place themselves on the same ground: they "tried to prove that the river in its wild state would produce more fish and tourists than the impounded river would produce butterfat."[62] But this was not true. Leopold acknowledged his mistake: "We should have claimed that a little gain in butterfat is less important to the state than a large loss in opportunity for a distinctive form of outdoor recreation."[63] Defending the protection of nature means defending the good of the community. And it is a duty of every farmer, as the farmer is a citizen, not just an economic agent: "A farmer has the same obligation to help within reason, to preserve the integrity of his community as he has, within reason, to preserve the culture which rests on it."[64] It is remarkable that it is around the word "culture" that the transition was made between two types of communities, biotic and social, and between the material relationship to the land (culture as in agriculture) and the spiritual relationship with the shared world (culture as a result of a collective spiritual creation).

The farmer "is the ultimate beneficiary of both," concluded Leopold.[65] And it can indeed be shown that in both cases, it is a question of a well-understood interest: each of us individually has an interest in the perpetuation of the State, in charge of the general, as well as of nature, in its entirety. But these benefits are indirect or

[60] Leopold 1999, p. 224.

[61] Leopold 1991, p. 344.

[62] Leopold 1991, p. 344.

[63] Leopold 1991, p. 344.

[64] Leopold 1999, p. 225.

[65] Leopold 1999, p. 225.

secondary benefits, and they can only be obtained if they remain secondary benefits, that is to say, if these collateral effects of an action do not become their principal aim. While it may be useful to understand that by devoting oneself to the common good, one is not sacrificing one's private interest, it is even more important to understand what makes it a primary goal and thus to show the values that are involved in nature conservation, as in the sense of the community: beauty, leisure, and gratuitousness. All these values which Leopold clearly states are not only economic but also "philosophical": "By value, I of course mean something far broader than mere economic value. I mean value in the philosophical sense."[66]

Generally critical of the recreational goal of the protection of nature, Leopold defended it when it linked beauty and gratuitousness. What is free has no price but has value. In order to regain a vital relationship with the earth and to find its own health, the hypochondriac society, which is concerned only with economic rationality, must regain the meaning of gratuitous: that of leisure and that of the beauty of nature. Shortly before he was assassinated in the attack on Charlie Hebdo, Bernard Maris, an economist who was criticizing the project of opening businesses on Sundays, was concerned about a society that no longer knows what a Sunday is. He continued Leopold's struggle for a society that was not only healthy but also happy.

Leopold is he who talks at the same time about the decisive importance of a "vital relationship" with the earth and praises gratuitousness. This may mean that life, and the health that assures its perpetuation, is not understood only in terms of functionalities. There are things in life that are free, to which beauty bears witness. That is why, whatever the objective criteria for understanding land health are, they are not sufficient. Health is also seen as a lived relationship, in beauty and happiness. It is not just a question of being healthy in a healthy environment but about being happy in the beauty of nature. This is appreciated only in the first person. But this first person can also speak in the plural which sheds new light on the gift we receive gratuitously from life and nature. We often hear that we only protect well that which has value. But do we have to do an economic evaluation for the value of what nature brings us, as the Millennium Ecosystem Assessment did by conducting a global monetary assessment of ecosystem services? It was not in this sense that Leopold emphasized making a distinction between value in the economic sense and in the philosophical sense. To appreciate the value of nature can be to understand what, in the relationship we have with it, escapes monetary valuation, to regain the sense of gratuitous, including in its political dimension.

References

Barbanti, Roberto, and Lorraine Verner, eds. 2016. *Les limites du vivant*. Paris: Dehors.
Bitbol, Michel. 2014. *La conscience a-t-elle une origine?* Paris: Flammarion.

[66] Leopold 1970, p. 261.

Buelle, Lawrence. 1996. *The Environmental Imagination. Thoreau, Nature Writing, and the Formation of American Culture*. New York: Bellknap.
Callicott, John Baird, ed. 1987. *Companion to a Sand County Almanac*. Madison: The University of Wisconsin Press.
———. 1989. *In Defense of the Land Ethic: Essays in Environmental Philosophy*. Albany: State University of New York Press.
———. 1994. The Land Aesthetic. In *Ecological Prospects: Scientific, Religious and Aesthetic Perspectives*, ed. C.K. Chapple, 169–183. Albany: SUNY Press.
———. 1999. *Beyond the Land Ethic, More Essays in Environmental Philosophy*. Albany: State University of New York Press.
———. 2013. *Thinking like a Planet. The Land Ethic and the Earth Ethic*. New York: Oxford University Press.
Larrère, Raphaël. 2013. L'esthétique de la nature. In *Le Multiculturalisme*, ed. Florence Burgat and Vanessa Nurock, 18–29. Marseille: Wildproject.
Le Camus, Antoine. 1753. *Médecine de l'esprit, où l'on traite des dispositions & des causes physiques qui, en conséquence de l'union de l'âme avec le corps, influent sur les opérations de l'esprit, & des moyens de maintenir ces opérations dans un bon état, ou de les corriger lorsqu'elles sont viciées*. Paris: Ganeau.
Leopold, Aldo. 1970. *A Sand County Almanach* [1949] *With Essays on Conservation from Round River* [1953]. New York: Ballantine Books.
———. 1991. In *The River of the Mother of God and Other Essays*, ed. Susan L. Flader and J. Baird Callicott. Madison: The University of Wisconsin Press.
———. 1999. In *For the Health of the Land, Previously Unpublished Essays and Other Writings*, ed. J. Baird Callicott and Eric T. Freyfogle. Washington, DC: Covelo, Ca, Island Press, Shearwater Books.
Roszac, Theodore. 2001 [1992]. *The Voice of the Earth, An Exploration of Ecopsychology*. New York: Phanes Press.
Rousseau, Jean-Jacques. 1959 [1782]. Les Rêveries du promeneur solitaire, *Œuvres complètes*. Paris: éditions Gallimard, bibliothèque de la Pléiade, t. I.
———. 1959 [1782, 1789]. *Les Confessions, Œuvres Complètes*. Paris: Gallimard, bibliothèque de la Pléiade, tome I.
———. 1964 [1761]. *La Nouvelle Héloïse*, IV, 11, *Œuvres Complètes*. Paris: Gallimard, bibliothèque de la Pléiade, tome II (trans. Eloisa: A Series of Original Letters, 1810, Harding and Wright, Printers, St John's Square, London).
———. 1966. *Œuvres politiques, Œuvres Complètes*. Paris: Gallimard, bibliothèque de la Pléiade, tome III.
Schoentjes, Pierre. 2015. *Ce qui a lieu, Essai d'écopoétique*. Wildproject: Marseille.

Chapter 11
Urban Space of the Living and Dead: The Conception of Environment and Death in Beijing from the Eighteenth Century to the Middle Twentieth Century

Xi Ju

Abstract In modern Beijing, urban space for the living and the dead is completely separate. However, before the 1940s, the separation was not that obvious. After the middle of the Qing dynasty (1644–1911), funerals of residents in Beijing were carried out according to "Master Zhu's Family Rituals" (*Zhuzi jiali* 朱子家禮, 12th). Typically there were three phases: the period in a mortuary, the farewell to the soul, and the burial. All three phases were ceremonies which clearly show that before the burial, and for a long period of time, the dead person was not separate from the living. During the first phase, the coffins were temporarily placed in a temple. The temples could also rent their rooms to the public and therefore served as hotels hosting the living and the dead at the same time. The second phase, which was the farewell ceremony to the soul of the dead, was typically held near water, especially large expanses of water in the heart of Beijing. Shi Cha Hai 什刹海, a lake in the center of Beijing, was thought to be the entrance to the other world, and thus, the space, in the city, reserved for the souls and ghosts. Moreover, the cemeteries were not only for the dead, they were the places for the living for recreation, to meet, and to present newborn infants to the ancestors and therefore embodied, in a sense, the true "home of the family". The way people treated death deeply influenced their knowledge of and feeling about the environment, which also impacted on their planning and usage of urban spaces. From the analysis of death rituals in late imperial Beijing, and of the spaces shared by the living and the dead, one can observe that death was not a polluting element for the living nor seen as a danger for the health. However, the transformation brought by the Japanese invasion, along with its 'hygienic modernity' completely overturned people's attitudes towards death and the dead body. Urban spaces for the dead and the living became strictly separate, which finally formed the modern urban landscape.

X. Ju (✉)
School of Sociology, Beijing Normal University, Beijing, China
e-mail: jessyca_ju@bnu.edu.cn

F. Bretelle-Establet et al. (eds.), *Making Sense of Health, Disease, and the Environment in Cross-Cultural History: The Arabic-Islamic World, China, Europe, and North America*, Boston Studies in the Philosophy and History of Science 333, https://doi.org/10.1007/978-3-030-19082-8_11

Keywords Urban space · Funeral rituals · Beijing · City landscape · Conception of environment

Separation is a basic principle of urban planning in modern cities. As David Harvey claims, "the system of production which capital established was founded on a physical separation between a place of work and a place of residence."[1] The notions of "health" and of "hygiene" have also played important roles in this process of separation and in the shaping of urban environments and landscape planning in modern Europe and in the rest of the world. For Ruth Rogaski, hygiene was a crucial element in the definition of Chinese modernity, and in the transformation of China this modernity was expected to bring, hygiene called for functional separation in the urban landscape. "Hygienic modernity," to borrow from Rogaski, truly determined which urban landscape could be seen and which potentially "contaminating activities" should not be seen. In this conception, death and corpses belonged strictly to the realm of what should remain invisible.[2] For Europeans at the end of the nineteenth century, "death no longer inspires fear solely because of its absolute negativity, it also turns our stomach, like any nauseating spectacle," it "becomes improper, like the biological acts of man, the secretions of the human body. It is indecent to let someone die in public."[3] The Chinese accepted this opinion from the mid-twentieth century and started to believe that death and the dead should be hidden because they were ugly and dirty. That is the crucial reason why Chinese cemeteries are isolated from residential areas in today's urban planning. However, before this time, death was not that disgusting scene with the spreading of germs, and dead bodies were not necessarily transferred to places that were hidden and removed from the eyes of the living. Until the 1940s, Beijing was a city where the living and the dead coexisted. As long as the funeral observances were carried out appropriately and ethically, the spaces in which the dead bodies and their souls were placed were considered safe and clean for those still alive. The analysis of the ceremonies, practices, and concepts related to death at that time allows us to see how this conception of death shaped the city landscape; it also allows us to understand what a "good and healthy environment" meant for Beijing residents.

This essay focuses on the time from the Qing dynasty (1644–1911) to the Republic period (1911–1949), with Beijing residents as main research subjects. It studies their funeral rituals and attitudes to death and discusses the way in which they used and imagined the urban space, in order to explain their environmental view at that time. Before stepping into this specific discussion, it is necessary to explain the concrete spatial scope, subjects, and main research sources.

---

[1] Harvey 2010, p. 33.

[2] Rogaski 2004, p. 2, pp. 193–195.

[3] Ariès 1977, trans. 1981, p. 569.

11.1 Research Scope and Materials

The space, landscape, and environment discussed in this essay are limited to Beijing city. Here, Beijing city refers to the area inside the old Beijing city wall. In the Yongle period (1403–1424) of the early Ming dynasty, the former city wall of Beijing built during the Yuan dynasty (1271–1368) was reshaped; the northern side of it was moved southward, forming the basis of Beijing city. This was called the Inner City (内城) after the Jiajing period (1507–1566). At that time, the Outer City (外城) was built beyond the south of the city wall, giving Beijing its final shape (see Picture 11.1). This city structure extended to the Qing dynasty (1644–1912). In the mid-seventeenth century, the Jurchen chieftain Nurhaci (努尔哈赤 1559–1626) controlled the northeast area outside the Great Wall. After having reorganized and unified various Jurchen tribes that he officially called "Manchu" (*manzhou* 满洲),

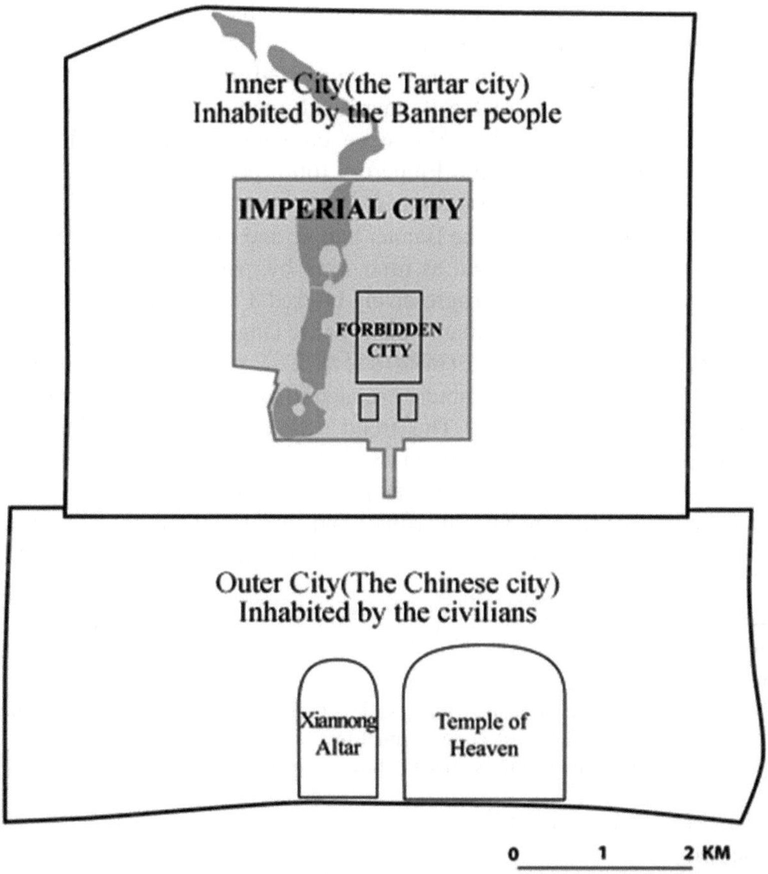

Picture 11.1 Beijing City in the early eighteenth to the middle twentieth century. (Drawn by the author)

he established the Manchu-led Eight Banners System, a military and administrative system which included Manchus, Mongols, and those Han Chinese and other ethnic groups who had capitulated, referred to as "Banner people" (旗人 *qiren*). "They [the Banner people]combined fighting skills with bureaucratic organization, agricultural surplus with monopolies on valuable trade items, and, by allocating followers to different 'banners', created a society-in-arms that expanded with each victory."[4]In 1644, Beijing was conquered by the Banner people and, in 1646, the Qing government ordered the separation of the Banner from the non-Banner people, who were referred to as "civilians" (民人*minren*), most of them being Han Chinese. The civilians were forced to evacuate their homes and move to the Outer City (the south part of Beijing or the Chinese City), while the Banner people were required to reside in the Inner City (the north part of Beijing or the Tartar City). However, this separation did not last for long. In 1785, the Qianlong Emperor (reign 1736–1796) could only say vaguely that the residents of the Inner City were "mostly" Banner people.[5]

The Manchu lifestyle became gradually influenced by that of the mostly Han Chinese civilians, especially as regards funeral customs, which were notably influenced by the Confucian funeral customs prescribed by the *Zhuzi Family Rites* (朱子家礼*Zhuzi jiali*).[6] The Manchu had originally practiced cremation of the dead. Before the eighteenth century, most of the Banner dead, including Emperor Nurhaci and Emperor Shunzhi (reign 1644–1662), were cremated. However, after Emperor Kangxi (reign 1662–1723), all the Emperors followed the Han custom and were interred. Emperor Qianlong even forbade the old custom of cremation.[7] In the eighteenth century, the law requiring the Banner people and the civilians to live in different parts of the city was lifted and as time went by, particularly in the nineteenth century, cultural identities were progressively blurred. For example, the Manchu language became more and more similar to the Chinese language. In *the Qing Dictionary Revised and Enlarged by Emperor* (御制增订清文鉴 *Yuzhi zengding qingwen jian*) published in 1771, most of the vocabulary related to funerals in the Manchu language was borrowed from the Chinese.[8] This trend was more obvious in the nineteenth century. In 1801, the scholar Suoning'an (索宁安) completed the *Compilation of the Manchu Four Rites* (满洲四礼集 *Manzhou sili ji*),[9] in which he brought together information on the traditional Manchu rituals and customs in order to provide a reference for later generations. While the worship of Heaven and God maintained the old

[4] Naquin 2000, p. 287.

[5] Ibid., p. 376, pp. 398–399.

[6] Compiled by a great Confucian scholar Zhu Xi (朱熹1130–1200) in the Song dynasty, the *Zhuzi Family Rites* later became the most popular ritual text which was to be followed by all of the society. The details of this book will be given below, see part 2, "Death Observances and Related Practices."

[7] *Qing Gaozong Shilu*, Vol 5, p. 241.

[8] See Guan 2010.

[9] Suoning'an (索宁安, 1747~?), Manchu, belonged to the Bordered Yellow Banner (镶黄旗, Xianghuangqi, one branch of the Eight Banners). The compilation of *Manchu Four Rites* was taken up by his grandfather, then continued by his father, and finally finished by himself.

Manchu customs, the three rites of weddings, funerals, and ancestor worship offered little difference from Han Chinese rites. Especially in the paragraph on funerals, the *Zhuzi Family Rites* was directly quoted to be the model for Manchu funerals.[10] Evidence from the diaries of some Banner people, such as the *Records of Leisurely Dreams under an Open Window* (闲窗录梦 *Xianchuang lumeng*),[11] testifies that the funeral rites practiced by Banner people who lived in the Inner City of Beijing were very similar to those of the Han Chinese, including dressing the dead, obituaries, encoffining, funeral wakes, carrying out the coffin, as well as the ethical values and corresponding opinions embodied in the process.[12] In general, from the end of the eighteenth century, the Banner people and civilians, or Manchu and Han Chinese, had few differences in dealing with death. They could be considered as a Beijing citizen group sharing the same lifestyle and same funerary observances. This group that we now refer to as "Beijing people" is the focus of this essay.

This essay concentrates on the Beijing people's concepts and practices regarding death; therefore, private historical sources, oral and written, have been favored. The most important materials are the ethnographic data that I collected in my fieldwork in the Inner City of Beijing during 2004 to 2015.[13] My research mainly focused on old people (over 60), who could still remember their personal experiences 50 years ago and accurately describe the changes in funeral customs since these years. Their memories can thus reflect the situation in Beijing from the 1930s to the present. The oral testimonies I was able to collect from the elderly are helpful in reconstructing and understanding what happened in this period of massive transformation. Different from written records, the oral materials from my fieldwork are more intimate. When the elderly in Beijing talk to me about death and the dead, they reveal their personal experience and their individual thoughts and feelings. These oral testimonies are particularly significant for understanding, on a personal (and individual) level, representations, and attitudes about death, the dead, and the environment.

This essay also makes use of some written materials, such as diaries written by Banner people in the Qing dynasty, notes or novels written by literati, and memoirs of the funeral experts in the Republic period. While some of these should be considered as rumors and folktales, they do reflect social values and popular opinions.[14] As the oral testimonies can only take us back to the 1940s, these written records can help fill some of the gaps and widen our knowledge of how Beijing people used to cope with death and death observances from the eighteenth century onward.

[10] Suoning'an 1801.

[11] The *Records of Dreams under Window of Leisure* is a Manchu diary manuscript, written by an Inner City Banner man, from 1828 to 1829. The author is Mu Qixian (穆齐贤). He was born in 1796, and he was a bondservant (包衣Baoyi) who belonged to the lower class in the Eight Banners System.

[12] On this question, see Ju 2017a, b.

[13] For the nature, content, and objects of this investigation, see Marianne Bujard 2011, pp. 1–17.

[14] Several scholars have discussed the historical value of notes and novels. For example, Robert Company thinks that mystery novels are not fiction, but the record of proven ideas. Edward Davis suggests that ancient Chinese stories are not only novels but also the document of private life concerning the subjective experience. Related discussion see Kang 2011, p. 11.

These materials have unparalleled value for the analysis of the thoughts and imaginings of the citizens. However, urban space distribution and landscape construction always result from the combination of private actions and public or governmental regulations. Therefore, in addition to the personal experiences, I could collect from fieldwork and documents; I also paid heed to government laws and regulations, especially the *Statutes of the Qing Dynasty* (大清会典 *Daqing huidian*) and governmental archives, notably those of the Republic period. In 1937, Japanese armies invaded Beijing and by 1939, Beijing was under Japanese control. On March 13th of that year, the Ministry of the Interior of the Provisional Government of the Republic of China (中华民国临时政府内政部) announced a ban on keeping encoffined corpses inside the city before burial,[15] but this ban was not enacted. In 1946, after the war, the Republic government itself introduced a ban on keeping encoffined corpses before burial and encouraged the construction of cemeteries; meanwhile, the government ordered the Peking Police Bureau and Bureau of Social Affairs to investigate all temples and public cemeteries for evidence that they were still used for keeping unburied coffins inside the city.[16] The People's Republic of China continued this policy. On May 30, 1949, the Peking Civil Affairs Bureau reported to Ye Jianying (叶剑英), the Mayor of Peking at that time, that they intended to move all the tombs and coffins to the suburbs. From 1951 to 1952, this work was carried out and fully completed in 1953. From then on, the custom of keeping coffins and burials inside the city vanished.[17] Therefore, the 1946 archive reflects the last days of the practice of burial inside the city. The situation recorded in these archives, from the practices involved in the funeral observances to the distribution of the temples where the coffins were kept, showed remarkable similarity with the oral evidence that we learned from the fieldwork. All the above sources help in addressing the issue at the core of this essay: how did the conception of death affect the repartition and the usage of urban space in Beijing, from the eighteenth century to the middle of the twentieth century?

11.2 Death Observances and Related Practices

Human activities are complex combinations of thoughts and actions, as were the death observances practiced in Beijing. While the overt actions, which included a highly stylized ritual process and non-ritual leisure activities, that we call, in this article, death observance practices, can be observed easily, the reasoning and comprehension of death behind these practices are less overt but can be revealed through discourse, anecdotes, and literature provided by the citizens themselves.

---

[15] *BA* 1939, J1-2-120.

[16] Eleven archive files preserved in Beijing Archive show the details of this investigation, such as J183-002-17547, J183-002-25256, J183-002-25922, etc.

[17] *Beijing zhi* 2003, pp. 465–466.

This essay will firstly explain the death observance practices while mainly focusing on the part specifically connected to the urban space.

From the time of the *Book of Etiquette and Ceremonial* (仪礼 *Yi Li*) and the *Book of Rites* (礼记 *Li Ji*), there were very detailed methods and procedures for dealing with corpses. It is generally acknowledged that the *Book of Etiquette and Ceremonial* was compiled by Confucius (511–479 BCE) in an effort to record the rites of the Zhou dynasty (1046–256 BCE), and the *Book of Rites* was the explanation of the former. These two books, along with the *Rites of the Zhou Dynasty* (周礼 *Zhou Li*), formed the *Three Rites*, which means "the three most important books of rites," and have become Confusion classics since the western Han dynasty (202 BCE-8 CE). As such, they formed the mainstream of behavioral standards for the whole of ancient China. In the Southern Song dynasty (1127–1279), Zhu Xi (朱熹 1130–1200) simplified and popularized the ritual practices prescribed in the *Three Rites* and composed a new text, the *Zhuzi Family Rites* (hereafter referred to as *Family Rites*). This book provided a sequence of ceremonies that needed to be held after the death. With its official endorsement (1385), it became the ritual text that had to be followed by the whole society. This situation changed in the Qing dynasty. As we said above, the emperors of the Qing dynasty were Manchu and not Han Chinese. The Manchu originally had their own funeral ceremonies, which gradually, however, as we have seen, became similar to the Han Chinese customs. While the *Compilation of the Manchu Four Rites* (later referred to as *Manchu Four Rites*) and the *Family Rites* are not exactly the same as regards the funeral observances, in both books the core of the observances relating to the corpse itself consisted in eight procedures: bathing, moving the dead body to a certain place, placing gold or rice into the mouth of the deceased, dressing up the dead, encoffining, condolences, taking out the coffin, and finally burial.[18] In reality, the observances may have varied from the rules prescribed in these two books, as some private written records attest. *Records of Leisurely Dreams under an Open Window*, a diary written by a Banner man in the mid-nineteenth century, shows that the funeral ceremonies/observances practiced by the author and his relatives differed slightly from the eight procedures described above. The *Red and White Joyous Event* (红白喜事 *Hongbai Xishi*), a book written by the funeral expert Chang Renchun (常人春1933–2015), who described the usual procedures for funerals in Beijing in the 1940s–1960s, also reveals that there were a few differences between the actual funerals and "rites guidelines." It seems however that the core of the ceremonies practiced by common people remained globally unchanged, especially as regards the ceremonial procedures relating to the corpse. These were mainly described as follows[19]:

① Changing bed. When the person was dying, an "auspicious board" was to be placed in the bedroom and located in the middle of the room; the person was to

[18] In addition, there are also several other procedures not listed here because they are not directly related to the corpse, such as the sacrificial practice held at noon of the day of burial (虞祭 Yuji).

[19] Mainly based on the two books *Records of Leisurely Dreams under an Open Window* and *Red and White Joyous Event*. Details see (Ju 2017b; Mu 2011).

be moved onto the board. This board could be rented from the coffin shop (funeral house), if this was not feasible, people also used a door plank instead.

② Dressing. Before the corpse was cold, it was bathed and dressed in mourning clothes. *Family Rites* and *Manchu Four Rites* both require that the sons (if the deceased was male) or daughters (if the deceased was female) were to do this work, but in reality, it was usually carried out by an "expert" from the funeral house.

③ Inviting monks or Daoist priests. The monks or Daoist priests were invited the very day of the death to chant the scriptures around the deathbed and held the religious ceremony. Common people generally invited 9–13 monks, while princes, high officials, or wealthy merchants felt honored through inviting monks, lamas (Tibetan monks), and Daoist priests to hold three kinds of ceremonies at the same time. There were even a few who were extremely rich that would add two groups of nuns and lay Buddhists. While the religious ceremonies were performed inside the house, professional funeral musicians (most of whom were not formally ordained monks) played funeral music outside the door. The authors of *Family Rites* and *Manchu Four Rites* were clearly opposed to the participation of monks and Daoist priests in the funeral. However, to the common people, monks and Daoist priests were important funeral "experts," whose understanding of death and the soul was deeply connected to the Confucian rites represented by the *Book of Etiquette and Ceremonial* and even by the *Family Rites*.

④ Encoffining and keeping the coffin. In general, on the second day after death, the body would be put into the coffin, that is "*guanlian*" (棺殓, encoffining), also called "*dalian*" (大殓, Grand Dress). Then the coffin would be kept in the house for several days for condolences, this period was called "keeping the coffin" (停灵 *tingling*). The coffin would be placed in different places depending on the familial position of the deceased. If the deceased was the elder in the family, the coffin would be placed in the principal room (atrium of the northern room), but if he/she was junior, the coffin would be placed in the east side of the yard. Around the coffin, a temporal mourning hall would be constructed; it was a white shed built in bamboo. In front of the coffin, there was a platform for offerings and objects; under the platform was the banquet for the relatives who came to pay their condolences, where they could eat and drink at any time. Seats and stages were placed around the banquet for Chinese monks, Daoist priests, and Tibetan lamas who were to perform the religious ceremony. The coffin would be kept in the house for at least 3 days and no more than 49 days. All the important ceremonies had to be performed on the odd days (the third, fifth, seventh, ninth day after the death).

⑤ "*jie san* and *song san* ceremony 接三-送三" or "receive on the third day and farewell on the third day." This ceremony was held on the third day after the death. It was considered the most significant ceremony in the funeral observances, but this was not mentioned in *Family Rites* and *Manchu Four Rites*. In Beijing, in the Qing dynasty, the ceremony of "receive on the third day" usually implied inviting monks, Daoist priests, lamas, nuns, and lay Buddhists (the first two were most common) to chant sutras all day long so as to release the deceased from suffering. If the ceremony was not held at day time, the monks had to be invited to perform "*fang yan kou*" (放焰口 the essential meaning of this ritual is to give food to starving ghosts[20]) at night. "Fang yan kou" was a ceremony held according to a Buddhist sutra *The Yoga Flaming-Mouth Ceremony* (瑜伽焰口施食全集 *Yuqie yankou shishi quanji*) with the worshipping of "king of the ghosts" (面燃大士 *Mianran dashi*, an incarnation of Avalokitesvara). After that, different kinds of funerary objects made of paper, usually including a carriage, horses, a sedan chair, a boat, and a barred box which represented treasure, were burned in the open air. That was the "farewell at the third day." The ceremony itself, which consisted of the act of chanting sutras and the act of burning paper objects, could be held every odd day during the coffin-keeping period. However, holding this ceremony in the third day after death had particularly special significance, because people believed that the dead person would look at his or her home from the platform called "*wang xiang tai*" (望乡台 a platform from which dead could see his/her hometown) or really go back home. As a result, the family members had to prepare some travel expenses and basic necessities for the deceased so as to offer a good reception and then send her/him away. Without the ceremony of the "farewell on the third day," the deceased soul would not leave her/his home. This was unacceptable to the citizens. Chang Renchun recorded that if poor people could not afford the fee for the ceremony of the third day, the house owner or the neighbors who had young children would come out to chastise them: "you should tell us you don't have money, we can club together to hold the ceremony, if the deceased soul does not leave and makes trouble in this courtyard, how can we live here? What can we do if it frightens the children?"[21] In a word, the ceremony of the third was regarded to be essential. Only when the living have invited "deities" to be the guides and prepared the vehicles and traveling fees for the soul of the deceased, and have secured the soul to be sent to the entrance of the Netherworld, can they enjoy peace and not fear being disturbed by ghosts. Although Chinese monks, Daoist priests, and Tibetan lamas have different descriptions of the afterlife, they might have been invited to do the salvation ceremonies at the same time. Giving food to starving ghosts was the most important ceremony on the third day, but the "king of the ghosts" who originally came from Tantra was not the only one who could guide the dead. In fact, Beijing citizens, like most Chinese people at that time, did not care about which god would

[20] If the deceased couldn't receive sufficient ritual offerings, his/her soul would become a starving ghost. This could be caused by two reasons: 1. If the deceased had no offspring or his/her descendants did not hold the sacrificial ceremony in time, his/her soul would be "starved" of ritual offerings which were supposed to be provided by his/her descendants regularly. I will give some details of this point in the part 3 of this article. 2. If the death was abnormal, caused by suicide, accident, or violence, the soul of the deceased then couldn't be guided by the deities to the Netherworld. It could only stay in our world and do harm to the living. To prevent damage from starving ghosts, the living had to offer them plenty of sacrificial food. In Buddhism, then in popular religious tradition, the ceremony of providing sacrificial food to starving ghosts is called Fang Yankou.

[21] Chang 1995, p. 282.

lead the deceased soul to what kind of world. They would rather invite "the gods and Buddha" (all the deities they knew) to offer help to guide the deceased, as long as the observances had been completed successfully.

⑥ Funeral wake. The night before the formal funeral, the relatives of the deceased were to sit beside the coffin in mourning. The funeral wake was the main day for memorials. The religious ceremony held on this day would surpass all the ceremonies previously performed.

⑦ *Fayin* (发引) or *Chusang* (出丧). It consisted of a ceremony during which the coffin would be carried to the cemetery, or to the temple where the coffin was to be kept longer. Keeping the coffin in a temple was called "temporary settlement" (暂厝 *zan cuo*). From the Qing dynasty until the Republic period, this ceremony was particularly splendid and luxurious. According to a collection of jottings *Miscellany of Yanjing (*燕京杂记 *Yanjing zaji)*, "the funeral procession in the capital city was especially famous for its ostentation and extravagance. The number of the people to carry the coffin was even up to a hundred, the least would also amount to several dozen. The length of the streamers was up to four to five Zhang (丈, about 44 to 55 feet), which needed several dozen people to carry them. These streamers were very expensive, being made of up to a hundred rolls of silk cloth. The other sacrificial objects were numerous and sparkling. The whole ceremony could employ nearly a thousand people and all the objects were leased from the funeral shop, and usually cost a lot of money for the half day usage."[22]

⑧ Burial. This was the last step of the funeral. The best place for burial was the family graveyard. Usually, the coffin of those who lived temporarily in Beijing, as visitors, would be returned to their hometown for burial. In the early Qing dynasty, all the Inner City Banner people originally came from the northeast area, north of the Great Wall. Emperor Kangxi permitted the Banner people in Beijing to take the coffin of a family member back to their hometown and to build houses in graveyards where they could live and guard the tomb for 3 years. However, most Banner people did not want to be buried in their homeland, but chose to build a new tomb in the suburbs of Beijing. As a result, the inhabitants of Beijing usually had their own family graveyards in the nearby suburb.[23] These graveyards often had a family temple and were surrounded by farmland. The family temple was used to make offerings to the ancestors and gods, and with the profits made from farming the land, the family could afford expenses such as repairs to the tomb and associated buildings, the annual ceremony of sacrifice for the ancestors, and the familial banquet after the sacrifice. If the deceased could not be buried in the family graveyard for various reasons, a place with good geomancy for burial was selected. Good geomancy (whose meaning is discussed in the following section) was not only to satisfy the deceased but also to guarantee good blessings for his/her offspring. The deceased were buried in public

[22] Anonymous 1925, p. 121.

[23] See Liu 2008, pp. 165–166.

cemeteries when no appropriate graveyard could be found due to poverty. The majority of these public cemeteries were located near the city wall or in the outer city and were constructed by those who came from the same province, worked in the same office, or dealt in the same business. They were called "Charity Land" (义地 *yi di*) by Beijing residents.

11.3 Conceptions of Death: Three Stages and Three Risks

The funeral process described above could be divided into three stages:

The first stage was wandering and farewell. At the time of the *Book of Etiquette and Rites* (about BCE 1046–476), a spirit medium (巫 *wu*) would "call back the spirit" (招魂 *zhaohun*) of the deceased, hoping that thc spirit would come back to the body so that it could be revived. In the *Family Rites*, Zhu Xi still recommended that the deceased should be put into the coffin on the third day. Before the deceased was encoffined, his/her relatives, the sons and grandsons in particular, were to be present and pay attention to the condition of the corpse, because there was the possibility that the deceased could be revived within 3 days. In Qing dynasty Beijing, there was a different explanation among the common people. According to the book *Red and White Joyous Event* by Chang Renchun, the deceased could not be put into the coffin immediately, on one hand because the preparation of the coffin and other funeral objects required much time, and, on the other hand, it was necessary to wait for a "favorable time." Before putting the corpse into the coffin, the relatives of the deceased were to stay around, all cats and dogs or other living animals were to be tied up, and it was necessary to keep the dead body from any disturbance. It was not because the death itself was considered contagious or the corpse could bring some bacteria, but because a dead person was yin (阴), while the living persons and animals were yang (阳). Yin should avoid meeting yang; otherwise, a "transformation of the corpse" (尸变 *shibian*) may arise, which meant that there was the risk of the deceased becoming a zombie (僵尸 *jiangshi*). For this reason, many people insisted that the corpse should be put into the coffin as soon as possible and not to wait for the third day as prescribed by the *Family Rites*. While *Family Rites* speaks of "revival of the dead" and the common people feared "transformation of the corpse," in both cases it was believed that, at this stage, the corpse was still in a period of transition which implied that the soul and the body had not totally separated.

Once the deceased was put into the coffin, the ceremony of "receive and farewell on the third day" was to be held. As mentioned above, the core significance of this ceremony was to send the deceased spirit up to the entrance of the Netherworld and let him/her go to the world of the afterlife peacefully. This ceremony was the final farewell to the deceased spirit, and when it was completed, the deceased had completed this transitional period, and thereafter his/her body and soul had separated permanently.

At the second stage, the deceased's body and soul had to be sent to different places. Before the formal burial, the coffin of the deceased was usually and temporarily placed in a temple for some time. This custom, called "temporarily keeping the coffin before burial," was linked to several contingencies.

Firstly, as the capital of the empire, Beijing had a very large population of people coming from outside. In addition to the great number of merchants, officials, and intellectuals who lived in Beijing but were not natives of the city, the imperial examinations brought about 40000 candidates every 4 years, each with their family members. If these visitors who were far from their hometown died in Beijing, their coffins would be kept temporarily in a temple, waiting for being returned to their hometown for burial.

Secondly, family members of deceased inhabitants of Beijing could also choose to keep the coffin in a temple temporarily. Some were waiting for the grave land with the best geomancy, some were waiting for the best time for burial, and some were waiting for a joint burial after the death of the parent who was still alive. Through fieldwork, I have found another important reason: for Beijing residents, a funeral was an opportunity for the deceased's whole family and his/her extended social network to meet, to make new alliances, and to reinforce their social relationships; it was also the opportunity to make visible the internal order of the family and then to solve any familial conflicts. As a result, the funeral had to maintain a certain scale and needed enough space for the large amount of people attending. If there was not enough space at home, the funeral had to be held in a temple with more space. Until all the arrangements were made the coffin would temporarily be placed in the temple.

Finally, the custom of temporarily keeping a coffin in a temple was related to the concept of "outside death"(外丧 *wai sang*). The so-called outside death referred to the fact of dying away from home. Someone who died accidently away from home or even died in hospital both belonged to "outside death," and the corpse could not be carried back to the family. In the Qing dynasty, the rule said that the coffins of officials or merchants who died outside were only allowed to enter the Inner City if they had a special permission from the Emperor. In this kind of situation, there was no other way than to hold the funeral in a temple.

When the coffin was kept temporarily in a temple, the deceased's spirit had already entered the Netherworld and become "Bardo" (中阴身 *zhongyin shen*, a concept from Lamaism, which, generally speaking, means the state of the soul between death and rebirth). People believed that the "Bardo" period lasted 49 days. During this period of time, if many Buddhist or Daoist salvation ceremonies were held, rewards would be accumulated to reincarnate the dead. Even after the period of "Bardo," the deceased spirit could not become an "ancestor" if the dead body had not been formally buried. In the second stage, while the soul of the deceased had already left the body, there was still some "connection" between the corpse and the deceased's spirit, so the religious ceremony was still needed to maintain the calm of the deceased's soul. Because of the importance of the ceremonies, it was the best choice to place the coffin temporarily in a temple in which monks and Daoist priests resided. From 1750 to 1950, the number of temples grew to 1500 in the Inner city of Beijing,[24] but only about 30 to 40 temples could keep coffins.[25] Various gods were worshipped in these temples and were not limited to Buddhism or Daoism. In fact, the most important thing for Beijing residents when choosing a temple to keep a coffin was not so much the kind of god that could offer protection as the fact that monks or Daoist priests could offer the most appropriate ceremonies.

In the third and final stage, the deceased became an ancestor. Not believing that the dead would stay in the Netherworld forever or be reincarnated as suggested by Buddhism or Daoism, Beijing citizens, like most Chinese people, apparently preferred the Confucian perspective: with formal burial, the deceased would become one of the ancestors, who could bring happiness and longevity to the descendants of the family as long as regular offerings could be provided. Those without proper offerings would turn into "ghosts" who were extraordinarily dangerous to humans because they were hungry and thirsty. Destined to wander among humans in search of food, they were the cause of disaster and illness.[26]

In addition to regular offerings, the deceased's descendants had to search for the most appropriate geomantic graveyard, which would definitely ensure that the ancestor would bless later generations. In a word, geomancy (风水 *fengshui*, lit. Wind-Water) was a long-standing set of ideas and practices related to the belief that the natural environment, such as mountains and rivers, had a deep influence on human life and fate. The geomancy of the ancestors' graveyard, in particular, was directly connected to the fate of the whole family.[27] The close attention paid to finding an appropriate geomantic place implied that the Banner people in the Qing dynasty spared no effort in searching for a good graveyard, regardless of price. In this essay, we will not go further on this phenomenon, since Liu Xiaomeng (刘小萌) has already discussed it in his research on the graveyards and sacrificial lands of the Banner people.[28] Nevertheless, geomancy did not only imply mountains, water, and topography but also involved other creatures in the natural environment. It was very auspicious if the deceased ancestors could harmoniously coexist with animals and plants, therefore, the creatures living here should not be disturbed by the construction of tombs. The anecdote reported in *Notes from Yuewei Cottage* (阅微草堂笔记 *Yuewei caotang biji*) attests to this belief:

"in the Qianlong period, there was a big python in the tomb of Mr. Zha, a cabinet bachelor. The python usually hung on the two pagoda trees. Later, Zha's wife died and needed to be buried in the same tomb. Zha's family first offered sacrifices, they saw hundreds of snakes slithering away led by the python, after that, the burial began. After the funeral, the snakes came back to the tomb".[29]

---

[24] According to the statistics of the China-Franco research group of "*Epigraphy and Oral Memory of Temples in Beijing: Social History of an imperial Capital*" in 2008.

[25] A detailed discussion on this number is in the following part.

[26] For the relationship between gods, ancestors, and ghosts, see Wolf 1974, pp. 131–182.

[27] There is a huge body of anthropological research on geomancy and tombs, for example, Freedman 1968, pp. 5–15, Ahem 1973, Feuchtwang 1974.

[28] See Liu 2008, pp. 163–200.

[29] Ji 1800, vol. 11, part 1, p. 184.

In this story, there seems to be an "agreement" between the snakes and human beings. Harmonious coexistence between the dead and animals had become the symbol of good geomancy. Another collection of folktales *Free Hand Notes of Night Words* (夜谭随录 *Yetan suilu*), written by a Banner man, He Bang'e (和邦额), from the Qianlong period, recorded a similar story: a merchant from Shanxi province (山西) died in Beijing and was buried under an old pagoda tree. After several decades, his son got rich through business and decided to move the tomb back to his hometown. An astrologer took a look at the old pagoda tree, and said: "This tomb was deeply connected to the wood, it cannot be moved." The tree had coexisted with the grave for a long time; it provided great protection and support to the grave. If the grave was moved, the descendants would be punished. The son of the merchant did not listen to the advice and insisted on moving the grave. He later fell from a horse and died.[30] These two stories showed that there was no one single objective standard for so-called good geomancy. Good geomancy was not only made of beautiful and well-known mountains and rivers. A place where graves and the natural environment could live side by side, with animals and plants not disturbed by human beings, was a good place for the deceased and could be considered as "good geomancy."

Generally speaking, from death to burial, the corpse and spirit would experience three stages: close connection, a process of separation, and eternal peace after the soul of the deceased finally found its position among the ancestors and the corpse reposed in an agreeable place with good geomancy. The period of time during which the soul and the corpse had not yet separated completely was considered to present great risks for both the deceased and the living. The deceased was at risk of becoming a zombie, while the soul risked becoming a ghost, both of which could do extreme harm to the living. When the separation between soul and corpse was fully achieved, the danger declined, the soul became an ancestor and the corpse rested peacefully. In order to achieve the separation between the soul and the corpse, the living were compelled to protect the deceased from danger by sending the soul and the corpse to the "correct places" for their own benefit. The only way to achieve this aim was to make sure that the ceremony was complete and effective and, in addition, to make sure that it was held in the right place and at the right time. That explains why the Beijing citizens spent a lot of time, money, and energy to hold the most appropriate funeral.

11.4 Urban Space Shared by the Dead and the Living

It is irrefutable that the "correct place" is extremely important in the thoughts and practices mentioned above. On the third day after death, the soul should be sent to the "correct place", the entrance of the Netherworld, so as to make sure the deceased spirit had definitely left the human world and gone to the afterlife. During the temporary period when the coffin was kept, the deceased should be kept in the

[30] Hebang'e 1791, p. 341.

"correct place," the temple, which, protected by gods, was the best place to ensure the calm of the deceased. Finally, the deceased should be buried in the right environment, that is, the one that benefited from the most flourishing geomancy. In short, the deceased would experience three different spaces before finding final peace. For the Beijing residents, these three different spaces were located physically in Beijing and were therefore constitutive parts of one kind of city landscape, which differed from the modern city landscape.

① The Way to the Netherworld

As mentioned above, the most important death ceremony for Beijing residents was the "farewell on the third day." After the religious ceremony, the monks or Daoist priests would lead the people carrying the paper carriage, horse, ox, and property for the deceased to an open air and burn them up. This was in order to prepare the travel necessities for the deceased so as to ease the passage from the world of the living. Chang Renchun recorded that all the sites for burning the paper objects were situated within the Inner city Beijing, including the South Square in the west of Desheng Bridge (德胜桥), Eight Bays Square (八道湾), Square in the west of the Drum Tower (鼓楼), back square of Bell Tower (钟楼), Icehouse Square at east side of Shichahai (什刹海), West Sidewalk Square in Inner Di'an Gate (地安门), South Brook Bank Square, Duck Temple, Guanyuan (官园) Ping'anli Square (平安里).[31] The following picture shows the location of these sites (Pictures 11.2).

From the above picture, we can see that all these sites were near bodies of water in the Inner City, along the Shichahai lakes (什刹海), Lake Jishuitan (积水潭), and the Nangouyan river (南沟沿). The squares of Ping'anli (平安里) and Guanyuan (官园) were a little more distant, but they were still not far from the Nangouyan river (南沟沿). The reason for these locations was that, because water moves mainly under the earth, it belonged to Yin in Chinese ancient cosmology and popular religion, and to send the deceased's spirit out of the living world was to send it to the Yin water.[32] During the Wei, Jin, and Southern and Northern dynasties (魏, 晋, 南北朝, 220–589), the Obon festival (盂兰盆节, "obon" is the transliteration of the Sanskrit "Ullam-bana," which means the salvation of those suffering) spread into China and was absorbed by Daoism. It finally formed the common ghost festival in Buddhism, Daoism, and popular religion – July 15th in the lunar calendar. In the Qing dynasty, on that day, Beijing citizens would burn funeral boats to help the ghosts cross the river to the Netherworld. Another example of the relationship between water and Netherworld lay in the decoration of coffins. Aquatic animals, like fishes and snakes, sometimes appeared on the coffin. "The Tale of the Green and the White Serpent" (*Qingshe baishe)*, from the collection of popular ballads *Miscellaneous Poems of the Capital* (都门杂咏 *Dumen zayong*) written by Yang Jingting (杨静亭) in the Daoguang period (1821–1850), also records that in the

[31] Chang 1995, p. 283, note 1.

[32] That is, to send them to Huangquan (黄泉 Yellow source). Since the earth is yellow in the systematic correlative thinking of Yin/Yang and five phases Yinyang Wuxing (阴阳五行), the place of the dead which is located under the earth is called "yellow source."

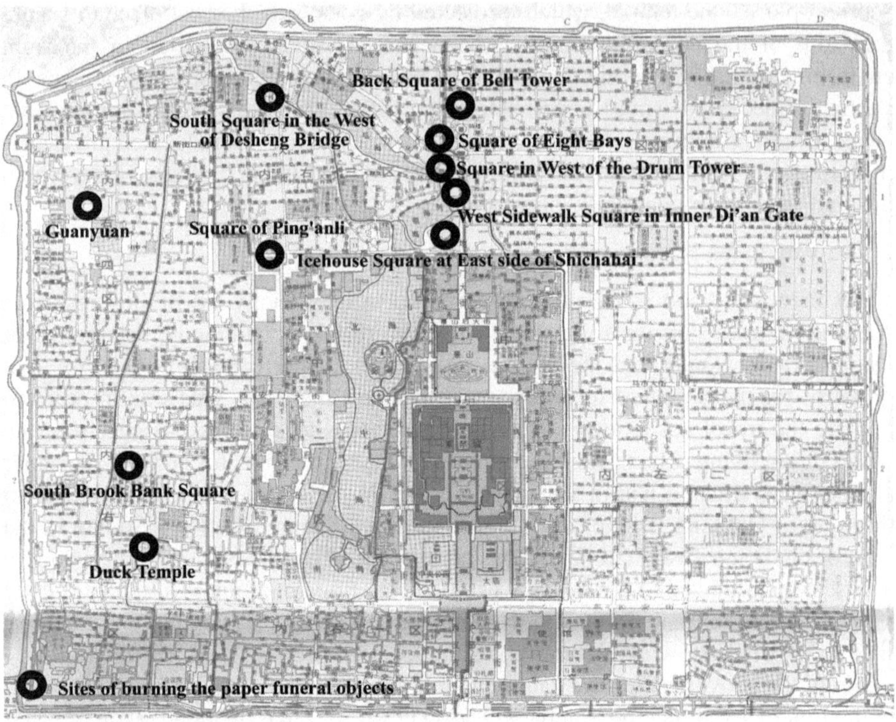

Picture 11.2 The distribution of the sites where people burned funerary objects in Inner City Beijing in 1930–1950 (the map is a part of the *Map of Beijing City in Republic Period (*民国北京城图 *Minguo Beijingcheng tu)* from the *Beijing History Atlas (*北京历史地图集 *Beijing lishi ditu ji*), (Hou 1997). The sites are marked by the author of this paper

Beijing funeral processions, there were usually two children preceding the coffin dressed up as a white snake and a green snake.[33] From the Qing dynasty to the Republic period, lakes Shichahai and Jishuitan were considered as the ghosts' entrance to the Netherworld.

In the Yuan dynasty (1271–1368), Guo Shoujing (郭守敬) built a canal from the Tonghui River (通惠河) to Beijing which formed the largest body of water in Inner City, including several connected lakes, Shichahai, Beihai (北海 "Northern Sea") and Zhongnanhai (中南海 "Middle and Southern Sea"). Before the Qing dynasty, the wharf of Jishuitan had been abandoned for a long time. Beihai and Zhongnanhai were in the imperial gardens and usually forbidden to ordinary people. However, Jishuitan and Shichahai were located in residential areas where the common people could freely go. They had become the most attractive water landscapes in the capital city. When talking about the Shichahai area, historians tend to emphasize that it was a bustling, well-known business area. This was indeed the case, but one thing has

[33] Quoted from Li 1937, p. 220.

always been ignored by researchers: the Shichahai and Jishuitan lakes had very long lakeshores, and while there were many restaurants and shops on the south shore and a part of north shore of Shichahai, other places along the shore were open ground. Ji Yun noted in his *Notes from Yuewei Cottage*, that, during the winter of the first year of Emperor Jiaqing (1796), he spent a night in the Shichahai Temple (located on the north shore of Shichahai, very near to the lake). He noted that the back hall had been abandoned for a long time and the building looked old and abandoned. When he took rest in the east corridor at night, he observed that "it was very cold, while there were many fires it was not hot, the candles produced a dim and green light." The other servants rested in the west corridor, and they held candles all through the night and could not fall asleep. Later a Daoist from another temple told him that the Shichahai Temple was haunted and that the ghosts had even killed people. Ji Yun agreed with that, and considered that in the closed rooms of the temple, there had to be a lot of ghosts.[34] This story shows that some parts along the shores of Shichahai were deserted at this time. In the minds of common residents, this place was actually the "world of ghosts" and not the "world of human beings." This perception lasted up to the twentieth century. From the fieldwork, we know that Shichahai and Jishuitan and the surrounding area were still the major locations in which to "make sacrifices to the ghosts" until the 1950s and 1960s. In addition to the "farewell on the third day," "Songku"(送库 "give and send a warehouse full of treasures to the Netherworld") and the burning of the offerings for ghosts, on July 15th every lunar year, floating river lanterns and sagebrush lanterns were always lit on the waterside to "illuminate the way for ghosts." One old resident said: "In our childhood, we used to light sagebrush lanterns on July fifteenth every lunar year. Since this place (Shichahai) was very deserted at that time, the sagebrush grew very high and it would wilt on July fifteenth, so we used to put incense and sagebrush together, make it like a tree. It was lit on July fifteenth to illuminate the way for ghosts."[35] Among other sources, the tabloids of modern Beijing collected in the *Compilation of Tabloids in Modern China* (中国近代各地小报汇刊 *Zhongguo jindai gedi xiaobao huikan*) show that much social news referring to Jishuitan was related to suicides, by drowning in the river, and murder. This phenomenon was related to the deserted environment, and perhaps it also had some connection with the opinion that this body of water had once been considered as the entrance to the Netherworld. Even now, residents nearby still hold this perception. A female resident told me that on the second day after the suicide of a university student in Shichahai, there were lots of water lanterns floating on the river, nobody knew where they came from and who had put them there.[36] Although Shichahai is today full of bars with endless parties, in the minds of many old people, it is still a place of water and favorable to ghosts.

---

[34] Ji 1800, Vol. 21, p. 404.

[35] Investigated: Wang Xiu (王岫), born in 1947. Investigator: Ju Xi. Investigating time: October 4th 2005. Location: Wang Xiu's home at Longtoujing (龙头井).

[36] Investigated: Sheng Xueying (盛雪莹), born in 1933. Investigator: Ju Xi. Investigating time: December 2003. Location: Sheng Xueying's home at Dongsi Shi'er Tiao (东四十二条).

② Temples for Keeping Coffins

The world inside the city did not only belong to the living but also to the newly deceased. In fact, the living and the deceased lived side by side as neighbors, in the houses and in the temples where coffins were kept. Houses were the places for the living; the newly deceased were wandering between life and death, so they could continue to stay in houses for some time until the deceased spirit had finally left. The temples where coffins could be kept were the places where gods resided, external to the "human world." The deceased could thus temporarily live there until a more suitable dwelling was found. If family members could not find a permanent "dwelling," the deceased could stay in the house of the gods forever. The mercy of gods not only benefited the deceased but the living as well, who, from the seventeenth century to the 1950s, could also stay in temples, temporarily or permanently, as "hotels" or cheap "apartments." In the residence of gods, the living and dead could live in harmony, without disturbing each other.

Due to a lack of information, it is difficult for us to know the exact number of temples which were both "temple housing" and "temporarily keeping coffins" at the same time during the Qing dynasty. In a report presented by the Imperial Household Department (内务府 *Neiwu Fu*) on May 22nd, in the 35th year of the reign of Emperor Qianlong (1770), 12 temples including Longchang Temple (隆长寺 Temple of Long Prosperity) in Xisi (西四) district of Beijing were reported as "renting their houses to others for dwelling so as to earn money from the house rent."[37] A Beijing travel guide book *Series Records of Cosmopolitan* (朝市丛载 *Chaoshi congzai*) at the end of the Qing dynasty (1886) had a chapter dedicated to "temple housing," which notably noted that Longchang Temple was well-known for "temple housing." The story of this temple is illustrative of these institutions that hosted the dead and the living under the same roof. From field research, we may infer that from 1918 to 1966, the tenants of Longchang Temple had all kinds of activities and identities including merchants, coal shop workers, etc. Meanwhile, Longchang Temple was also well-known for keeping coffins and holding funerals. According to Peking Police Bureau investigation records in 1944, six coffins including that of Tang Erhe (汤尔和), the supervisor of the Board of Education during the Japanese occupation, were kept in Longchang Temple. One of the coffins had been kept there for more than 10 years.[38] When we investigated the older residents who lived in the houses of this temple from the 1930s, they could still remember that coffins were kept in the main hall and the back hall until the 1950s, and they felt that there was nothing inappropriate. Again, in the Wenchang Zitong Temple (文昌梓潼庙 Temple of God of Literature) in Mao'er Hutong (帽儿胡同), 95 coffins were kept there when an official investigation was carried out in 1946.[39] Meanwhile, the monks also used to

[37] *FHA* 1770, 05-0277-031.

[38] See: (*BA* 1944, J183-002-25922).

[39] See: (*BA* 1936, J183-002-40250).

"rent the houses to make a living." Up to the 1960s, tenants lived with coffins as neighbors.[40]

In addition to being inexpensive, the temple accommodation was also considered to be safe and clean, which were leading reasons for tenants to stay there. Ms. Wang, who was born in 1950 and lived in the Wenchang Zitong Temple (The Temple of God of Literature mentioned above), clearly remembered that "Originally, (the corpses) were just placed in the surroundings of the back yard, they were in the coffins and placed on the ground." While these more than 90 coffins were only separated from her home by a simple partition, she did not feel anything wrong with that and did not smell any odor either, because "the coffin boards were especially thick at that time." She told her own experience to us in order to demonstrate that the corpses did not harm the environment: her aunt who had passed away several decades previously was put secretly into a coffin for burial to avoid the then obligatory cremation (after 1950 burials were no longer allowed). Several months before our investigation, the family members opened the coffin for relocation. They found the dead body of her aunt was in "extremely good" condition, only the belly was not that good.[41] On telling us her own experience, she tried to convince us that the corpses were not dreadful, and the living were not disturbed by corpses in well-sealed coffins, even at a very close distance.

Jiuding Niangniang miao (九顶娘娘庙 Temple of Our Lady of Nine Peaks) in Lama Temple Avenue (雍和宫大街) was another temple well-known for keeping coffins. It still kept 23 coffins in 1946, and the older residents who lived there did not feel disturbed but rather proud, they said: "Gods protect anyone living here, nobody was sick, neither adult nor child."[42] Bao'an Temple (保安寺 Temple of Bringing Safety) in Di'an Gate West Avenue (地安门西大街) also used to keep coffins, some with dead bodies, and others empty. The grandmother of Mr. You (游) who lived in Bao'an Temple had deposited an empty coffin here. Mr. You said: "The old people could sleep several nights in the coffin when they wanted, that's the habit, if the coffin fits then they feel at ease. My grandma did so, she often spent whole nights in her own coffin when she was alive."[43] Much fieldwork proves that the temples were not only for the living but also for the dead. Living in the house of gods, whether one was alive or dead, one could feel to be well protected and cared for. At the turn of the twentieth centuries, there is no doubt that many temples were both "temple housing" and "temporarily keeping coffins" at the same time.

[40] This is what Shanguo (善果), the Abbot of Wenchang Zitong Temples said. See: (*BA* 1930-1939 J2-8-114).

[41] Investigated: Wang Shuzhen (王淑珍), female, born in 1950. Investigator: Ju Xi. Investigating Time: November 6, 2014. Location: Inside Zitong Wenchang Temple in Mao'er Hutong.

[42] Investigated: Ms. Li (李), female, born in 1930. Investigator: Ju Xi. Investigating Time: September 2006. Location: inside Jiuding niangniang miao, No. 175, Yonghegong Avenue.

[43] Investigated: Mr. You, born in 1952. Investigator: Ju Xi. Investigating Time: March 15, 2015 Location: Baoan Temple in Dianmen West Avenue.

In 1946, there were 26 temples registered as “temporarily keeping coffin” by the Peking Police Bureau in the Inner and the Outer city.[44] By examining the temple property registry, we find that all of them also had rooms to rent out. The number 26 is certainly wrong and does not take into account many other temples that were used for keeping coffins. For instance, in 1939, Wu Yushuai (吴玉双) died and his coffin was kept in Nianhua Temple (拈花寺 Temple of Picking Flower),[45] which was not registered in the 1946 official statistical table. In addition, according to *Red and White Joyous Event*, the number of famous temples for keeping coffins was up to about 31, and there were 25 in the Inner and the Outer city.[46]

The following picture is the distribution map of the temples which used to keep coffins depicted according to the Police Bureau registry, the *Red and White Joyous Event* by Chang Renchun and my own fieldwork. There are 38 temples in total. If we take into account temples in the outskirts but close to the city wall and used by city residents, the number possibly amounts to 100.[47]

As this picture shows, the distribution of the temples was homogeneous (Picture 11.3). Only places which concentrated official buildings, that is, the administrative center, did not have temples for keeping coffins. It is reasonable to consider that Beijing citizens organized their religious community around these temples even if further study is still needed. The existence of these temples within the heart of the city shows that Beijing was a space where the living and the dead could actually live as neighbors.

③ Graveyards

Today, Shichahai is no longer desolate and gloomy – on the contrary it is a place for lay life, par excellence: consumption, tourism, performance, and excessive commercial development, in sharp contrast with its situation in the Republic period. Conversely, today’s Beijing cemetery, Babao Mountain public cemetery (八宝山公墓 Babaoshan Gongmu), is a quiet and desolate place, also in sharp contrast with the graveyards for Beijing citizens from the seventeenth century to the 1950s.

As previously mentioned, for Beijing citizens, the order of preference for burial was the following: firstly was the family cemetery; secondly was the private cemetery with good geomancy, which would later become the family cemetery; and last was the collective cemetery, usually for the lowest social classes who were unable to afford the price of a private cemetery. There were two types of collective cemetery, the first were known as the “land of charity,” which were managed by charitable associations and sometimes made a small charge. The other were the “collective graves,” which were for the poorest or unclaimed corpses and were totally free of charge.

---

[44] BA1946, J181-016-02958, pp. 28–39.

[45] BA 1939, J183-002-27402.

[46] Chang 1995, pp. 441–447.

[47] The author of the book *Beijing Funeral History* (北京丧葬史 *Beijing Sangzang Shi*) thinks the number of temples for keeping coffins was 73, the number for the public cemeteries was 130, but he does not mention his sources. (Zhou 2002).

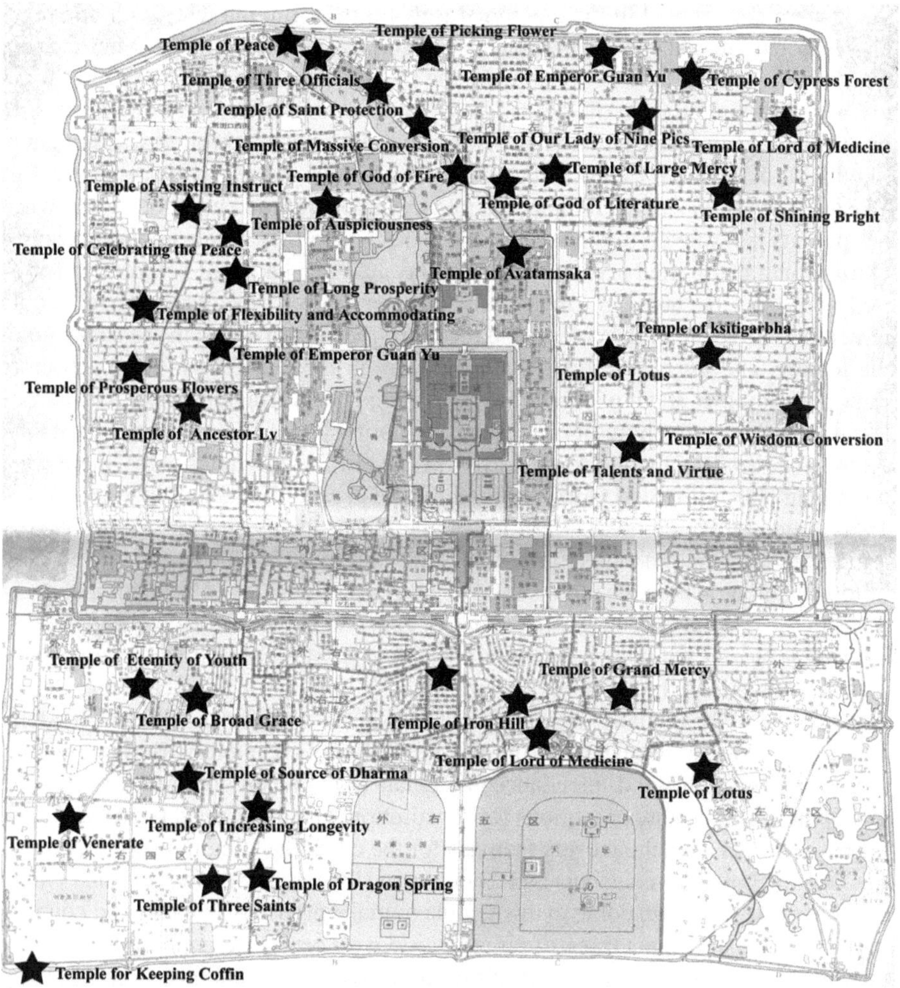

Picture 11.3 The distribution of temples for keeping coffin in Beijing in 1930–1950. ((Hou 1997) The temples are marked by the author of this article)

As we have already stated, the Inner and Outer city residents shared the same conception of funerals and shared the same observances. However, the distribution of cemeteries differed between the two parts of Beijing city. In Beijing, most of the family cemeteries and public cemeteries were located in the Outer City and in the suburbs. The main reason for this was that the Inner City was overcrowded. In the sixth year of the reign of Emperor Qianlong (1741), there was nowhere to bury the unclaimed dead since the Inner City was densely populated with limited land. As a result, the government rented land in the Outer City to construct some public graveyards. After consultation between the governor of the imperial prefecture (顺天府尹 *Shuntianfu Yin*) and the Ministry of Revenue (户部 *Hu Bu*), seven public graveyards were constructed in the outer city at that time.[48] But the temples in the Inner City which used to keep coffins could also become a cemetery. In fact, when a coffin was unclaimed for a long period of time, the monks would use bricks to build a tomb on the ground in the form of "*Qiuzi*"(坵子 "little tomb like a small hill") in order to protect the coffin from the wind and rain. Consequently, the ground was gradually filled with these small tombs and actually bore no difference from a cemetery. Ms. Wang who lived in Zitong Temple was told by her parents and her brother that when her parents had moved to the temple, the room where they dwelt was just beside a Qiuzi which had buried a young lady who because she had died before being married could not be buried in the family cemetery. Even Ms. Wang herself could remember the scene of the coffins and the Qiuzi that were packed in the backyard of the temple, and they were finally moved away in the late 1950s.[49] A much more famous example is Jiaxing Temple (嘉兴寺 Temple of Auspiciousness). As a temple located on the south shore of Shichahai, the north side of West Imperial City Wall Avenue (西皇城根, the most prosperous avenue in Beijing Inner City), there were more than 500 coffins and Qiuzi in Jiaxing Temple, according to our investigation of 1946 Police Office archives. Until today, the old residents nearby can still clearly remember the situation at that time, when the open space behind the temple was full of tombs among which the monks used to cultivate vegetables. A collection of anecdotes in the late Qing dynasty *Miscellany of court and commonalty since the period of Daoguang and Xianfeng (1821–1861)* (道咸以来朝野杂记*Daoxian yilai chaoye zaji)* also mentions the Long'an Temple (隆安寺 Temple of Prosperity and Peace) and Jiangnan Chenghuang Temple (江南城隍庙 Temple for the God of the City south of the Yangzi River) as having this kind of public cemetery. Both these temples were big with good attendance. Jiangnan Chenghuang Temple was close to Qianmen (前门), an area notorious for prostitution, and many dead prostitutes were buried there. As a result, at every festivals for the dead like Qingming (清明, around the 5th April of solar calendar) and Zhongyuan (中元 the 15th July of lunar calendar), prostitutes used to come to this temple to make offerings to their dead counterparts. This became something of an attraction with people scrambling to meet the prostitutes who were making offerings to their deceased sisters.[50]

As the abovementioned examples show, a few temples in the Inner City became cemeteries when they kept many unclaimed coffins, but most of the graveyards were located in the outer city and suburbs. This was primarily because the Inner City had many people but little land; however, the underlying conception of death still suggested that the graveyard should be far away from the noisy human living areas and be in contact with animals, plants, mountains, and water. This was the standard for a "good" graveyard. Only the poor who were unable to choose the good graveyards would be buried inside the city with the living all around. This type of graveyard was considered "bad" because its position, among the living, would do

[48] *Daqing huidian zeli* 1986, vol149, Court of Sensors (都察院 Ducha Yuan).

[49] Investigated: Wang Shuzhen, born in 1950. Investigator: Ju Xi. Investigating Time: November 6, 2014. Location: Inside Wenchang Temple in Dianmen.

[50] Chongyi 1982, p. 26.

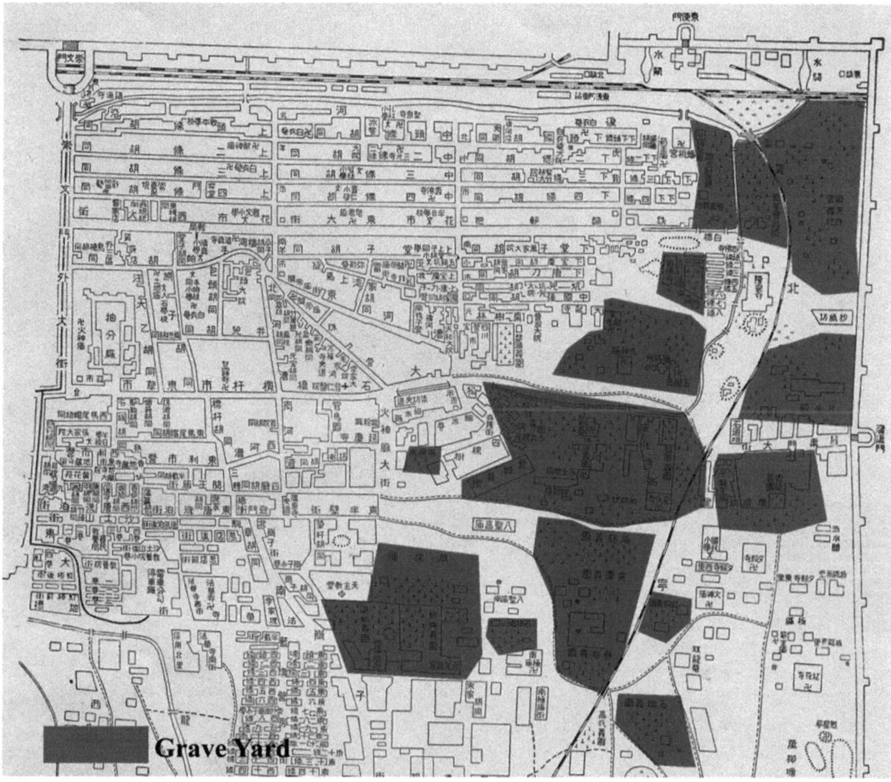

Picture 11.4 The graveyards in Outer City in 1937 (the western part) (Based on the map *Beiping jiequ fangxiang zhi gailue* 1937–1938.) The blue marks were added by the author of this article

harm to the deceased. As we have already shown, the fact that some graveyards were located in the most flourishing streets did not make people uncomfortable. On the contrary, some graveyards could even become places to visit (as in the case of the Jiangnan Chenghuang Temple). In fact, there were plenty of cemeteries in the densely populated area of the Outer City. As Pictures 11.4 and 11.5 show, based on the *Outline of Peking Street and Alley* (北平街衢坊巷之概略 *Beiping Jiequ Fangxiang Zhi Gailue*) mapped by the Peking Special Office (北平特别市公署) in 1937–1938, graveyards (in blue) were close to the residential areas, some even being located in the central area of the Outer City.

In fact, graveyards, and in particular family graveyards, were also places for entertainment for Beijing citizens. The family cemetery was not only the place for the ancestors but also the "root" of the whole family. It was the space for the living to rest, to gather as well as temporarily escape from daily life and return to the ancestors so as to "be reborn." In other words, the cemetery was not only a "place for the dead" but also a place for "new life." One of the important reasons attributed to the transformation from the graveyard to the land of new life was filial piety and ancestor worship, which were highly rewarded by the Emperors of the Qing dynasty.

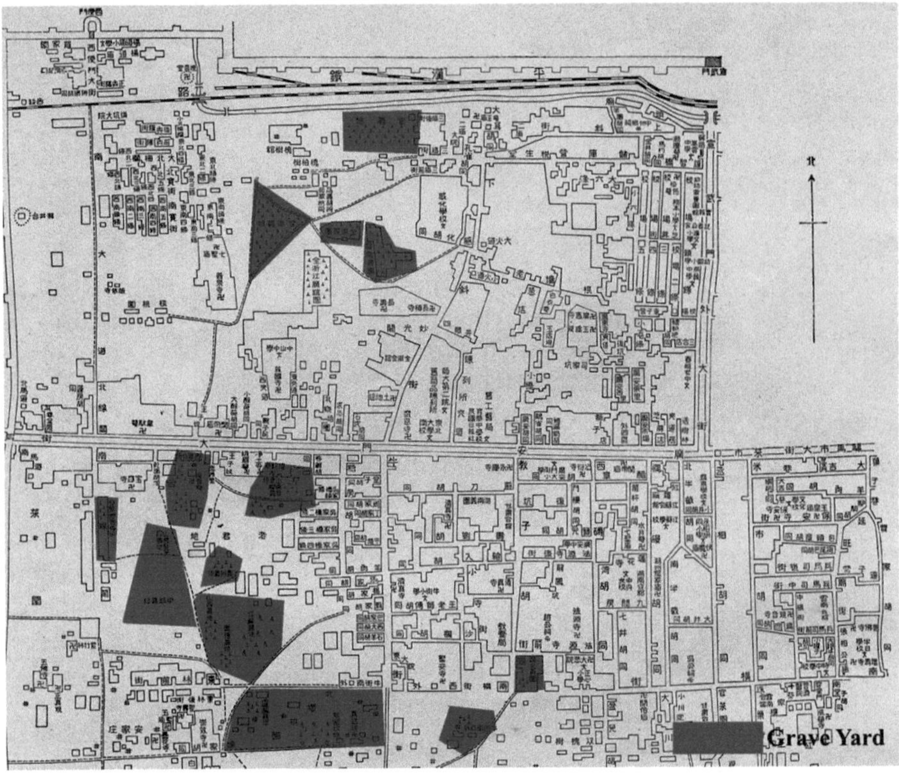

Picture 11.5 The graveyards in Outer City in 1937 (the eastern part)

According to the laws of the Qing dynasty, even when a criminal family was punished by confiscation, while all their belongings were confiscated, the graveyards with its surrounding land were not included.[51] So, people from different classes devoted themselves to looking after graveyard, cultivating the land and constructing houses, so as to prevent the family's descendants from being homeless if they ever committed a crime. The graves of the ancestors were the spiritual home and were the last shelter for the citizens. For instance, according to research by Liu Xiaomeng, many Banner people chose to return to graveyards when they faced poverty or a career setback. On the one hand, they could rest in a place that was harmonious with nature. On the other, they could easily manage the graveyard and earn a living with farm work.[52] Moreover, for the Beijing citizens at that time, visiting graves was a leisure activity. Tomb Sweeping (扫墓 *saomu*) was the main part of Qingming Festival together with the "spring walk" (春游). *Sketch of Scenery in the Imperial Capital* (帝京景物略 *Dijing jingwu lüe*), a very beautiful prose work recorded in 1635: "the day of Qingming in March (lunar calendar), men and women swept the

[51] Liu 2008, p. 187.

[52] Ibid., p. 187–189.

tomb [...] after crying, they walk among the beautiful trees and gardens, sit down and drink [...] on that day, people wear the leaves and branches of willow as a head-dress and visit Gaoliang Bridge (高粱桥), or so called spring outing."[53] This custom had existed for a long time, as *Kangxi Wanping County Chronicles* (康熙宛平县志 *Kangxi wanping xianzhi*) recalls: "On the day of Qingming, men and women go out with willow branches on the head to sweep the tomb [...] then walk among beautiful trees, choose a garden, sit together to eat something, and return after that."[54] A poem in *Miscellaneous Poems of the Capital* also said: people usually ate and drank in the fragrant woods and gardens after ancestor worship; they would soon turn to leisure and happiness after crying in front of the tomb.[55] Graves, whose sites had been carefully chosen, were places enjoying "good geomancy," people paid heed to whether the animals and plants could grow freely in the graveyard, and the beauty of the landscape had a double meaning to the living and the dead. For the deceased, it was the symbol of the harmonious coexistence with nature and earth; for the living, it was also a place where they could return to nature, relax, and find spiritual comfort.

Today in Beijing, the cemeteries are strictly separated from the areas for the living, but (at least) from the seventeenth century to the early twentieth century, the living areas and the cemeteries were close to each other. In the Outer City, just next to the busy streets and the residential houses, there were plenty of public and familial graveyards. And in the densely populated Inner City, the custom of keeping coffins in temples led to the creation of cemeteries around some of these sites. A burial near the bustling street was not a good choice, because the dead needed a calm place with good geomancy where plants and animals grew harmoniously, so most of the cemeteries inside the city walls were public, and the familial graveyards far away from the city were agreable places for both the dead and their descendants, they were not only the Netherworld for the dead, but also heaven for the living. In a word, the cemeteries were not places with a potential for contaminating the living and which should be removed from the view of citizens at that time. They made up an indispensable part of the city landscape.

11.5 Conclusion

From the seventeenth century to the 1910s, before the western concept of hygiene flowed into China, the period from death to burial was considered as unfolding in three stages, during which, the deceased could face different dangers. The living were responsible for helping the deceased to get through these dangerous stages and successfully become an "ancestor," instead of a wandering, starving ghost or a zombie. In return, and only if they had successfully completed this journey, the

---

[53] Liu, Yu 1635, p. 102.

[54] Wang 1684, p. 8.

[55] Li 1937, p. 77.

deceased would bring safety and happiness instead of disasters to the living. What was most important lies in the fact that, each stage in this journey had a real incarnation in a physical place. The dead spirit had to be sent to the entrance of the Netherworld, their corpse had to be placed firstly in the house of gods, then in the land where the ancestors rested, these three places, respectively, represented by three different forms: water, temple, and tomb.

With these conceptions of death and burial observances, Beijing city was a place shared by the living and the dead and at no time was the natural environment invested by death suspected of being unhealthy or dreadful for the living. Bodies of water were not only "nature" or "environment"; they were full of metaphors and religious sense. Although Shichahai, the biggest body of water in the city, was located in the core area of citizens' daily life, in people's mind, it was still the entrance to the Netherworld, the route for ghosts and the place for holding funeral ceremonies. Correspondingly, along the shore of Shichahai, there were numerous temples for keeping coffins (see Picture 11.3). Before the collapse of these conceptions and cosmology, due to the respect and awe toward the Netherworld, Shichahai still maintained its wildness and natural aspect and was not totally taken over by commerce, even though it was located in the center of the most prosperous area. Temples were important places for maintaining the safety of life. Gods and their servants – clergies headed by monks and Daoist priests – protected the order of the afterlife and helped the deceased spirit to leave peacefully. A funeral under their auspices helped the living overcome the crisis, deal with the difficulties of death and reclaim their social connections. These temples were community oriented, they served the people nearby and were familiar to the surrounding residents and served a pivotal role between the living and the dead. Death was not unfamiliar for the citizens. Under the protection of gods, the world of the living and the dead was stable and in good order; the city was considered safe, clean, and healthy. The graveyards were not a world of doom; on the contrary, they were the places that could bring vitality to the living. Good geomancy, undisturbed animals and plants, well-cared tombs: all these features implied that the graves were not a cold otherworld but were within our invigorating living world.

At the end of the Qing dynasty, western science expanded in China and, widely supported by Manchu and Chinese authorities, progressively reduced the indigenous cultures. Temples gradually became the byword for "superstition," many of them were converted into schools or official offices. However, temples for keeping coffins which were deeply rooted in the daily life of the residents were not concerned by this change. The real turning point began in 1937 after the Japanese occupation of Beijing. In order to carry out "hygienic modernity," Japanese occupiers began to crack down on the custom of temporally keeping coffins; meanwhile, they prepared to construct some modern cemeteries and bury all the coffins all together.[56] This regulation was greatly resisted until the end of World War II. In 1945, a large number of the temples for keeping coffins had not been destroyed. Nevertheless, knowl-

[56] See (*BA* 1939, J001-002-00120).

edge of "bacteria" and "hygiene" had already become popular, some modern intellectuals kept demanding that the coffins be moved out of the temples and all the graveyards in the city demolished.[57] From the beginning of 1946, the Beijing government of the Republic began to register and remove the coffins from the temples. In 1948, it finally decided to continue the Japanese policy toward cemeteries, by prohibiting keeping coffins and holding funerals in the temples.[58] From then on, temples were progressively reformed or transformed into schools or administrative buildings until the Cultural Revolution (1966–1976) which prohibited all ancient funeral ceremonies in the name of anti-superstition. Funerals were to be administered in one and only one way, the familial "graveyard" no longer existed, and the place where it was possible to offer a quiet environment to the dead and to allow the living be reborn were erased from the surface of Beijing. When the popular religion was totally banned in the name of "superstition," Shichahai as "the way to the Netherworld" soon faded away and was progressively transformed into the commercial, leisure area we know today.

References

Primary Sources (Ordered by Original Publication Date)[59]

(a) *Diaries, Collections of Folktales, Customs and Rituals*

Zhu, Xi朱熹. *Jiali* 家礼 (Family Rites), 1169 or 1170 [2002]. Shanghai: Shanghai Ancient Book Press.

Liu, Tong 刘侗, Yu Yizheng 于奕正. *Dijing jingwu lue* 帝京景物略 (Sketch of Scenery in the Imperial Capital). 1635[2001]. Shanghai: Shanghai Ancient Book Press.

Hebang'e 和邦额. 1791[1993]. *Yetan suilu* 夜谈随录 (Free Hand Notes of Night Words). Zhengzhou: Zhongzhou Ancient Book Press.

Ji, Yun 纪昀. 1800[1998]. *Yuewei caotang biji* 阅微草堂笔记 (Notes from Yuewei Cottage). Hangzhou: Zhejiang Ancient Press.

Suoning'an 索宁安. 1801. *Manzhou sili ji* 满洲四礼集 (Manchus Four Rites). Place unknown: Xingfeitang.

Mu, Qixian 穆齐贤. 1828–1830. Zhao, Lingzhi 赵令志, Guan, Kang 关康 (trans.). 2011. *Xianchuang lumeng quanyi* 闲窗录梦全译 (Translation of the book *Records of Leisurely Dreams under an Open Window*). Beijing: Press of Minzu University of China.

Anonymous. 1925 [1986]. *Yanjing zaji* 燕京杂记 (Miscellany of Yanjing). Beijing: Beijing Ancient Book Press.

Li, Jiarui 李家瑞 (ed.). 1937 [2010]. *Beiping Fengsu Leizheng* 北平风俗类征 (Systematization of Peking Custom). Beijing: Beijing Press.

[57] See (*BA* 1944, J183-002-28754).

[58] See (*BA*1948, J181-016-02568).

[59] The original publication date, when known, follows the title. The edition consulted is in square brackets.

Chongyi 崇彝. 1982. *Daoxian yilai chaoye zaji* 道咸以来朝野杂记 (Miscellany of court and commonalty since the period of Daoguang and Xianfeng 1821–1861). Beijing: Beijing Ancient Book Press.

Chang, Renchun 常人春. 1995. *Hongbai xishi: Jiujing hunsang lisu* 红白喜事 *(Red and White Joyous Event: Custom of Wedding and Funeral in the Old Capital)*. Beijing: Beijing Yanshan Press.

(b) *Gazetteers, Chronicles and Collection of Regulations*

Qing Gaozong shilu 清高宗实录 (Chronicle of the Reign of Qing Gaozong), 1986. Beijing, Zhonghua shuju.

Beijing zhi · Zhengwu juan 北京志·政务卷 (Beijing Chronicles Government Affairs). 2003. Beijing Chronicles Compilation Committee (comp.). Beijing: Beijing Press.

Daqing huidian zeli 大清会典则例 (Instances of Qing Collected Regulations). 1986. 四库全书版 Siku quanshu edition.

Wang Yanglian 王养濂 (comp.). 1684 [2007] *Kangxi wanping xianzhi* 康熙宛平县志 (Wanping County Chronicles in 1621–1722). Beijing: Beijing Yanshan Press.

(c) *Archives Kept in Beijing Archive Collection (Abbreviated to BA in the Text):*

内五区文昌梓潼庙僧人善果登记庙产和法魁接充住持的呈文及社会局的批示 *Report and Reply of Bureau of Social Affairs on Temple Property Registration by Monk Shanguo and Fakui Succeed to be the Abbot of Wenchang Zitong Temple in Inner Five Districts*. 1930–1939. Archive Number: J2-8-114.

北平市内外城各寺停灵柩调查表 *Survey of the coffins kept in the temples in Inner and Outer City of Peking*. 1936. Archive Number: J183-002-40250.

内政部公布公墓条例及取缔停柩暂行章程 *Temporary Rules for Beijing Cemetery Regulations and Ban on Keeping Coffin before Burial Issued by Interior Bureau.* 1939a. Archive number: J1-2-120.

———*Public Cemetery Provisions and Temporary Regulation of the Ban of Keeping Coffin Published by Ministry of the Interior.* 1939b. Archive Number: J001-002-00120.

北平市警察局内五分局包括兴亚纪念周报告书、侨居英国人迁移住址、钱方车家迁入德国人、拈花寺停厝情形的呈*Report of Inner Five District Substation of Peking Police Bureau for Thriving Asia Week, Moving Location of the British Immigrants, Germans Move in Qianfangchejia, Ban for Keeping Coffin in Nianhua Temple*. 1939. Archive number: J183-002-27402.

北平市警察局内四区关于界内停灵处所调查表及领回尸体移葬的呈 *Peking Police Bureau Report on the places for Keeping Coffins in Inner Four District and Notice for Taking the Corpses Out for Burial.* 1944. Archive number: J183-002-25922.

北平市警察局内五分局关于请调查农盛号制碱商内部设备、查设福民胰皂土碱工厂、嘉兴寺和尚任意停灵、店铺交接等的呈报 *Peking Police Bureau Inner Five Substation Report on the Internal Equipment Inspection of Nongshenghao Alkali Factory, Establishment of Fumin Soap and Alkali Factory, Arbitrarily Keeping Coffin in Jiaxing Temple, Shop Handover, etc.* 1944. Archive Number: J183-002-28754

北平市警察局关于禁止各庙宇停放灵柩的训令 *Peking Police Bureau Order for Temples about the Ban of Keeping Coffin*, in Beijing Archives Collections. 1946. Archive Number: J181-016-02958.
北平市警察局关于取缔庙宇停柩的呈报 *Peking Police Bureau Report on the Ban of Keeping coffin in temples*. 1948. Archive Number: J181-016-02568.

(d) *Archives Kept in the First Historical Archives of China (Abbreviated to FHA in the Text)*

Memorial to the Throne of Imperial Household Department. 1770. 奏为会查粘修三官庙等处官管寺庙殿宇房间数目事*Memorial to the Throne about the Investigation of the Number of Halls and Rooms in San Guan Temple and other other official temples*. In the. Archive number: 05-0277-031.

(e) *Maps*

Hou, Renzhi 侯仁之 (ed.). 1997. *Beijing lishi dituji* 北京历史地图集 *(Beijing History Atlas)*. Beijing: Beijing Press.
Beiping jiequ fangxiang zhi gailue 北平街衢坊巷之概略 (the *Outline of Peking Street and Alley*). 1937–1938. Peking Special Office.

Secondary Sources

Ahem, Emily Martin. 1973. *The Cult of the Dead in a Chinese Village*. Berkeley/Los Angeles: Stanford University Press.
Ariès, Philippe. 1977. *L'homme devant la Mort*. Paris: Editions du Seuil. Translated by Helen Weaver *The Hour of Our Death* (New York, Alfred A Knopf, 1981).
Bujard, Marianne (ed.). Xu Minglong, Ju Xi (trans.). 2011. *Temples and Steles of Beijing: Preface*. Beijing: National Library Press.
Feuchtwang, Stephan. 1974. *An Anthropological Analysis of Chinese Geomancy*. Taipei: Vithagna.
Freedman, Maurice. 1968. Geomancy. *Proceedings of the Royal Anthropological Institue of Great Britain and Ireland*, 1968: 5-15.
Guan, Xiaojing 关笑晶. 2010. "清代满族的丧葬习俗——从《御制增订清文鉴》谈起" (Funeral Custom of Manchu in Qing Dynasty: From the Qing Dictionary Revised and Enlarged by Emperor). 满语研究 *Manchu Studies* 1: 91–103.
Harvey, David. 2010. The Urban Process under Capitalism: A Framework for Analysis. In *The Blackwell City Reader*, ed. G. Bridge and S. Watson, 2nd ed. Chichester: Wiley-Blackwell.
Ju, Xi 鞠熙. 2017a. 北京的下层寺庙与社区公共空间——以西四北大街双关帝庙的碑文和仪式为例 (Lower Temples and Public Space of Beijing: An example of the Inscriptions and Ceremonies of Shuang Guandi Temple in Xisi North Street), in Marianne Bujard, Luca Gabbiani (ed.), *Temple and its actors*. Beijing: China Citic Press.
———. 2017b. 礼仪的文献记载与生活实践——以清代北京旗人丧葬礼俗为例 (The Funeral Rites of Banner-men in Local Chronicle Records and Daily Life Practices). 徐州工程学院学报 *Journal of Xuzhou Institute of Technology (Social Sciences Edition)*, 1: 1–8.

KANG, Xiaofei. 2006. *The Cult of the Fox: Power, Gender, and Popular Religion in Late Imperial and Modern China*. New York: Columbia University Press.

Liu, Xiaomeng 刘小萌. 2008. *Qingdai Beijing qiren shehui* 清代北京旗人社会 *(Beijing Banner People Society in Qing Dynasty)*. Beijing: Chinese Social Science Press.

Naquin, Susan. 2000. *Peking: Temples and City Life, 1400-1900*. Berkeley/Los Angeles: University of California Press.

Rogaski, Ruth. 2004. *Hygienic Modernity: Meanings of Health and Disease in Treaty-Port China*. Berkeley/Los Angeles: University of California Press.

Wolf, Arthur P. 1974. Gods, Ghosts, and Ancestors. In *Religion and Ritual in Chinese Society*, ed. Arthur Wolf. Berkeley/Los Angeles: Stanford University Press.

Zhou, Jiping 周吉平. 2002. *Beijing binzang shihua*北京殡葬史话 *(Beijing Funeral History)*. Beijing: Beijing Yanshan Press.

Part IV
Healthy or Unhealthy Environments: Techniques, Tools, and Concepts to Reconsider the Relationship Between the Environment, Disease and Health

Chapter 12
Urban Nature: (The) Good and (The) Bad

Nathalie Blanc

Abstract The article discusses the evolution of the relationship between nature and the urban environment during the nineteenth and twentieth centuries in Europe and the United States, a relationship which was mainly framed in terms of public health issues. From the first hygienist utopias, it examines how urban ecologies developed through the analysis of the great figures of urbanism, who are the source of urban theories and utopias. Ultimately, it explores how environments today are redesigned and standardized for health ideals enrolled in the tradition of urban planning. The article argues the following thesis: spatial and urban planning have long been the tools of biopolitics, i.e. political engagement in all dimensions of life in order to monitor, preserve and control it. Public health becomes the objective of biopolitics as the way we tried, from the eighteenth century onwards, to rationalize the problems posed to government practice by the phenomena resulting from an assembly of living beings making up a population: health, hygiene, birth, life and race relations. It was then that some key concepts were shaped (such as population), forms of knowledge (such as statistics and demography) and practices (hygiene, public safety, etc.). These notions are all linked with the development of modern urban planning conceived as a discipline. To act on the territory is to act on the population, categorised as bodies.

Keywords Urban health · Nature in the city · Spatial planning · Biocontrol

12.1 Introduction

The French government's 3rd National Environmental Health Plan (PNSE3) emphasizes the importance of nature for the health and well-being of the French population. Thus, in presenting the PNSE3 to the Council of Ministers on November 12, Minister of Ecology, Sustainable Development and Energy, Ségolène Royal, stated:

N. Blanc (✉)
Université de Paris, LADYSS, UMR 7533, CNRS-Université de Paris, Paris, France
e-mail: nathali.blanc@wanadoo.fr

F. Bretelle-Establet et al. (eds.), *Making Sense of Health, Disease, and the Environment in Cross-Cultural History: The Arabic-Islamic World, China, Europe, and North America*, Boston Studies in the Philosophy and History of Science 333, https://doi.org/10.1007/978-3-030-19082-8_12

"La nature peut être notre premier outil de prévention et notre premier médicament, comme le montrent les bienfaits pour la santé de sa présence en ville et les bénéfices thérapeutiques des jardins en milieu hospitalier (Nature can be our primary tool in prevention and our primary medicine, as demonstrated by the health benefits of its presence in cities and the therapeutic benefits of gardens in hospitals)".[1] And later: "Le Ministère de l'Ecologie soutient, en coopération avec les collectivités territoriales, l'implantation et la valorisation des espaces naturels en ville. Un bilan des connaissances sur ce sujet permettra d'identifier les pistes les plus prometteuses d'aménagements d'espaces de nature au bénéfice de la santé (The Ministry of Ecology supports, in co-operation with local and regional authorities, the establishment and enhancement of natural areas in cities. A review of knowledge on this subject will identify the most promising avenues for the development of natural spaces for the benefit of health)". However, there is no consensus currently on the nature of the relationship between health status and well-being, particularly among urban populations: the issue is complex due in particular to the difficulty in reaching agreement on what we call health and what we call urban.[2] There is also a wide variety of ways of understanding what is meant by nature in urban areas.[3]

The issues raised in the present article concern the relationship between "health" and urban areas, as articulated and qualified as a public health issue in the nineteenth and twentieth centuries in Europe and the United States. It examines how an urban ecology has evolved since the first hygienist utopias, through the analysis of great figures of urban planning, who were responsible for theories and urban utopias. Ultimately, it explores how today's redeveloped and standardized environments are composed for the ideals of health included in this tradition of urban planning. To begin, we will summarize a relationship between nature and urban areas by considering the characterization of public health issues in the nineteenth century and then in the twentieth century. Afterwards we will discuss how the issue of ecology has modified the perception of nature in towns, from a hindrance to a benefit.

12.2 The First Urban Hygienist Utopias

Until the end of the nineteenth century, the causes of disease and infection were looked for in the unhealthy city air, in the urban environment corrupted by putrid exhalations carrying miasmas resulting from the decomposition of organic matter

---

[1] See the website of the l'Association Nationale des Villes pour le Développement de la Santé Publique: http://www.espt.asso.fr/activites/participation-au-debat-public/seminaires-colloques-et-rencontres/357-5e-congres-national-sante-environnement-organise-par-la-sfse.html. Accessed on December 4th 2017.

[2] The collective work by Roderick Lawrence and Gilles Barbey (2015) shows the difficulty of assessing quality of life and health in the environment. See also Lawrence 2006.

[3] I presented the great variability of the perceptions of nature in my thesis *Nature in the town* defended in 1996.

and waste.[4] Writers testify to the catastrophic situation of the working class. Indeed, as shown by Jules Verne (1828–1905) in *Les Cinq cent millions de la Begum* (1879) and Victor Hugo (1802–1885) in *Les Misérables*, in certain districts, human beings, often from the working classes, lived in destitution, overcrowded and enduring the insalubrity of the housing. But the criticism of the physical environment also addressed the behaviour of the inhabitants. Those from the countryside were seen as wise and virtuous, those from towns as dangerous. The appeal of the countryside was therefore linked to the rejection of the town.[5] To a diagnosis implying bad management of putrid areas, the remedy would be spatial: it was necessary to act on the pathogenic areas, to find technical ways to make them healthier. The mobilization of space for therapeutic purposes continues today.

Numerous nineteenth century thinkers reflected on the question. Robert Owen (1771–1858) advocated the creation of gardens and communities with a reasonable population density. It was a matter of favouring man's nature, of placing people "under those [circumstances] which, being congenial to the natural constitution of man and the wellbeing of society, cannot fail to produce that amelioration in their condition which all classes have so great an interest in promoting".[6] For the same reasons, industry would be separated from dwellings by open spaces, "gardens bordered by roads", thus isolating the different activities. In 1822 Charles Fourier also proposed a community in which open spaces were important, notably for reasons of hygiene: "Autant la police de propreté est difficile dans les maisons resserrées et obstruées, comme celles de nos capitales, autant elle est facile dans un édifice où les espaces vacants maintiennent les courants d'air. On éviterait donc ici les vices d'insalubrité, avantage de haute importance (Much as the policy of cleanliness is difficult in the tightly-packed and obstructed houses such as those in our capitals, it is easy in a building where empty space preserves a flow of air. Here, then, we should avoid the vices of insalubrity, a benefit of great importance)".[7] These city projects highlight the need for hygiene in urban areas, thus incorporating numerous proposals concerning the practices of urban cleanliness. In 1840, Étienne Cabet (1788–1856) developed "la vision d'un socialisme étatique (the vision of state socialism)" in *Voyage en Icarie*. Icara, a model town proposed by E. Cabet, is organized according to principles of rationalization of living, which one finds in other utopias and in this case includes the interior of the house – since the author writes that the design of the furniture is a model design, "le plus parfait, sous tous les rap-

[4] Albert Lévy (2012) presents the various aspects of this question.

[5] This is particularly evident in Rousseau, who even assigns value to the most isolated and uncultivated places (the sea, mountains, etc.). According to Jean Viard (1990), and also Keith Thomas (1985), it took the development of cities and urban life to develop a special taste for the countryside and nature. It is thus in a setting away from the countryside, rural nature, which had been created beforehand at the cost of daily work and through a popular knowledge of the natural environment, that this taste for nature, rural and untamed, was built. It was a setting aside of "natural" space for the benefit of town dwellers, a setting aside that would reach its greatest visibility with the creation of the natural parks.

[6] Owen 1972, pp. 59–60.

[7] Fourier 1822, p.566. (The translation into English is by the editors).

ports de la commodité, de la simplicité, de l'économie de temps et de matières (The most perfect one (was chosen) based on the criteria of convenience, comfort, simplicity, saving of time and materials, and finally elegance and charm)"[8] – and hygiene. The measures taken for healthiness, for the free circulation of air, for the preservation of its purity and even for the cleaning of Icara are dealt with at length. This question of hygiene requires a maintained, binding cleanliness, to the point of obsession: "Jamais je ne pourrais te répéter toutes les précautions prises pour la propreté des rues. Que les trottoirs soient balayés et lavés tous les matins, et toujours parfaitement propres, c'est tout simple : mais les rues sont tellement pavées ou construites que les eaux n'y séjournent jamais, trouvant à chaque pas des ouvertures pour s'échapper dans les canaux souterrains. Non seulement la boue, ramassée et balayée à l'aide d'instruments ingénieux et commodes, disparaît entraînée dans les mêmes canaux par les eaux de fontaines, mais tous les moyens que tu pourrais concevoir sont employés pour qu'il se forme le moins de boue et de poussière que cela est possible (I will never be able to give you an idea of all the precautions taken to ensure the cleanliness of the streets. Not only are the sidewalks swept and washed every morning and kept perfectly clean, but the streets are paved or constructed in such a way that water never collects on them, for there are frequent openings through which the water goes into underground canals. Not only does mud, collected and raked up by ingenious, useful instruments, disappear into the same canals, pushed along by water from the fountains, but all imaginable means are used so that the least possible mud and dust collect in the first place)".[9]

Benjamin Ward Richardson (1828–1896), an English physician who inventoried ways of fighting against the deplorable sanitary state of large towns in what was initially a communication to the 1875 Congress of the Social Science Association, imagined a town, Hygeia, that even personified urban hygiene. It is a town whose hygiene is guaranteed "by the character of the houses, which ensures an equal distribution of the population".[10] These houses leave the streets sunny. The streets are considered in relation to air circulation and are planted with trees on both sides. The public facilities are "surrounded with garden space and add not only to the beauty but to the healthiness of the city".[11] In fact all the elements that are part of the construction of the town are considered from a health perspective. The kitchens are installed "immediately beneath the garden roofs". They are "lighted to perfection, so that all uncleanliness is at once detected. The smell which arises from cooking is never disseminated through the other rooms of the house".[12] Unlike the other utopias, the Hygeia model was essentially to demonstrate the layout of a healthy town. Urban hygiene is the sole value that determines the location and design of private homes and public buildings.

[8] Cabet 2003, p. 58.
[9] Cabet 2003, p. 36.
[10] Richardson 1876, p. 19.
[11] Richardson 1876, p. 20.
[12] Richardson 1876, p. 24.

Jean-Baptiste Godin (1819–1888) also writes: "Dans le palais social, la lumière doit pénétrer partout avec abondance : pas de cabinets noirs, pas d'endroits obscurs ; la clarté et l'espace sont les premières conditions de l'hygiène. Aussi, tout est largement éclairé au Familistère, comme tout est largement pourvu d'air et d'eau. L'espace consacré aux communs, la grandeur des cours, les jardins et les promenades qui entourent ce palais, tout concourt à donner libre-accès partout à l'air et à la lumière (In the social palace, light must penetrate everywhere in abundance: no gloomy rooms, no dark corners, clarity and space are the foremost conditions for hygiene. Equally, everything is comfortably lit at the Familistère, as everywhere is adequately provided with air and water. The shared space, the size of the courtyards, the gardens and the promenades that surround this palace, all contribute to giving free access everywhere to air and to light)".[13] The organization of this Palace, and the principles of hygiene that were put into effect, offered the inhabitants the possibility of having a useful life. In 1874, Jean-Baptiste Godin even wrote that "le palais social au contraire, appelle ses habitants à la vie utile, parce que leur activité est directement productive (the social palace, on the contrary, calls its inhabitants to a useful life, because their activity is directly productive)".[14]

As State intervention in the field of urban policies developed progressively, especially from the eighteenth century onwards, policies for the control of nature were put into place.[15] All these interventions fall within an intention of public utility and effectiveness, "of prophylaxis of the urban environment".[16] We ventilate, construct wide roads, pave the streets and allow sunlight to penetrate the streets and houses; we create planted walkways and gardens and move cemeteries and slaughterhouses out of town; we open the town to its rural environment, and the old fortifications are planted.[17] The air, the water and the sun circulate and purify the town. The vacant areas, the gardens, the woods and the forests separating the built features are an opening for the circulation of air and the penetration of rays of sunlight and correlatively arrange the spatial organization of the city. However, they are also places for walking and relaxation. Finally, the separation of the buildings by these same natural elements is a means of assigning specific functions (factories on one side and dwellings on the other, etc.) to areas. Between the lines of the Utopian texts, lies the idea—which has its continuation in our epoch in the diagnosis of the crisis in large housing developments—that space produces a social order, even more well-behaved behaviours. The utopians thus sought to produce a healthy society by developing space as a vector for healthiness. The introduction of urban hygiene was also a way

[13] Quoted by Choay (Choay 1965, pp. 142–144) and by Roncayolo & Paquot (Roncayolo and Paquot 1990, pp. 81–89). (Translation by the editors.).

[14] Quoted by Choay 1965, p. 142.

[15] According to Jean-Claude Perrot (translated by the editors), "from the functional perspective, it is a question of trimming, grafting, and modifying nature. Thus a populationist policy through the control of empiricists (midwives, surgeons, apothecaries), the surveillance of epidemics, the birth of social medicine (free consultations)" (Roncayolo and Paquot 1990, p. 39).

[16] Luginbuhl 1992, p. 28.

[17] Guillerme 1983

of restoring productive energy to the workers. On the other hand, society, and more particularly the city, underwent a medicalization from the end of the eighteenth century.[18] This is reflected in the use of medical terms or those borrowed from the natural sciences in the analysis of urban functions. This practice led certain thinkers and utopians, in particular Benjamin Ward Richardson, to develop urban analyses that led to understanding the city as a living being and giving it a nature (1876; 2004). Unlike the first urban utopias (e.g. Thomas More's *Utopia*), these texts have little interest in the exploitation of the surrounding nature in the town. For these thinkers, it was a question of elaborating the selection of the healthy natural elements from those that are unhealthy and of constituting an urban model for the control of natural hazards. There was thus a change in perspective in the way of considering the relationship between nature and the town in the utopias of the nineteenth century compared to the first urban utopias: from the exploitation of nature and its integration into the production process of the town, we move to a selection and an effort to remove the disadvantages of nature by means of the development of techniques in order to create a town that is in keeping with the needs of man and capable of ensuring social progress.

12.3 Nature Between Utopia and Technology

Of the authors of the nineteenth-century utopias, very few succeeded in realizing their ideal cities. The true control of the effects of nature in urban areas combined with a technical implementation employing the city as a network occurred in large towns around the 1850s. In France, this movement would take place, notably with the transformation of Paris under Haussmann and then prefect of the Seine department (1853–1870). The transformation of urban areas would allow the development and penetration of urban techniques that promoted not only comfort and hygiene but also the introduction of a controlled nature. According to André Guillerme, "les

[18] Jean-Claude Perrot notes in his analysis of urban history that "la perception des fonctions urbaines résulte d'un glissement analogique depuis des domaines de connaissance déjà constitués. Le bagage épistémologique provient des sciences de la nature animée. Avant 1750, la philosophie des sciences était cartésienne, puis newtonienne; après 1750, elle devient vitaliste. Elle porte à saisir la ville comme un être vivant et lui donne une nature (the perception of urban functions is the result of an analogical shift from already constituted areas of knowledge. The epistemological background comes from the animated natural sciences. Before 1750, the philosophy of science was Cartesian, then Newtonian; after 1750, it became vitalistic. It led to understanding the town as a living being and giving it a nature)". The author adds that these analogies are still common today: "sous nos yeux, l'analyse des fonctions est en train de céder lentement la place à celle des structures et le vocabulaire biologique (artère, cœur, croissance) à l'outillage de la mécanique astrale (axes de circulation, pôles de développement, ville satellite, nébuleuse urbaine..) (before our eyes, the analysis of functions is slowly giving way to that of structures and biological vocabulary (artery, heart, growth) to the tools of astral mechanics (axes of circulation, centres of development, satellite town, urban nebula, etc.)". (Roncayolo and Paquot 1990, p. 38, translation by the editors).

techniques permettent de soustraire la ville aux effets des éléments naturels : contre le feu de l'incendie, contre l'eau de l'inondation, contre l'excès de chaleur et de froid, contre la pluie et le brouillard, contre la poussière, les mauvaises odeurs et les miasmes, le génie urbain débarrasse la ville de sa mauvaise nature, la "désaisonne" (the techniques make it possible to remove the town from the effects of the natural elements: from the flames of the fire, from the waters of the flood, from the excess of heat and cold, from the rain and the fog, from dust, bad odours and miasmas, urban engineering rids the town of its bad nature, 'de-seasons' it)".[19] The introduction of natural vegetation, at the instigation of Napoleon III, contributed to this technical implementation of the city in order to control urban areas and improve urban well-being. Haussmann, in charge of the application of the Imperial scheme, entrusted the engineer Adolphe Alphand with the task of setting up a large Promenades Department.[20] In 1867, Alphand reported on his mission in a work that is still a reference today: *Les Promenades de Paris*. While the implantation of this vegetation was part of an aesthetic vision for urban areas, it also fell within a public health strategy.[21] This was assured at the same time by the establishment of technical networks, which would ensure the conditions for urban growth.[22] It was not only urban hygiene or aesthetics but also the foundation of true urban comfort. The woods and parks are there as places for walking and for relaxation. The implementation of technical and urban operations that took shape during the transformation of major towns between 1850 and 1890 would contribute decisively to the rise of urban planning.

The utopias, which also contributed to the elaboration of urban planning as a body of practice and public policy, developed between the end of the nineteenth and the beginning of the twentieth century, still focusing on urban hygiene and the myth of a town reconciled with man and nature. This was especially true in England, with its garden cities mixing urban and social reform. These utopias, unlike those described above, would result in urban realizations but not as envisioned by the originator of the garden city, Ebenezer Howard.[23] This thinker proposed a limited

---

[19] Guillerme 1994. Translation by the editors.

[20] Stefulesco 1993

[21] Luginbuhl (1992, p. 43, translated by the editors) explains that "l'hygiénisme et l'esthétisme contenus dans l'urbanisme du second empire, empreints du sens moral déjà souligné, ne sont pas pour autant dénués d'un sens social: la ville s'organise en effet dans une différenciation socio-spatiale, comprise également dans des espaces verts (...) Si le besoin de beauté et de santé, c'est à dire de nature, est effectivement un des arguments de l'organisation formelle et sociale de la ville, il reste dans les pratiques et les discours fortement teinté d'une idéologie paternaliste et ségrégative (the hygienism and aestheticism contained in the urban planning of the second empire, marked by an already emphasized moral sensibility, are nevertheless not devoid of a social sense: The town organizes itself in a socio-spatial differentiation, included also in the green spaces (...) If the need for beauty and health, that is to say nature, is indeed one of the arguments for the formal and social organization of the town, it remains so in practices and discourses strongly tinged with a paternalistic and segregative ideology)".

[22] Roncayolo 1990

[23] Choay 1965, pp. 277–290; Roncayolo and Paquot 1990, pp. 252–255; Ebenezer 1946

urban area in a new relationship with nature and thus fits into the history of the English taste for nature in towns.[24] With the garden cities, he launched a major theme in contemporary urban planning. He was self-taught, militant and socialist. In 1898, he published *To-morrow: A Peaceful Path to Real Reform*,[25] a new utopia in which the theory of the Garden city was presented and which was to become reality from 1903. Garden cities are the spatial implementation of relationships between town and nature. They are, as in older towns, enclosed, demarcated, limited and understandable geographically and therefore contrary to nature. They constitute the meeting of the town and the countryside in terms of urban planning and also the ways of life that they permit. This mixture of town and countryside can also be seen in the so-called progressive utopias of the nineteenth century, and in some of the later developments of modern urban planning, especially the new towns. They therefore include in the urban design, in the elaboration of its morphology, the forms of nature: trees are an indispensable element of the garden city, along with hedges, lawns, verdant tranquillity and climbing plants, which unite with the stone and also those very cultivated forms of private nature, gardens. The garden city includes a conception of the town that maintains a nature/culture dialectic, a very assertive relationship between what is the town and what is not.[26] It is thus a question of asserting the harmony between nature and the town.[27]

12.4 Urban Planning in the Twentieth Century: From the Respect of the Site to an Abstract Nature

Urban planning took shape at the confluence of these two trends, the social utopias on one hand and on the other the technical processing of large cities towards the end of the nineteenth and the beginning of the twentieth century. The term *urbanisme* entered the French language around 1900–1910.[28] The individualization of urban functions and the zoning of areas were bound together at the heart of the urban

---

[24] Thomas 1985, p. 277.

[25] Re-published as *Garden-Cities of Tomorrow*, new ed. with prefaces by Sir F. Osborn & Prof. L. Mumford, Faber & Faber, London, 1946. Translated into French by L. E. Crepelet: *Villes-jardins de demain*, Tietsin press limited, China, 1902, pp. 15–26, 83–84, 77–79, 81, 128, 134.

[26] IAURIF 1978

[27] Roncayolo 1983, p. 149. But the principles of garden cities as described here would not be respected when they were realized. In England, these garden cities would be suburban cities quickly adsorbed into a continuous urban fabric. While in France, under the pressure of various constraints, "la cité-jardin n'est qu'un quartier de l'agglomération, une réponse à la question du logement et de l'habitat salubre (the garden city is only a district of the agglomeration, an answer to the question of clean and healthy housing)", gradually what was a project of individual dwellings then became a series of collective buildings. We drift then "du mythe de la nature vers l'équilibre des formes et des masses (from the myth of nature towards the balance of forms and masses)".

[28] Roncayolo and Paquot 1990, p. 229.

project and of the urban modelling of the modern town in space. From a geographical perspective, zoning allows certain areas to be set aside, in particular those that have to be preserved as such, a particular allocation of certain areas (residential, industrial, etc.), and also to preserve free spaces planned for the purposes of urban hygiene, relaxation and leisure activities.

Zoning and its implementation would allow construction in urban areas to be regulated according to principles of urban hygiene. It was then a matter of regulating the distance between the buildings and the allocation of urban areas. Some urban planners would divide up the urban space to allow the sun and the air to penetrate. Aesthetic concerns were of enormous importance, as the architects, like the urban planners, had mostly trained as artists. Most urban planners, close to or coming from the modernist trend advocated by the CIAM (Congrès Internationaux d'Architecture Moderne) that Le Corbusier founded in 1928 in Switzerland, sought to organize the urban elements. Critical of towns,[29] these thinkers wanted to establish a space that enhanced the conditions of human existence by applying aesthetic principles that take into account these urban hygiene needs and also the interest there is for urban comfort to rationalize the production of buildings and the town. Walter Gropius (1883–1969), who had an influence comparable to that of Le Corbusier on architecture and contemporary town planning, and who had similar conceptions of architecture between 1920 and 1930, explained that free space allows the air and the sun to reach the apartment: "[The ground floor windows] command a clear view of the sky over the broad expanses of grass and trees which separate the blocks and serve as playgrounds for the children".[30] While Le Corbusier (1887–1965), for whom architecture and urban planning were inseparable, represented a vision of the town covered in greenery: "Au lieu de tracer des villes en massifs quadrangulaires avec l'étroite rigole des rues cantonnées par les sept étages d'immeubles à pic sur la chaussée et encerclant des cours malsaines, sentines sans air et sans soleil, on tracerait en occupant les mêmes superficies, et avec la même densité de population, des massifs de maison à redents successifs serpentant le long d'avenues axiales. Plus de cours mais des appartements ouvrant sur toutes les faces à l'air et à la lumière, en donnant non pas sur les arbres malingres du boulevard actuel mais sur des pelouses, des terrains de jeux et des plantations abondantes (Instead of laying out cities in quadrangular blocks with the narrow channels of the streets constricted by the seven-story apartment buildings that drop right down to the pavement and wrap around the unhealthy wells of airless and sunless courtyards we will lay out (in the same surface area and with the same surface areas) apartment blocks with successive indents snaking along axial avenues. No more courtyards, but apartments opening on every side to air and light, and overlooking not on the sickly trees of the today's boulevards but lawns, playgrounds and luxuriant

[29] Hygienism has inaugurated a unitary view of the town as a space for the generation of diseases: the urban territory appears as a medical/health unit (Murard and Zylberman 1985).

[30] Gropius 1935, p. 102.

vegetation)".[31] Thus, according to F. Choay,[32] nature has again been taken into account. Instead of becoming a barren stonescape, the town is a vast park. The urban agglomeration is treated as a green town with sun, space and greenery. The buildings are laid out in the town behind a filigree of trees. The pact is signed with nature.[33]

This image, where nature restricted to certain elements of climate, air and sunlight, provides hygienic living conditions, fails to address the historical and social conditions for the production of unhealthiness. It is no longer in response to a specific urban situation, as for the first urban utopias, but abstractly, for themselves, that air and light are desired. In addition, through all these texts on urban hygiene, the repetition of terms such as "air", "sun" and "greenery" does not hide the absence of the diversities of the climate in favour of a sort of climatic standard, which would meet an urban hygiene standard. Moreover, these thinkers' urban planning was essentially architectural. They included all the functions of the town in a single very tall building, leaving the urban space outside with no function other than to be a healthy space, with the additional remit of traffic. This, when it was automobiles, would be realized on roads that would join the buildings together. Other flows would be possible at other levels of the town.

The buildings would be very tall and thus free up a large amount of ground while maintaining high population densities. They were no longer buildings but residential units set up in relation to the sun in a leafy park. This land, which was assured beforehand to be flat, because "wherever civilization intensifies, flat ground provides the usual solutions",[34] was the plan that supports buildings where "from the fourteenth floor, the calm is absolute and the air pure". There is therefore a distancing from the ground, which remains only as a green lawn. According to Jean Castex, Jean-Charles Depaule and Philippe Panerai, "le terrain n'est qu'un plateau de représentation pour un objet machine-sculpture, déterminé abstraitement (the terrain is only a stage for an object, an abstractly determined machine-sculpture)".[35] Tony Garnier, Walter Gropius and Le Corbusier worked in the same perspective: to create a "dwelling house", "dwelling model" or "housing unit" corresponding to the needs of human beings. These needs are identical from one individual to another and do not take into account cultural or sociological differences. They are essentially understood from the body and from its conformity. They are medical. This conception of a standard creation included all the urban elements and had to organize the building as well as the interior of the dwelling. Le Corbusier was in our opinion the most explicit: "rechercher l'échelle humaine, la fonction humaine, c'est définir les besoins humains. Ils sont peu nombreux ; ils sont très identiques entre

[31] Le Corbusier 1986, pp. 61–63.

[32] Choay 1965, pp. 233–249.

[33] Le Corbusier's urbanistic work is of little significance from the point of view of realizations, including never executed blueprints. However many of his books had a great influence on a whole generation of architects and town planners.

[34] Choay 1965, p. 243.

[35] Castex, Depaule, Panerai 1980, p. 140.

tous les hommes, les hommes étant tous fait sur le même moule depuis les époques les plus lointaines que nous connaissons. Le Larousse chargé de nous donner une définition de l'homme nous donne trois images pour démonter celui-ci sous nos yeux : toute la machine est là, carcasse, système nerveux, système sanguine (to search on the human scale, for human function is to define human needs. They are not very numerous; since man had been made out of the same mould from the earliest times known to us. The Larousse (dictionary), whose task it is to furnish us with the definition of man, gives us three images to dismantle him before our eyes: the whole machine is there, carcass, nervous system, and circulatory system)".[36] It is then simply a question of improving the human tools, to have a typical house tool: as "une maison est une machine à habiter. Bains, soleil, eau chaude, eau froide, température à volonté, conservation des mets, hygiène, beauté par proportion (a house is a machine for living in. Baths, sun, hot-water, cold-water, warmth at will, conservation of food, hygiene, beauty in the sense of good proportion)".[37]

The ideas of the American antiurban trend crystallized into a new movement that borrowed partially from European urban planning and its evolution.[38] Too radically utopian to have lent itself to realization but called on however to mark the thinking of a section of American sociologists and *town planners*, this model was elaborated under the name *Broadacre City* by the great American architect F. L. Wright in three successive publications in 1932, 1945 and 1958.[39] This model established nature against artifice. For F. L. Wright, large towns alienate human beings by their artifices, and only nature can contribute to the harmonious development of the "entire person". This relationship with nature is mentioned in special terms: "Man exchanged his original trade with rivers, woods, fields and animals, for the constant agitation, the taint of carbon monoxide rising to a rented aggregate of cells upended on hard pavements".[40] The result "The citizen if properly citified has lost sight of the true aims of human existence and accepted substitute aims. His unnaturally gregarious life tends toward the promiscuous, blind adventure of a crafty animal; some form of graft, the febrile pursuit of sex as 'relief' from factual routine in prevailing uproar of mechanical conflict".[41]

Other research has attempted to evaluate urban planning and the place of nature in urban areas, by means of psycho-ecological studies. These studies caused a revolution in how to take mental health into account in towns (Levy 2012). The central

---

[36] Translation in Butler and Parr 2005, p. 32.

[37] Le Corbusier 1986, p. 95.

[38] Many American thinkers on urban planning travelled to Europe as part of their training. However, beyond the transfer of ideas and similarities in the relationship between nature and the town, it must be borne in mind that the place of nature in Europe and North America and the history of European and American towns are very different. European nature is the pastoral, bucolic and possibly cultivated nature, whereas the American nature emerges from untamed nature, the wilderness.

[39] Choay 1965, pp. 297–311.

[40] Wright 1945, p. 3.

[41] Wright 1932, p. 25.

concept here is that of mental hygiene. It has emerged that public health and a rational division of urban space cannot ensure the pleasurable living necessary for mental health and therefore physical health. Leonard Duhl, a psychiatrist, and "activist for a global approach to urban planning", included "the study of diverse environmental factors likely to affect behaviour" in his psychiatric approach.[42] He criticized, in particular, garden cities in 1963 as an example of humanist urban planning: "But not all will find a garden city a welcome relief from the chaos of a modern city. For some people the design of a clean new city may mean not peace and serenity but boredom, a lack of colour and vitality that old cities have provided in chaos".[43]

The study of these texts that are critical of contemporary urban planning leads us to a reflection on how to envisage the development of the urban living environment and, furthermore, the way in which the conversation on the theme of nature and its role in towns highlights a particular vocabulary. We notice that the nature to which it makes reference is thus solely plant and climatic. Animals are absent from urban areas. Urban hygiene was probably one of the normative principles of these rules which probably reduced the place of animals and increased that of plants. Indeed, from the eighteenth and nineteenth centuries, animals generating insalubrity and excrement were pushed out of town and ultimately only remain in domestic space: "La respiration des animaux, les fermentations, les combustions, et enfin les effluves de toutes espèces corrompraient bientôt l'air de l'atmosphère et le rendraient mortel à tous les animaux si la nature n'avait pas trouvé un moyen de ramener l'air corrompu à l'état d'air commun : c'est la végétation des plantes (Animal respiration, fermentation, combustion, in short, the effluvia of every kind, would infect the air of the atmosphere, and render it mortal to all animals, if nature had not a method of reducing the corrupted air to the state of common air: the vegetation of plants)", wrote Lavoisier.[44] "Dans l'ordre du vivant, le règne végétal supplée la neutralité du minéral pour réduire les nuisances du règne animal. D'où les plantations le long des boulevards qui parent la ville de « pompes aspirantes et foulantes mues par l'air », régulant ainsi la circulation verticale des vapeurs » (In the order of living things, the vegetable kingdom supplements the neutrality of the mineral in order to reduce the nuisance of the animal kingdom. Hence the plantations along the boulevards which adorn the town with "air driving pumps", thus regulating the vertical circulation of the vapours)".[45] We see, therefore, that in the very development of the urban environment, there are specific concerns in the life sciences which led to the application of technical arrangements likely to later transform urban life.

At the same time as cities were being developed as a result of technical implementation from the eighteenth century onwards, the relationship with nature also evolved, with nature being perceived as playing two principal and opposing roles: the first as a vector of disorder and insalubrity and the second beneficial and as a vector of hygiene and purity. The positive side of nature contributes to the creation

[42] Choay 1965, pp. 379–384.

[43] Duhl 2011, p. 137.

[44] Lavoisier 1776, p. 147.

[45] Guillerme 1983, p. 202.

of the idea of the city, in the sense of an exemplary socio-spatial organization, of a model town, whereas urban engineering constructed the town by removing the negative side of nature from it. In fact, it is not the inhabitants who would apply the rules of urban hygiene. It is the buildings, the way in which they were placed in the space that would compel them to do so and also allow them to have opportunities to ensure the hygiene of their home. The nature/town relationship thus affects the general organization of the town, the layout of the streets, the density of the buildings and their height and also the dwellings down to personal space.[46] While the opening of the town to the countryside, which was advocated in many town planning projects, corresponded to the implementation of urban public health, it also consisted in reconciling the advantages of rural life with those of urban life. The aim was to make the benefits of the countryside available to urban dwellers. The organization of the workers' housing with, in the case of garden cities, the provision of natural areas for everyone can be paralleled with the establishment of protected sites and nature reserves: it was a question of building places of peace and contemplation in the very heart of the town and urban affairs.[47] In general, this urban hygiene policy targeted a particular category of inhabitant, those who did not know how to "live cleanly", especially the poor and the workers. All the urban models, whether they emanate from industrialist utopias or twentieth-century architectural projects, and those of the CIAM in particular, aimed to teach people to live through the affirmation of a positive urban space. It is therefore a materiality that would allow harmonious social development, fruitful from the point of view of the productivity of the worker.[48] These urban hygiene policies were part of an ideology of progress. It is therefore not surprising that these modelling texts have largely contributed, on the ideal level, to the development of social housing. The implementation of nature in urban areas is part of the idea of a moral order. Whether garden cities, industrialist utopias or urban projects of the Second French Empire, these projects or realizations link social order, urban hygiene, nature and beauty in the same urban plan.[49] Finally, the development and realization of these urban models would establish standards, more precisely specified in the second half of the twentieth century and notably in the construction of ZUPs (priority urban development zones) in France in the 1970s. The implementation of these standards implies that the introduction of nature into towns can be based on a combination of natural spaces quantified according to the number of inhabitants, their sex, their age and their professional activity.[50] The delimitation of the different functions by zone would allow space to be given to

[46] See Special Issue 3 "Machines au foyer" of *Culture technique* published in September 1980 by the Centre de recherche sur la Culture technique.

[47] Luginbuhl 1992, p. 44.

[48] It is still a logic of the milieu, used by urban planners. Detached from a global reflection on society, they base social changes on the context of the built environment that constitutes their only field of intervention.

[49] Luginbuhl explains that Haussmann's achievements in particular are tainted by a segregative and paternalistic ideology (Luginbuhl 1992, p. 43).

[50] IAURIF 1979

the green areas which "previously had a strong visual meaning and had to offer a domesticated image of nature" and "must now compensate for the lack of untamed nature and free space". Nature was introduced into urban areas in a dual ideological and material environmental dimension.

In short, spatial and urban planning have long been the tools of biopolitics that is governing the relationships of living bodies with one another and with themselves. As Michel Foucault said, from the eighteenth century onwards, there was "une sorte d'étatisation du biologique (a kind of biological nationalization)", that is, a political investment in life in all its dimensions in order to monitor, preserve and regulate it. Public health thus becomes the objective of biopolitics defined by the author as "la manière dont on a essayé, depuis le XVIIIème, de rationaliser les problèmes posés à la pratique gouvernementale par les phénomènes propres à un ensemble de vivants constitués en population : santé, hygiène, natalité, longévité, races…(the way in which attempts have, since the eighteenth century, been made to rationalize the problems posed for governmental practice by phenomena characteristic of a group of living beings constituted as a population: health, hygiene, natality, longevity, races …)".[51] The relationship of opposition to the State of some thinkers on alternative utopias notwithstanding, Michel Foucault's reflections remain relevant to understanding these urban utopias as playing a role in regulating the living body, even if not in the way that the State may wish. The origins of this position date back to the second half of the eighteenth century, with the formation of key notions (such as those for population), on forms of knowledge (such as statistics and demography), and on practices (urban hygiene, public safety, etc.).[52] These notions are all connected with the formation of modern urban planning conceived of as a discipline. Acting on the territory means acting on populations categorized as bodies.

12.5 Towards Ecological Urban Planning

There are many health risks for twenty-first-century populations that are no longer the products of a miasmatic environment or decay but are the results of a morphology and a materiality (biological, physical and chemical), from the level of the town buildings to the region. A reversal is produced: public health awareness, the doctrine of modern urban planning, is no longer in line with the idea of progress. A "clean" town is no longer even a healthy town. Excess cleanliness in urban areas is accompanied by the proliferation of polluted soils, significant air pollution and sometimes open-air garbage dumps.

The principles which have been at the foundation of the "modern" town have brought about major dysfunctions. Functionalist town planning has neglected the quality of daily life and the quality of architectural and urban landscapes, feeding on

[51] Foucault 2004a, p.24, (and for the translation of this excerpt into English, Garrett, Jotterand, Ralston 2012, p.101).

[52] Foucault 2004b

strong patrimonial reactions. Cultural identities, from which the progressive movement has sought to free itself, are once again a central feature of the town and its policies, as well as its competitiveness.[53]

Although the profession of urban planning was not completely acquainted with this view, changes of perspective asserted themselves during the 1990s.[54] Several European initiatives took shape during the first *European Conference on Sustainable Cities & Towns* in 1994, and the text of the Aalborg Charter is contrary in every respect to that of the Athens Charter.[55] The Aalborg Charter calls for transversal urban policies that include the environmental, ecological and social impact of development on several levels. It emphasizes the potential and power of towns to solve a number of environmental and social problems. It invites them to fully reappropriate local political space that of urban democracy.

Ecological town planning is part of the renovation of the remit and procedures of town planning, inviting nature into its very heart on numerous occasions for equally diverse reasons. It is indeed a question of integrating the modern town and the productivist project on the basis of a redefinition of the urban landscape in artificial components as well as in natural ones. Lately the great temptation has been to replace civil engineering with the implementation of ecological engineering and to replace technical urban systems with quasi-natural ones. The urban landscape has to become as efficient from an ecological point of view as rural areas. Consequently, although urban planning, which was born a century ago, is beginning to emerge, in the Western context of course, from the grip of hygienism and functionalism, when it considers the environmental and social question, it remains difficult to think of inhabiting it other than in the frameworks provided for by modern urban planning.

Recently, towns have been conceived of and built by professionals in relation to their functions and uses, ignoring the fact that they are an environment for living and fulfilment, especially for human beings who are not only functional but alive and sensitive. Urban citizens need to live their perceptions and sensitivities, to enter the world beyond the productive sphere alone. Urban gardening, escaping to suburban environments, and green consumerism speak volumes about the frustrations that life can generate in towns. In an era of eco-technologies, is it possible to reflect an ecology of existence, including those dimensions that are immediate, biologically sensitive, but also imaginative and thus permeated by anxieties about global ecological dead ends? Beyond the challenges of environmental rationalization to which ecological issues are often reduced, ecology raises identity and social issues. The beauty of life or, conversely, the multiple degradations of the environment that affect everyday life are central to the construction of identities, whether urban or individual. Among these initiatives, which call on urban planning to transform itself

---

[53] See, for example, the *Green Paper on the Urban Environment*, a Communication from the Commission to the Council and Parliament, Brussels, 1990.

[54] The white paper by the Architects' Council of Europe: *Europe and Architecture Tomorrow.*

[55] The Charter of the Society of European Spatial Planners, 2003. Accessed online on December 4th 2017: http://www.demo.ba.itc.cnr.it/RE/Documenti/The%20New%20Charter%20of%20Athens%202003.htm.

and to take account of issues of health and well-being, the question of gardening has become a major issue at the crossroads of multiple legitimacies (food, biodiversity, social bonding, permeabilization of the ground, etc.) at different levels allowing another relationship to the land and another town (town-nature) to be imagined.

In this new context, where the environment has become problematic, issues in terms of health and care have evolved. Practices are emerging in hospital settings that more generally reflect a change in the relationship with nature. Ecopsychology, for example, refers to a discourse on the psyche in relation to the Earth and requires us to look at the human-nature relationship.[56] In 1960, psychiatrist and psychoanalyst Harold Searles published *The Nonhuman Environment*, which focuses on the influence of the environment, neglected in psychoanalysis. Psychotics are very sensitive to even small changes in the environment. Harold Searles deduced that "the non-human element of the human environment represents one of the most fundamental constituents of psychic life. (…) there is within the human individual a sense, whether at a conscious or unconscious level, of relatedness to his nonhuman environment (…) it is a source of ambivalent feelings to him, and that, finally, if he tries to ignore its importance to himself, he does so at peril to his psychological well-being".[57] In 1963, Robert Greenway, a psychologist, published an essay on the human-nature relationship, seeking to unify the fields of ecology and psychology and set up nature immersion workshops with his students. While psychiatric facilities have, on a practical level, encouraged patients to visit and become active in their surroundings, sometimes being equipped with farms, these practices were not theorized. These first considerations have enriched theoretical developments in psychoanalysis, initially centred on the intrapsychic, and then open to a conception of an increasingly broad attachment. The term "ecopsychology", however, was invented by Theodore Roszak in 1992: a historian and writer, he was involved in a discussion group including psychologists at the University of Berkeley and later published *The Voice of the Earth: An Exploration of Ecopsychology*, a second collective work was published in 1995. Ecopsychology is associated with the ecological movement, which is based on the ecosystemic analysis of dynamic interdependencies between forms of life and the environment. It is also linked to ecofeminism and the educational movement that promotes immersion in nature in order to restore authentic relationships. It is equally linked to the antipsychiatric movement, which considers people's suffering to be the symptoms of dysfunctions in their environment.

Among the social practices associated with the development of the links between psychology, psychoanalysis and the environment, hortitherapy, included in therapeutic protocols for people suffering from physical or mental health problems, is interested in the action of gardening on the body, the intellect, the psyche and the mind. Hortitherapy was pioneered in Canada and America. It then spread to the

[56] The following observations are the product of a workshop organized by LADYSS in the context of a reflection on cultural ecosystem services and concerning the relationship between health and environment: The relationship to living beings as a condition of habitability, June 4, 2015, National Museum of Natural History (MNHN).

[57] Searles 1960, p. 6.

United Kingdom and Japan and is growing in Europe. For the Canadian Horticultural Therapy Association (CHTA) and the American Horticultural Therapy Association (AHTA), the definition of hortitherapy dates from 2008: “The use of plants by a trained professional as a means of achieving pre-established clinical objectives”. For example, the Centre Hospitalier Universitaire de Nancy has a garden devoted to memory. For hospitals, such as the Centre François Baclesse in Caen (where French hortitherapy was born), the therapeutic garden can also be a way of making the hospital more hospitable. The garden in Caen, still under construction, includes seven different settings: a forest, zen, aquatic, one for children’s games, one for relaxing, one for sunbathing and another more shaded with a pergola. Depending on mood, the season and the weather people are free to go where they feel best, either alone or with someone else. This garden is in response to a request from the patients and the healthcare team.

These developments correspond with a great deal of research, notably on the well-being of the residents, quality of life and the place of nature.[58] The perceptions and practices of the latter highlight singular elements of language in the context of a survey carried out in the 1980s in the large housing developments in Rennes. On the basis of numerous surveys, it appears that the description of nature in the town is often a description of so-called natural and plant elements (such as birds and animals or plants: grass or trees) and even urban areas such as parks and play parks. Nature is a resource against the torments of daily life. It is a practical resource: “il n’y a rien qui m’évoque la nature (en ville), la seule chose que je veux, c’est d’aller de ce côté là (l’enquêtée désigne l’autre côté de l’appartement) et regarder la campagne mais sinon je ne regarde pas là-bas (elle désigne la ZUP) (There is nothing that reminds me of nature (in the town), the only thing I want is to go that way (the respondent indicates the other side of the apartment) and look at the countryside but I don’t look there (he means the ZUP)” or “ pour s’évader un peu il n’y a ici que la télé et les jeux, alors il faut partir dans la nature, c’est autre chose, on ne peut pas rester ici enfermé dans les tours, je suis capable de faire des kilomètres en marchant dans la nature et après je suis ressourcée (to escape a little there is only the TV and games, so you have to go out into nature, it’s something else, you can’t stay here locked up in the tower blocks, I can walk miles in the countryside and afterwards I’m revitalised) or “dès qu’il y a un peu d’herbe, c’est déjà la nature, on peut s’oxygéner (as soon as there is a bit of grass, you have nature, you can breathe)”. Nature is isolation from others “la nature, c’est vivre à la campagne avec les animaux, et pas de problèmes avec les autres habitants (nature is living in the countryside with animals, and no problems with the other residents)” because here “on est rejetés (you are left alone)”, or it is another way of living either in the countryside or in a “lotissement neuf (new housing estate)” where “chacun est bien chez soi (everyone is comfortable at home” with greenery. We agree with J. Palmade’s analysis (1977), which explains that the only spaces that can be appropriated by this population, nature and housing are always located in the divide between other

[58] From the 1980s, geographers developed many works related to quality of life and well-being (Bailly and Racine 1988).

spaces and the town. This affirmation is reinforced by the appreciation that certain inhabitants have of their neighbourhood compared to the rest of the town: "Je ne connais pas la ville, la ZUP ce n'est pas la ville, la ville, c'est le centre (I don't know the town, the ZUP isn't the town, the town is the centre)".

In interviews about their relationship with nature, individuals often talk about the issue of well-being and talk about revitalisation, thinking that plants protect them from pollution defined here as what is not in its place. In other words, the inhabitants use the natural spaces with a view to their well-being, i.e. to satisfy their requirements for a good life. It is an ethic/aesthetic. The countryside thus brings the perception of nature. Perceptions of the town are negative and draw one towards evil, while perceptions of nature and the countryside are positive and draw one towards good. The sensation of well-being experienced in nature, of which Ernst Bloch speaks in particular, could well refer to the representations of the Garden of Eden and thus consecrate an age-old concept.[59] Until the end of the eighteenth century, this relationship with nature was constant and unaltered.[60] Jean Viard, commenting on nature in Jean-Jacques Rousseau, and later in the Romantics, writes: "L'homme n'a pas à être soumis au rythme de la technique et de la science, il se doit à ses rythmes propres, aux rythmes de la nature. Et la nature elle-même n'a pas à être totalement soumise à l'homme, car l'homme n'y est pas extérieur. Il est nature lui-même (Man does not have to be subject to the rhythm of technology and science,

[59] Bloch 1995

[60] Robert Lenoble states that two periods of thinking on nature can be distinguished in the eighteenth century: "l'une qui étend les principes du mécanisme à tous les détails de la nature, mais surtout invente une métaphysique mécaniste de l'homme, où la raison pure, réduite d'ailleurs à une mécanique de sensations, suffit à assurer le bonheur; puis à partir de 1750, une réaction violente de l'affectivité, qui éclate chez Rousseau dans une apologie totale du sentiment (one which extends mechanistic principles to all the details of nature, but above all invents a mechanistic metaphysics for man, in which pure reason, reduced, moreover, to a mechanism of sensations, is sufficient to ensure happiness; then from 1750, a violent emotional reaction, which bursts out in Rousseau in a complete apologia of the sentiment)". Rousseau's relationship with nature and then that of the Romantics is not characterized in the same way as in the text of the Encyclopédie de d'Alembert and Diderot. The definition of nature in this very Encyclopaedia clearly shows an effacement of the symbolic force of the idea of nature in the first half of the XVIII century. The article refers to nature in other terms such as the system of the world, cause, essence and being. Nature in itself is nothing. We refer to God and Providence in the articles, if we understand by Nature "l'action de la providence, le principe de toutes choses, c'est à dire cette puissance ou être spirituel qui agit et opère sur tous les corps, pour leur donner certaines propriétés et y produire certains effets (the action of providence, the principle of all things, that is to say, that power or spiritual being who acts and operates on all bodies, to give certain properties and produce certain effects)". And when we speak of the action of nature, in the Encyclopaedia, "on n'entend point autre chose que l'action des corps les uns sur les autres, conformément aux lois du mouvement établies par le créateur. C'est en cela que consiste tout le sens de ce mot, qui n'est qu'une façon abrégée d'exprimer l'action des corps et qu'on exprimerait peut être mieux par le mot mécanisme des corps. (we understand nothing but the action of bodies on each other, according to the laws of motion established by the Creator. It is in this that the whole meaning of this word, which is but an abridged way of expressing the action of bodies, and which may be better expressed by the term mechanism of the bodies)" (Lenoble 1969, pp. 342–344).

he needs to live to his own rhythms, to the rhythms of nature. And nature itself does not have to be totally submissive to man, for man is not external to it. He is nature himself)".[61] These representations illustrate the difficulties that man encounters in living in towns. According to the philosopher G. Hottois (1989), the development of technology results in "ruiner l'habitabilité des villes réduites à des espaces de déplacement fonctionnels (ruining the habitability of cities reduced to spaces of functional displacement)".

Today, the reintroduction of nature into towns has led to two currents of thinking. The first, which identifies living things as an element of urban well-being and an important element of an urban ecology, consists of an economic valuation of the natural elements (vegetation, water, etc.) and a technical definition of the objects considered natural. The second can be to assimilate nature into the formal dimension of the living environment. The decorative logic that tends to reduce the aesthetics of the elements of nature to their visual aspects transforms nature into environmental amenities identifiable as marketable goods. In short, the aesthetic logic of the urban fabric tends to transform the production of environments into abstract problems of their ecosystematic processes. However, it is indeed a very concrete challenge, in all its bio-physico-chemical dimensions, that is at the centre of urban health.

Moreover, it is not only a question of nature external to the body. In recent years, chronic diseases that developed significantly in the twentieth and twenty first centuries have become the leading causes of death ahead of infectious diseases. These diseases cannot be cured once and for all but have to be lived with, day after day, month after month and year after year, in the uncertainty of one's condition, and the difficulty in expressing oneself, especially in public. Chronic disease requires the patient to adapt constantly to a state that is unpredictable from 1 day to the next. Moreover, the origin of the disease is often indeterminate and leaves room for the patient's narrative, especially if the disease is rare. To measure the importance of this unforeseeable nature, and even unpredictability, is to take into account an "outside of us", a natural otherness that medical science would try to define. One can then hypothesize that this kind of disease does not make sense except in terms that it is nature and takes its meaning outside of us human beings.[62] The difference with other natures is that it is our body, our nature, that every day, day after day, takes precedence over any other definition of ourselves so pressing is the need for meaning. From then on, it is about well-being and also about sickness, and the town is as much our environment as our body is.

[61] Viard 1990, p. 115.

[62] Maurice Merleau-Ponty in his lectures at the Collège de France published in 1995 defines nature in these terms.

12.6 Conclusion: Promethean and Orphic Attitudes

The relationship between "health" and urban areas falls within attitudes towards nature that historically have developed as being opposites according to the philosopher Pierre Hadot.[63] According to Hadot, the "Promethean" attitude consists in wanting to discover the secrets of nature, especially through science, in order to bring progress and comfort to humanity. The "Orphic" attitude, on the other hand, is wary of revealing the secrets of nature which could be taken as a forcing of the latter and refers to a more humble attitude towards nature and being attentive to its messages. Criticism of the "Promethean" attitude concerns the depletion of resources which is characteristic of indiscriminate industrial development, which endangers nature and human relationships with it: the Promethean attitude thus involves both benefits and risks. Furthermore, the Orphic attitude is not without danger due to the risk of loss of individuality, which is the result of a blind fusion with nature, sometimes likened to romantic impulses. Indeed, from the end of the eighteenth century, thinkers reacted against violence against nature, notably through the romantic movement, the work of Rousseau and the *wilderness* movement in the United States. In the nineteenth century, scientific positivism consecrating nature as an area to be administered, however, asserted itself and formalized the victory of the "Promethean" approach. Will recent developments linking nature and well-being through horti-therapeutic gardens lead to a renewal of the Orphic attitude?

References

Bailly, Antoine, and Racine Jean-Bernard. 1988. Qualité de la vie, bien-être, indicateurs sociaux territoriaux: l'homo geographicus entre choix et contraintes. *Espace géographique* 17 (3): 232–240.

Blanc, Nathalie. 1996. *La Nature dans la Cité*. Doctorat de l'Université de Paris 1 – Sorbonne.

Bloch, Ernst. 1995. *The Principle of Hope*. Trans. Neiville Place, Stephen Plaice and Paul Knight.

Butler, Ruth, and Hester Parr. 2005. *Mind and Body Spaces: Geographies of Illness. Impairment and Disability*. London: Routledge. Cambridge MA: MIT Press, Reprint edition, Studies in Contemporary German Social Thought.

Cabet, Etienne. 1840. *Voyage en Icarie. Travels in Icaria.* 2003. Trans. Leslie J. Roberts. Syracuse: Syracuse University Press, Series Utopianism and Communitarianism.

Castex, Jean, Jean-Charles Depaule, and Philippe Panerai. 1980. *Formes urbaines: de l'îlot à la barre*. Paris: Dunod, coll. Aspects de l'urbanisme.

Choay, Françoise, ed. 1965. *L'urbanisme, utopies et réalités*. Paris: Seuil collection Points.

Duhl, Leonard. 2011. The Human Measure: Man and Family in the Megalopolis. In 1963. *Cities and Space: The Future Use of Urban Land*,ed. Lowdon, Wingo Jr. New-York: Earthscan (First Published Baltimore: John Hopkins University Press).

Foucault, Michel. 2004a. *Naissance du biopolitique*. Paris: Gallimard/Le Seuil. (English translation of some excerpts are in: *The Development of Bioethics in the United States*, published

[63] Hadot considers nature as a hidden entity, guardian of secrets (Hadot 2004).

by R. Garrett, Jeremy, Fabrice Jotterand, and D. Christopher Ralston, Springer Science & Business Media, 2012).

———. 2004b. *Sécurité, territoire, population. Cours au Collège de France (1977–1978)*. Paris: Le Seuil-Gallimard.

Fourier, Charles. 1822. *Traité de l'association domestique-agricole*. Paris: Bossange père & Mongié P.

Gropius, Walter. 1935. *The New Architecture and the Bauhaus*. Trans. P.M. Shand and F. Pick. London: Faber and Faber.

Guillerme, André. 1983. *Le temps de l'eau, la cité, l'eau et les techniques*. Seyssel: ChampVallon.

———. 1994. *La disparition des saisons dans la ville. Les Annales de la recherche urbaine, "Les saisons dans la ville"*. Vol. 61, 9–15.

Hadot, Pierre. 2004. *Le Voile d'Isis. Essai sur l'histoire de l'idée de nature*. Paris: Gallimard NRF.

Hottois, Gilbert. 1989. Le technocosme urbain. La ville comme thème de philosophie de la technique. In *Penser la ville, Choix de textes philosophiques*, ed. Pierre Ansay and Roger Schoonbrodt. Bruxelles: Ed. AAM.

Howard, Ebenezer. 1946. *Tomorrow: a peaceful path to social reform*. Re-issued as *Garden-cities of tomorrow*, new ed. with prefaces by Sir F. Osborn & Prof. L. Mumford, London: Faber & Faber, (Translaled into French by L.E. Crepelet: *Villes-jardins de demain*, Tietsin press limited, China, 1902, pp. 15–26, 83–84, 77–79, 81, 128, 134).

IAURIF. 1978. Les cités-jardins de la région Ile-de-France. Recherche documentaire: les cités jardins en France et à l'étranger. *Cahiers de l'IAURIF* 51.

———. 1979. *Étude économique des espaces verts: rapport de synthèse*. Étude réalisée pour le compte de Région Ile-de-France, Établissement public d'aménagement de la ville nouvelle de Marne-la-Vallée, Ministère de l'environnement et du cadre de vie, Secrétariat général des villes nouvelles, par IAURIF, Institut d'aménagement et d'urbanisme de la Région d'Ile-de-France – Paris: IAURIF.

Lavoisier, Antoine Laurent. 1776. *Essays Physical and Chemical: By M. Lavoisier*. Trans. Thomas Henry. London: printed for Joseph Johnson.

Lawrence, Roderick. 2006. Representations sociales de l'habitat: la pertinence de l'écologie humaine. In *Psychologie sociale de l'environnement*, ed. Karine Weiss and Dorothée Marchand, 35–47. Rennes: Presses universitaires de Rennes.

Lawrence, Roderick, and Gilles Barbey. 2015. *Repenser l'habitat: donner un sens au logement*. Paris: InFolio, Archigraphy.

Le Corbusier, Edouard. 1986. *Toward a New Architecture*. Trans. Frederick Etchells. New-York: Dover Publications Inc (Originally *Vers une architecture*. 1923).

Lenoble, Robert. 1969. *Esquisse d'une histoire de l'idée de nature*. Paris: Albin Michel, coll. L'évolution de l'humanité.

Lévy, Albert. 2012. *Ville, urbanisme et santé: Les Trois révolutions*. Paris: Pascal, coll. Santé et société.

Luginbuhl, Yves. 1992. Nature, paysage, environnement, obscurs objets du désir de totalité. In *Du milieu à l'environnement*. Paris.

Murard, Lionel and Patrick Zylberman. 1985. L'ordre et la règle. L'hygiénisme en France dans l'entre deux-guerres. *Les cahiers de la recherche architecturale,* n°15 à 17.

Owen, Robert. 1818 Reprinted 1972. *New View of Society*. Report to the Committee of the Association for the Relief of the manufacturing and Labouring Poor. AMS Press Inc.

Palmade, Jacqueline. 1977. *Système symbolique et idéologique de l'habiter*. Paris: CSTB.

Panerai, Philippe, Jean Castex, Jean-Charles Depaule, and Ivor Samuels. 2004. *Urban Forms: The Death and Life of the Urban Block. Translated by Olga Vitale Samuels*. London: Routledge.

Richardson, Benjamin Ward. 1876. *Hygeia: A City of Health*. London: Macmillan.

———. 2004. *Early Urban Planning, Volume 1. Hygeia: A City of Health*, ed. Richard LeGates. Reprint Taylor & Francis.

Roncayolo, Marcel. 1983. *L'urbanisme, la guerre, la crise. Histoire de la France urbaine, t. 4*. Paris: Seuil.

———. 1990. *La ville et ses territoires*. Paris: Seuil.
Roncayolo, Marcel, and Paquot Thierry, eds. 1990. *Villes et Civilisations urbaines-XVIIIe-XXe siècle*. Paris: Larousse, coll. Textes essentiels.
Searles, Harold Frederic. 1960. *The Nonhuman Environment: In Normal Development and Schizophrenia*. New-York: International Universities Press Inc.
Stefulesco, Caroline. 1993. *L'urbanisme vegetal*. Paris: IDF, coll. mission du paysage.
Thomas, Keith. 1985. *Dans le jardin de la nature*. Paris: Gallimard, Bibliothèque des histoires.
Viard, Jean. 1990. *Le tiers espace, essai sur la nature*. Paris: Librairie des Méridiens Klincksieck.
Wright, Frank Lloyd. 1932. *The Disappearing City*. New-York: William Farquhar Payson. Reissued by the University of Michigan.
———. 1945. *When Democracy Builds*. Chicago: University of Chicago Press.

Chapter 13
Health and the Environment in Ecological Transition: The Case of the Permaculture Movement

Laura Centemeri

Abstract In this contribution my aim is to discuss how the permaculture movement promotes, through its concepts and practices, an understanding of human health as inseparable from the health of the environment – primarily intended as the health of the soil – and strictly dependent on the re-grounding of human subsistence activities within the environment of proximity. This process of re-grounding should not be conceived simply as the increase in self-reliance and self-sufficiency in providing for basic needs. It is better understood as a more engaged process of "reinhabitation," meaning by this term a normative orientation of all life activities toward doing what is best for the long-term health and viability of the area where one lives.

In this article, I analyze the recovery of a sense of place, through developing bonds to a specific spot on the earth we can know intimately, which is crucial to the vision of permaculture. I argue that this recovery can be more precisely understood if we take into account that there are various ways of valuing the environment. One of these, we may describe as "emplaced," develops in the "conscious design" of human permaculture settlements. In permaculture thinking, humans are seen as responsible ecosystem managers within, rather than separate from nature, and their ways of cohabiting with the multiplicity of beings (human and other) inhabiting their environments must be oriented by care. As a concluding remark, I will discuss the political implications of the permaculture vision of human health and flourishing.

Keywords Permaculture · Valuation · Design activism · Ecological transition

L. Centemeri (✉)
CEMS, CNRS-EHESS-Inserm, Paris, France
e-mail: laura.centemeri@ehess.fr

F. Bretelle-Establet et al. (eds.), *Making Sense of Health, Disease, and the Environment in Cross-Cultural History: The Arabic-Islamic World, China, Europe, and North America*, Boston Studies in the Philosophy and History of Science 333, https://doi.org/10.1007/978-3-030-19082-8_13

13.1 Introduction

In this contribution my aim is to discuss how the permaculture movement promotes, through its concepts and practice, an understanding of human health as inseparable from the health of the environment – primarily intended as the health of the soil – and strictly dependent on *the re-grounding of human subsistence activities within the environment of proximity.*[1] This process of re-grounding should not be mistaken for self-sufficiency in providing for basic needs. Individual self-sufficiency is repeatedly defined as a pointless goal in the most influential writings on permaculture.[2] Permaculturists are neither survivalists nor "peakists."[3] The re-grounding of individual subsistence activities within the environment of proximity has the aim of sustaining the emergence of *self-reliant communities*. Moreover, this practical re-grounding should be combined with a more engaging individual and collective process of "reinhabitation." By this term, introduced by American bioregional thinkers, permaculturists refer to a normative orientation of all life activities toward doing what is best for the long-term health and viability of the area where one lives.[4] More precisely, following Berg and Dasman, "reinhabitation means learning to live-in-place in an area that has been disrupted and injured through past exploitation. It involves becoming native to a place through becoming aware of the particular ecological relationships that operate within and around it. It means understanding activities and evolving social behavior that will enrich the life of that place, restore its life-supporting systems, and establish an ecologically and socially sustainable pattern of existence within it. Simply stated it involves applying for membership in a biotic community and ceasing to be its exploiter."[5]

The idea of "living in place" through developing bonds to a specific "spot on the earth we can know intimately"[6] can be more precisely understood in terms of the importance attributed to what I am going to discuss as *emplaced modes of valuing the environment* in orienting the "conscious design" of permacultural human

---

[1] This contribution discusses some of the issues I am currently exploring in an ongoing research program funded by the French ANR (SYMBIOS – *Social Movements For The Transition Towards A Frugal Society*, ANR-14-CE03–0005-01) and directed by Gildas Renou (University of Strasbourg).

[2] See on this point the analysis of Suh (2014a: 90) and the critique of the "Myth of Self-Reliance" as formulated by Toby Hemenway in his blog (http://tobyhemenway.com/107-the-myth-of-self-reliance/). Hemenway is an influential permaculturist and the author of *Gaia's Garden: A Guide to Home-Scale Permaculture*, a best-selling permaculture publication.

[3] On "peakism" as the "ideology of peak oil believers," see Schneider-Mayerson (2015).

[4] Bioregionalism is an intellectual and political movement that stresses the importance of the place-based dimension of social life. Bioregions are usually defined on the basis of physical and environmental characteristics accounting for a form of territorial coherence (watershed boundaries, soil characteristics, etc.). On the theoretical and practical dimensions of bioregionalism, see McGinnis (1999).

[5] Quoted in Aberley 1999: 23.

[6] Starhawk 2002: 163.

settlements.[7] Through the concept of emplaced modes of valuation that I introduce here as an analytical tool, I want to point to those capacities of evaluation that rest on a corporeal, sensorial, and affective understanding of what is "good" (or valuable) in the human-environment relationship. More in general through the lens of modes of valuation, it is possible to go beyond a simple phenomenological understanding of the sense of place, to explore the process of reinhabiting and living in place in terms of the recovery of actual evaluative capacities and practices guiding the way people engage with social and natural environments.[8]

My contribution is organized as follows. I first briefly present permaculture, as a concept and as a movement, its history, and main traits. I then discuss the permacultural understanding of health, stressing the fact that the health of the person is conceived of as being dependent upon what permaculture should provide: the possibility of regaining a form of *control* over one's own "lifestyle and future."[9] In permaculture, climate change, pollution, energy depletion, soil destruction, erosion, and impoverishment are presented as sources of increasing uncertainty concerning the future. The permacultural response is to actively design our environments so as to create the conditions for authentic human flourishing, which is authentic in as far as it contributes to the well-being of the biotic community. The ecological transition is thus seen, potentially, as an opportunity to rethink human subsistence as a fundamental locus of "commoning," challenging the capitalist way of dealing with human needs, including human health.

The idea of "controlling" the environment originates directly from a systemic thinking approach to environmental issues, which can prompt a utilitarian understanding of human-nature relationships, quite close to the one supporting the mainstream vision of nature as "capital" providing "services." In the original framework of permaculture, however, utility to human beings and human control over nature are not intended as detached from a "sense of place" but, on the contrary, are grounded in the reflexive mobilization of capacities of "contemplation, reflection and experimentation."[10] These capacities, I argue, imply recognizing the importance of the emplaced experience of the environment as a source of knowledge and specific normative orientations.

In this framework, humans are seen as "responsible ecosystem managers" *within*, rather than separate from nature. The goal of the permaculture movement is not nature preservation, but the active creation, by design, of the conditions for a perennial human-environment coevolution. To manage means here "(to) have a way, make do and work with nature."[11] In classic permaculturist writings, this practical

---

[7] A permacultural human settlement is meant to yield "an abundance of food, fiber, and energy for provision of local needs" with limited use of energy and natural resources. See Holmgren 2011: XIX.

[8] See Tuan (1977) and Casey (1997) for a philosophical history of the concept of place.

[9] Mollison and Holmgren 1978

[10] Smith 2011: XI.

[11] The etymology of the word "management" is from the old French word "ménagement," from the Latin "manus agere," which means literally to lead by the hand ("manus" being the Latin for hand).

wisdom is expressed in terms of ethical principles of *care*, both *of* the people and *of* the environment, *fair share*, and *personal responsibility*. These principles are considered as fundamental in order to have a "realistic rather than romantic understanding of what it means to live with and from nature."[12] As a concluding remark, I will discuss the political implications of the permacultural vision of human health and human flourishing.

13.2 The Permaculture Movement: A Composite Ethical Framework for Ecological Design Activism in a World of Energy Descent

Before being a movement, permaculture is, first of all, a *holistic design system* for the creation of *sustainable human settlements*. In other words, permaculture is "a practical *in* situ approach to creating collectively sustainable human settlements."[13] From the ecological vantage point, permacultural "consciously designed landscapes" *mimic* the *patterns* and *relationships* found in nature. They draw inspiration from traditional models of ecological organization as well.[14] The idea is to "work with nature," and not against it, in order to create environments that are "healthful and nurturing" for humans and other species.

Permaculture is about "our relationships with, and the design and redesign of, natural resource management systems, so that they may support the health and well-being of all present and future generations," in a world considered to have declining energy and resource availability and increasingly vulnerable to catastrophic events.[15] It is based on "assembling conceptual, material and strategic components in a pattern which functions *to benefit life in all its forms*."[16]

Concretely, in its original formulation, a permacultural design results in the creation of an integrated and evolutionary agroforestry system that includes a variety of species (plant, animal, etc.) while being perennial (or auto-perpetuated)

In particular, Olivier de Serres, who is considered a founding father of French agroecology (Tassin 2011), published a book in 1600, *Le théâtre d'agriculture et mesnage des champs*, in which the concept of "mesnage" (management) is related to a practical philosophy of cultivating land through imitating nature. Catherine and Raphaël Larrère have introduced the concept of "have a way, make do, and work with nature" (in French, *faire-avec la nature*) to qualify the arts of "managing" nature (in French, *piloter*) as opposed to the demiurgic techniques of exploiting nature.

[12] Holmgren 2011: 61.

[13] Suh 2014a: 76.

[14] Holmgren 2011: XVII. As reported by Suh (2014a: 79), Bill Mollison, one of permaculture's founding fathers, travelled extensively in the 1970s across India, Southwest Asia, and peasant Europe where he could observe the organization of traditional farming systems that were thousands of years old.

[15] Hill 2011: V; Holmgren 2002: XVI.

[16] Mollison 1988: 69 (my emphasis).

and *beneficial to human beings and their biotic community*. Animals can be explicitly included in this design, and they are always considered from the perspective of the multiple functions they can provide and never reduced to that of only providing food.[17]

It is important to stress, however, that permaculture is not simply reducible to a set of ecological engineering techniques, since permaculture tries not to separate the ecological and cultural dimensions implied in the design of sustainable human settlements. This means that permaculture is equally concerned with the design of dimensions of collective life pertaining to cultural and political domains, such as education, health, collective decision-making, etc. In this sense permaculture is a holistic design system since ecological solutions are devised and developed in parallel with cultural and political experimentation that accompany them and vice versa.[18]

Bill Mollison (1928–2016), a bio-agronomist, and David Holmgren (1955), his student, initiated permaculture in Tasmania (Australia), in the 1970s, in a sociohistorical context marked by an increase in ecological struggles and countercultural movements, locally and globally.[19] New concepts such as "bioethics" were emerging in the intellectual sphere. In the original idea of Van Rensselaer Potter, who first introduced it, bioethics implied an ecological understanding of the conditions of human well-being. A similar understanding of health is identifiable in permacultural thinking, as I am going to show.[20] Both Mollison and Holmgren were involved in the intellectual and sociopolitical effervescence of their time, and permaculture was initially intended mainly as a support for the many "back to the landers" who, in Tasmania as elsewhere, were moving from urban to rural areas to develop subsistence agriculture, often being totally unfamiliar with farming.

The word "permaculture" originates from the contraction of the expression "permanent agriculture" that appeared in the 1910s, in US agricultural science. As reported by Suh, the idea of "permanent agriculture" at this stage was not explicitly defined by the first American authors who introduced it: by this expression, they wanted to praise Oriental and Far Eastern agricultural systems in opposition to the kind of exploitative agriculture promoted by the US Department of Agriculture.[21] As reported by Fergusson and Lovell, in the 1920s, the expression "permanent

---

[17] An example is the introduction of ducks in rice paddies as discussed by Suh (2014b): ducks feed on insects and weeds in paddies and fertilize rice plants.

[18] In this sense, to focus exclusively on the agroecological dimension of permaculture, and to assess the permaculture movement's relevance in terms of agricultural productivity, is somehow very partial. This is the approach primarily promoted, in France, by the INRA, which is not the one I am interested in developing in my own work.

[19] Tasmania has been the cradle of the first ecological political party, the United Tasmania Group, formed in March 1972. The constitution of this political subject is related to a movement of protest for the preservation of Lake Pedder (in the southwest part of the island) against the construction of a dam. The publication of reports such as *The Limits to Growth* and *A Blueprint for Survival* inspired the movement (Wall 1998; Mulligan and Hill 2001).

[20] For a discussion of Van Rensselaer Potter's conception of bioethics, see Gaille (2011).

[21] Suh 2014a: 77.

agriculture" was used in a somewhat different meaning in the USA, to connote agricultural systems incorporating a high proportion of perennial species.[22]

Mollison and Holmgren's interest in both these declinations of "permanent agriculture" was motivated by the search for a low-energy and high-yielding "perennial agriculture" to sustain "perennial culture," based on village self-reliance, in which there would be direct personal responsibility in activities of providing for basic needs: "to empower the powerless and create 'a million villages' to replace nation-states is the only safe future for the preservation of the biosphere."[23] In this sense, "perennial" is not simply a synonym of "sustainable": perennial points to a more radical change in the social, economic, and political organization that is needed to face the trajectory of energy descent, considered as unavoidable. It is for this reason that its founders consider permaculture "much more than a form of organic gardening": permaculture is a "positivistic" response to an environmental crisis that threatens "the well-being and even survival of the world's expanding population."[24] It is important to highlight that permaculture is meant by Mollison and Holmgren as a necessarily collective endeavor, grounded, however, in a process of personal transformation involving the habits of everyday life.

Influenced by the reading of the political ecologist Ivan Illich, Mollison and Holmgren are critical of anonymous institutions taking control of basic needs. The main reason for their emphasis on relying on a network of close connections is the conviction of the unsoundness of the patterns of energy consumption of the industrial-capitalist model of social organization and the inevitable collapse to be expected. The goal of permaculture is not overall self-sufficiency but "the reduction of dependence on the wider industrial system" to create "inter-dependence within a locality."[25]

Although permaculture emerged inside academia, it has been viewed with suspicion by academics because of its strong interdisciplinary perspective and because it has been considered as utopian, from the point of view of mainstream agriculture. The divorce from academia over the last 40 years has brought accusations of permaculture being a "pseudoscience," because it has not embraced more recent developments in ecological theory. Another accusation has been that of the idiosyncratic use of terminology.[26] Highlighting the lack of a substantive assessment, in peer-reviewed literature, of permaculture as an agroecological alternative, Ferguson and Lovell suggest considering permaculture as a form of "folk science," meaning by this that success in practice has not always been followed by in-depth scrutiny.[27] The same

---

[22] Fergusson and Lovell 2014. A perennial plant, or simply "perennial," is a plant that lives for more than 2 years.

[23] Mollison 1988: IX.

[24] Holmgren 2011: XI, XV.

[25] Mollison and Holmgren 1978: 12.

[26] According to Ferguson and Lovell (2014), this is the case with the use, in permaculture literature, of the term "guild" to express mutually beneficial plant combinations, while the same concept means exactly the opposite in the vocabulary.

[27] Ferguson and Lovell 2014. It should be noted, however, that starting from 2012, on the initiative of the British-based Permaculture Association (existing since 1983), a Permaculture International Research Network has been created in order to widen "the knowledge base of permaculture

authors also stress the "populist orientation" of permaculture literature, based on books written by nonacademics and intended for a "popular" audience to whom the solutions to environmental and social crises are portrayed as both "simple and known."[28]

This orientation toward lowering the access barriers to knowledge on the design of self-reliable human settlements is an important aspect of permaculture, as a practice oriented toward *empowering* people to take direct action (and responsibility) concerning a variety of everyday activities: permaculture "is about what we want to do and can do, rather than what we oppose and want others to change. This response is both ethical and pragmatic, philosophical and technical."[29] The transformation of everyday activities prompted by permaculture is, for Mollison and Holmgren, a pillar of broader social change seen as based, predominantly, on *individual personal responsibility* and *voluntary action* more than on the capacity to influence the policy-making process: permaculture aims for a "sustainable future" in which "empowered individuals" move "from being dependent consumers to becoming responsible and productive citizens."[30]

Permaculture "pioneers," mostly itinerant permaculture teacher-activists, have been disseminating permaculture through teaching and training, first in Australia and then worldwide. Today, permaculture is "one of Australia's most significant intellectual exports."[31] The Permaculture Design Course (PDC), whose curriculum was codified in 1984, has, since then, been the prime vehicle for the global dissemination of permaculture.[32] As a transnational movement, permaculture has quite a loose form of organization, mostly structured through certified local/regional teaching and training institutions, grounded in local contexts. International and bioregional permaculture conferences (called convergences) have been held regularly since the 1980s. Beside the organization of these public gatherings and the offering of training opportunities, permaculture has been promoted via the creation of "demo sites" (short for "demonstration sites"). Demo sites are places in which the principles of permaculture have been applied to the design of human settlements that can range from simple gardens to ecovillages. These places are devoted to

research," as part of an effort to increase links with universities and other academic institutions and policy makers in order to conduct high-quality research projects. See the website https://pirn.permaculture.org.uk/. I want to stress again that the agroecological dimension is just one dimension of permaculture and that it is somewhat limited to consider permaculture as simply an alternative way to practice agriculture.

[28] Ferguson and Lovell 2014

[29] Holmgren 2011: XV. On permaculture and empowerment, see Smith 2002.

[30] Holmgren 2011: XIX.

[31] Holmgren 2011: XX. See for the global dimension of the movement the website http://permacultureglobal.org/projects.

[32] Mollison copyrighted the word permaculture "for the purpose of consistent education": "The word 'permaculture' can be used by anyone adhering to the ethics and principles expressed herein. The only restriction on use is that of teaching; only graduates of a Permaculture Institute can teach 'permaculture,' and they adhere to agreed-on curricula developed by the College of Graduates of the Institutes of Permaculture" (Mollison 1988: IX).

socio-ecological experimentation and permaculture teaching activities, and they provide practical examples of applications of permacultural design principles. Permaculture activists do not valorize proselytism as a tool of dissemination but instead the small-scale realizations of the type of world they envision. This concrete "prefiguration" (Yates 2015) of permacultural ways of organizing human activities is considered of the highest importance to sustaining the movement's diffusion.

It is important, however, to distinguish the logic of demonstration from that of exemplarity, which is predominant in many "Lifestyle Movements (LMs)."[33] In this respect similar to religious movements, many LMs highly valorize moral coherence between actions and ethical principles. The articulation between principles and actions is usually guaranteed by precepts, for example, the elimination of animal products from one's diet in veganism. The permaculture movement too is grounded in ethical principles. These principles, however, are not translated into precepts: they are complemented by guidelines meant to orient problem-solving design activities in specific situations, thus accounting for a variety of in situ practical permacultural realizations. This implies that there is no singular, identifiable permaculture lifestyle; the way in which a permaculturist lives can vary according to the specific local conditions of her place of living.

This "fluidity" is coherent with the bioregionalistic imaginary that inspires permaculture. As a "fluid object," the permaculture movement "isn't too rigorously bounded," "doesn't impose itself but tries to serve," and "is adaptable, flexible and responsive."[34] English is the "lingua franca" of the movement, and a basic knowledge of English is an important skill for a permaculturist who wants to be actively involved in the transnational organization of the movement. National permaculture associations today ensure access to learning in local languages. Voluntary initiatives have started to take place in order to create a dictionary in which the main concepts and terms of permaculture are translated by permaculturists all over the world into more than 15 different languages. To my knowledge, however, this work has not been accompanied by discussion concerning the cultural implications of different ways of translating permaculture concepts and especially its ethical principles.

Whether permaculture should be considered as a social movement – and of which kind – is an object for debate inside the self-same community of permaculture practitioners.[35] All depends on how one conceives of not only social movements but also political action. Permaculture can be understood as a *practice-based form of activism*, which can sustain a variety of political projects, from the more "radical" to the more "reformist."[36] This accounts for a certain degree of political incongruity

---

[33] On LMs see Haenfler et al. 2012; on the logic of public demonstration, see Rosental 2013.

[34] The reference is here to de Laet and Mol (2000: 226) and their analysis of the Zimbabwe Bush Pump as "fluid technology." I wish to thank Alexandre Gaudin for suggesting to me the category of fluidity as pertinent to describing the permaculture movement.

[35] See, for an example, the post "What permaculture isn't –and is" in Toby Hemenway's blog: http://tobyhemenway.com/668-what-permaculture-isnt-and-is/.

[36] As noted by Chatterton in the case of UK autonomous urban social center activists, permaculturists too share the same condition of "dwell(ing) both in the hoped-for and actual world": actions and

that is observable in the various initiatives referring to permaculture as a major influence and that I suggest understanding as an expression of the fluidity of permaculture as social movement. This fluidity results from the composite ethical framework, inspiring permaculture, which represents, in my interpretation, the most original aspect of this movement. This ethical framework is based on *care* (of the earth and of people), *distributive justice* ("fair share"), and *personal responsibility for choice* (notably choice of procreation and consumption).

The acknowledged influence, at least in Holmgren's thinking, of Stewart Brand's *Whole Earth Catalog* points to the importance, in permaculture, of the vision of the self as the ultimate driver of social change, in a world conceived of as an interconnected system of energy flows and information, in which technology is a potential countercultural force.[37] According to this vision, political transformation is the result not of agonistic politics but of empowered individuals becoming "comprehensive designers" who are able to see the "whole picture." In this sense, permaculture can be considered as an expression of "design activism," as theorized by the "design revolutionary" Victor Papanek for whom design "could revolutionize the living world" through "changing the relationship between humans and nature" and "help to heal social and ecological relations."[38]

Openly acknowledging Papanek's intellectual influence, David Holmgren defines design as "the *conscious* and *intuitive* effort to impose meaningful order" in complexity, thus reducing entropy. According to him: "design is not simply the result of rational, analytical and reductionist thinking, but also depends on our *intuitive* and *integrative* capabilities."[39]

---

living examples, not demonstrations and other forms of propaganda, are their main tools to "inspire people" to change. This implies paying attention to a "more complex and subtle understanding of anti-capitalist practice as not actually just 'anti-' but also 'post-' and 'despite-' capitalist" Chatterton 2010: 1216; 1221. An example of a radical critique of capitalism, expressed through applying permacultural thinking and practice, is that promoted by the activist group known as the Laboratory of Insurrectionary Imagination. See http://www.labofii.net/.

[37] I do not develop this point, since it is not central to the argument of the present essay. I just want to point out the proximity of permacultural thinking to some aspects of what Fred Turner (2006) describes as the peculiar political culture and worldview (based on personal integrity, individualism, collaborative sociability, and soft technology) having given rise to "digital utopianism." Stewart Brand (born in 1938), a Californian journalist and entrepreneur, is, according to Turner, the key figure of junction, in the 1970s, between cybernetic counterculture and the emerging digital technologies hub in the Silicon Valley. The *Whole Earth Catalog*, published between 1968 and 1972, was founded by Brand "in order to help those heading back to the land find the tools they would need to build their new communities" (p. 5). One of his most famous quotes exemplifies the controversial aspects of his vision: "We are as gods and might as well get good at it."

[38] Anker 2010: 81. Victor Papanek (1928–1998) is the author of *Design For The Real World: Human Ecology and Social Change* (New York: Pantheon Books, 1971).

[39] Holmgren 2011: 14 (my emphasis). In this sense, we can say that permaculture inherits from the long history of relationship of design and ecological sciences, which dates back to early 1930s, and to the Bauhaus movement, when many biologists saw in modernist design a way to improve social health and the environment through the unifying of art and science. On this history see Anker 2010. See also the "mesological" analysis of eco-design developed by Petit 2015.

The importance given to "intuition" and "integrative capabilities" based on extended in situ observation counteracts the considerable influence exerted on permaculture's vision by the thermodynamic perspective of Howard T. Odum's ecosystem ecology.[40] Most notably, permaculture keeps a "humanistic" approach to "whole systems design," based on the taking into account of the plurality of emotional and spiritual needs necessary for human flourishing, besides their biological and environmental needs.[41] Accordingly, nature is considered as "comprehensible only as *sensation* and *system*."[42] As I already pointed out, the apparently contradictory couple of principles such as "care and autonomy," "sensation and system," "strategy and harmony,"[43]and "rationality and intuition" are omnipresent in the writings of the most influential permaculturists. The effort not to separate dimensions of reality and experience that the modern understanding of science, ethics, and politics has disconnected can partially explain the difficulties that permaculture has been facing in academia while accounting for the practical success of the permaculture message. As Holmgren remarked: "for most people on the planet, the spiritual and rational still coexist in some fashion."[44]

In my understanding, the innovative idea of permaculture is then to think about the environment as a network of dynamic, interacting and overlapping systems, while being, at the same time, what a person experiences as her place of living.

It is possible to intervene in such a complex web of interdependencies with design tools based on the imitation of nature and its observable patterns meant to increase and support *collective resilience* and *self-sustainability*.[45] Of paramount importance then is the *protracted and thoughtful observation* of those "patches" in nature that are "wealthy" and "vibrant," in order to emulate them in landscape

[40] Howard T. Odum (1924–2002) together with his brother Eugene P. Odum was a pioneer in the introduction of energetic systems theory to ecology. The approach to environmental issues they promoted was based on diagramming the flows of energy in the natural world as input and output circuits in a cybernetic ecosystem. All biological life (including human behavior) could thus be reduced to charts of energy circuits. This reductionist understanding of biological phenomena can justify an approach of techno-scientific management to governing societies, which is rather far from permaculture. In fact, where Odum conceived the ecosystem manager as situated outside the system, in permaculture, on the contrary, she is considered as part of the system. For a critique of the technocratic approach of Odum, see Taylor 1988.

[41] While Odum's approach was technocratic and heavily influenced by ecology applied to the recreation of ecosystems for the purpose of the sought-after dream of space colonization (very influential in the 1950s and 1960s), permaculture's founders openly reject as meaningless these visions of enclosed and artificial life.

[42] Mollison 1988: 71 (my emphasis).

[43] This is the motto of the Permaculture Council for Europe (formerly European Permaculture Institute).

[44] Holmgren 2011:3.

[45] Mollison and Holmgren 1978: 4. This attempt to base design on biology points to the very core of the modernist heritage of Bauhaus. See on this point Anker (2010).

designs and repair the damage that has been done to natural systems by industrialized agriculture.[46]

The theoretical assumption behind the idea of "emulating nature" and "working with nature" is that, in nature, there are not only competitive and predatory relationships but "coexistence with no competition" and *cooperative* and *symbiotic* relationships improving the environment for each of the beings involved.[47] Human beings can design their activities so as to live in such a symbiotic, noncompetitive, and mutually benefitting way with their environments. In permaculture, human flourishing is not conceived of as conflicting with the environment: on the contrary, true human well-being, oriented by the consciousness of what we can call the "mesological condition" of human experience,[48] entails healthy and biodiverse environments.

In this sense, permaculture shows the limits of a simplistic understanding of anthropocentrism, as an ethical posture opposed to ecocentrism. The specific ethics and normativity that guide permaculture are clearly anthropo*genic* but not exclusively anthropo*centric*, since they are deep-rooted in an understanding of human beings as being in a symbiotic relationship with the environment they inhabit.[49] As stated by Holmgren, permaculture is "unashamedly human-centered," since human beings have "the power and intelligence" to affect their own situation and, conversely, the responsibility for their own situation "as far as possible."[50] At the same time, human beings are conceived of as actors in a complex environment whose ways of functioning are beyond the capacities of rational human understanding.[51] This irreducible uncertainty requires the development of skills of empathic observation and intuition, together with small-scale and local experimentation. As stressed by Holmgren: "separating the useful from the useless in complex biological and human systems is (…) difficult."[52] In fact, "in attempting to fix any system, we may

[46] Michael and Meachan 2001. Permaculture design is organized according to concentric zones (from the house, the zone 0, to the wilderness, the zone 5): the last zone is kept as a learning space, of no active human interference and just observation of spontaneous dynamics.

[47] Peter Kropotkin's work and his refutation of social Darwinism in *The Mutual Aid: A Factor of Evolution* (1902) had a strong influence on David Holmgren's early thinking. See Holmgren 2011:55.

[48] See Berque 2000.

[49] Where environment is intended in the sense of the French *milieu*. For a discussion of environmental ethics as necessarily anthropogenic, without being necessarily anthropocentric, see Maris 2010.

[50] Holmgren 2011: 6.

[51] The example mobilized in Permaculture Design Courses to illustrate this irreducible uncertainty of the extent of the system and mode of functioning is that of the mycorrhizas. Mycorrhizas are beneficial fungi, growing in association with plant roots that greatly increase the absorptive area of a plant, acting as extensions to the root system. The extent of this area covered by the fungi is largely unknown. This implies that hidden associations can be at work in a natural system. Mycorrhizas are less frequent in cultivated soils, since the laboring of the soil destroys them: their beneficial action is replaced by the use of fertilizers.

[52] Holmgren 2011: 57.

damage another that is working perfectly well."[53] This implies that the emphasis on the need to take action and assume responsibility is, to a certain extent, always balanced (or contained), in permaculture, by the need to observe and let the systems develop, accepting feedback. Activity and passivity are both crucial.

According to this vision, what appears to us as a disease could be, in fact, a healthy response to some other disequilibrium in the system we are not aware of. This emphasis on "accepting feedback" and "letting the system evolve" has an influence on the way to conceive of human health as well, and especially, the practice of medicine.

13.3 An Ecological Approach to Human Health: "Healing" the Soil for the Healthy Development of Human Societies

In line with the broader systemic understanding of nature, permaculture conceives of human health holistically: "we depend on good health in all systems for our survival."[54] It is as important to "accept feedback" concerning human health as it is in nature. One should try to understand (mobilizing both systemic thinking and intuition) what a symptom points to, not to eliminate the symptom, as allopathic medicine is accused of doing. In other words, it is better not to interfere too much with natural processes. At the same time, active interference in the environment is necessary, in order to design environments that can evolve so as to guarantee, permanently, both the health of the person and that of the environment. In this systemic understanding, health depends, first of all, upon not only eating but also growing "healthy plants and animals," avoiding "biocides and pollutants," and drinking "clean water or beverages."[55] The link between the health of the environment and human health is primarily, but not exclusively, through food. According to Mollison: "Both dangers and benefits arise from our food. Natural levels of soil antibiotics may sustain us, and natural resistance to disease is in great part transmitted to us via food."[56] However there are no food restrictions in permaculture, besides the general guideline to adapt food choices as much as possible to local conditions of permacultural food production. In this sense, practices of killing and eating animals (farm or wild animals) are possible if justified by local systemic dynamics.

The need to adapt food choices to local conditions of production implies that a change in how people eat becomes necessary, together with a change in how food is produced. For example, in order to increase local food self-reliance, it may become necessary to shift to a diet less dependent on carbohydrates as well as to use

[53] Holmgren 2011: 241.

[54] Mollison 1988: 3.

[55] Mollison 1988: 185.

[56] Mollison 1988: 184.

alternative ways to access proteins. It is important to notice that these changes in the way of eating are not, for permaculturists, simply a lifestyle question but, primarily, a form of ecological adaptation. To conceive of this adaptation simply as a lifestyle choice is somehow correct but superficial. This adaptation and the acceptance of the constraints imposed by the objective of increasing local food self-reliance are not considered as harmful to health, quite the contrary. Permaculturists consider the current epidemic of chronic pathologies (cancers, diabetes, etc.) in the Western world as largely due to the huge deregulation of our eating habits, over the last 40 years. They consider the industrialization of agriculture and livestock farming as the driver of this change.[57]

Local self-reliance also implies rediscovering, learning, and sharing competences in the use of herbs for healing purposes. It implies, equally, de-hospitalizing birth and death.[58] More generally, permaculture is sympathetic to an approach to health in which preventive measures (hygiene of life) and capacities of self-care are of crucial importance. *Self-care* is, in fact, the "soil" in which people care and earth care can grow: according to Holmgren, "to be able to contribute to a wider good, one must be healthy and secure."[59] In this respect, permaculture's approach to health emphasizes, first of all, individual responsibility in self-care, as well as the importance of local, self-organized networks of health support. As I have already pointed out, permaculture shares, with other movements that are critical of both the industri alization/standardization and the marketization of basic needs (health care included), a critical approach toward large, anonymous institutions (such as the market or the state) as providers for basic needs. Some authors argue that through reinforcing a vision of self-disciplined individuals and self-reliant ordered communities, while denouncing the state as incapable (or corrupted), these movements reproduce neoliberal subjectivities and practices, and they can easily be co-opted in projects of dismantling public health programs, with the consequential contribution to the further increase of social inequality. At the same time, other authors underline the possibility of strategic alliances with other actors, engaged in struggles against the inequalities produced by capitalist social relations, as a way to create and "feed" alternative "imaginaries" (notably through the framework of the "commons").[60] Truth is that public actors often consider these alternative approaches to health as a

---

[57] Mollison and Holmgren 1978. It is important to stress that today the epidemic of chronic pathologies is not confined to the north of the planet but it has progressively become a plague in the south as well, due to the deregulation of local eating habits caused by the invasion of cheap "junk" food in local markets. See Saligram and Nugent (2016) and Coste (2014).

[58] In Holmgren's "Permaculture Flower" – a diagram showing synthetically the key domains that require transformation to create a sustainable culture – health is conceived together with "spiritual well-being" as requiring the exercise of yoga or other body/mind/spirit disciplines, the practice of holistic medicine, home birth, and "dying with dignity." See Holmgren 2011: XX.

[59] Holmgren 2011: 7.

[60] See on this point the analysis that Calvário and Otero (2015) develop concerning, more in general, "back-to-the-landers." There are various examples of such alliances, as the initiatives reclaiming themselves from a perspective of "liberation permaculture," the "Black Permaculture Network," and the previously mentioned "Laboratory of Insurrectionary Imagination."

promotion of a "do-it-yourself" health judged as potentially harmful to the person and the collectivity, since it jeopardizes the very idea of public health itself.[61]

Reclaiming a holistic approach to health, in which the link between health and environment is crucial, is not particularly specific to permaculture, as compared to other movements supporting alternative visions of health and education (e.g., like anthroposophy). What is specific to permaculture, I argue, is the idea of promoting health through providing the possibility of living *a meaningful life*, starting from the opportunity to carry out meaningful work. Gardening for self-production should not be considered simply as a way of increasing self-reliance: according to Mollison, gardening is a form of "meditation."[62] More generally, to produce one's own food and fibers, according to permaculture principles, provides a meaningful experience of work, based on a sense of wholeness: "rather than the menial and repetitive labour of sowing, ploughing and reaping in a labour-intensive annual crop system, work in the permaculture system usually involves observation and control rather than power functions."[63]

This contemplative – one could say passive – dimension that is highly valued in permaculture (as part of observation and as a result of a humble attitude toward the complexity of nature) is always balanced by a vision of permaculture as "empowering" people and helping them to take action. As expressed by Mollison: "I sometimes think that the only real purpose of an initial design is to evolve some sort of plan to get one started in an otherwise confusing and complex situation. If so, a design has a value for this reason alone, for as soon as we decide to start doing we learn how to proceed."[64]

According to its founders, permaculture is beneficial especially since it provides the tools to engage in direct action to address basic problems, thus giving the opportunity to recover a sense of control over one's own life and escaping the paralysis of "paranoia," which can be induced by challenges apparently beyond our control such as climate change, energy transition, etc.: "perhaps the most valuable product of a city devoted to permaculture would be peace of mind; a paranoia pervades cities and it is a product of helplessness in the face of approaching energy shortages and uncertain futures."[65]

It is especially at the urban level that permaculture is considered as an activity that can have a decisive impact on health: "a city or community involved in permaculture is taking an important step towards controlling its destiny; consultancy

[61] For example, when alternative visions on health lead to the refusal of vaccinations. Moreover, the alternative health practices promoted as holistic approaches are denounced, sometimes, as the place for potentially sectarian "deviations" which require a form of monitoring. See, in particular concerning France, the activity of the Interministerial Mission against Sectarian Deviations (Miviludes, http://www.derives-sectes.gouv.fr/,) that monitor alternative therapeutic practices in health, in order to repress sectarian deviations and inform the public on the existence of this risk.

[62] Mollison 1988: 183.

[63] Mollison and Holmgren 1978: 9. On the use of agriculture for therapeutic and "empowering purposes and its connection with countercultural movements, see Doidy (2012).

[64] Mollison 1988: 47.

[65] Mollison and Holmgren 1978: 90.

groups can advise and monitor dangerous substances in the environment; irresponsible industry and public authority is thus identified and can be regulated or replaced. People, seeing a food resource develop nearby, are relieved of a great deal of their anxiety about their future, and can become actively involved in constructive work which assists community survival. It cannot be too much stressed that this factor alone will greatly assist community health."[66]

The refusal to separate "human survival" from the "survival of natural systems" implies the development of "general earthcare ethics" because all life is "our family."[67] This "general earthcare ethics" is rooted in the care of soil. In Holmgren's words "care for the earth can be taken to mean caring for living soil as the source of (terrestrial) life and for which we have the greatest responsibility."[68] The question of soil is considered from "viewpoints or sets of values" that are quite different if compared to the current ones. The state of the soil becomes "the best measure of the future health and well-being of society."[69] That is the reason why permaculture is not concerned exclusively with "preserving" but, more broadly, with the "healing" and regeneration of long-term soil fertility. Rehabilitative work to heal damage to ecosystems is what permaculturists are meant to do, as a form of stewardship "in which one leaves the land in better shape than it was when one received it." This "veneration of humus" is about increasing "the biological capacity of soil for the benefit of future generations."[70] The almost religious tone that shows through these quotations concerning the soil reveals the influence that the environmental philosophies of Taoism and Zen Buddhism had on the formulation of the original concepts of permaculture. This also explains the many similarities between *feng shui* and permaculture. As remarked by Suh: "the proponents of both feng-shui and permaculture consider soil to be the interface between non-living mineral earth and the atmosphere, and thus the most important site of interaction for all terrestrial life, inclusive of humankind."[71]

As pointed out by Mollison, the only places where the soil is conserved or increased, apart from "uncut forests" and "under the quiet water of lakes and ponds," are "the prairies and meadows of permanent plants" and "where we grow plants with mulched or non-tillage systems."[72] However, how to care for the soil remains partially open to interpretation, since "soils defy precise treatment," being the result of a "very long and varied history."[73] According to Holmgren, to care for the soil, as a way to care for the earth, means "looking after home, place, country, or territory,

[66] Mollison and Holmgren 1978: 94. The initiatives known today as "transition towns" have been firstly promoted, in the UK, by a permaculture designer, Rob Hopkins. See Hopkins 2011. Transition towns have created a network: see https://www.transitionnetwork.org/.

[67] Mollison 1988: 3.

[68] Holmgren 2011: 5.

[69] Holmgren 2011: 5.

[70] Holmgren 2011: 10.

[71] Suh 2014a: 81.

[72] Mollison 1988: 183.

[73] Mollison 1988: 182.

as understood by indigenous cultures and, more recently, through the concept of *bioregionalism*."[74] This is why I argue that at the core of the permacultural understanding of health, conceived as inextricably human and environmental flourishing, there is the process of *reinhabiting*. I propose to understand and investigate this process in terms of the recovering, or developing, at the personal and collective levels, of emplaced modes of valuing our relationship to the environment.

13.4 Recovering a "Sense of Place" Through Learning and Teaching Emplaced Modes of Valuing the Environment

As Lockyer and Veteto highlight, bioregionalism, permaculture, and ecovillages can be considered as "three interrelated and often overlapping ecotopian social movements."[75] The aspect I am interested in is the convergence of permaculture, bioregionalism, and ecovillages in stressing the importance of *reinhabiting*: "Reinhabitation entails a process whereby individuals and communities decide to commit themselves to a particular bioregion and live 'as if' their descendants will be living there thousands of years into the future. (…) Bioregionalists often take the indigenous societies of their bioregions as models of long-term inhabitation and sustainability, but work within their own cultural traditions, with a sense of dynamism that does not reify or essentialize traditional place-based cultures."[76] To quote Holmgren: "Wherever we live, we must become new indigenes."[77]

An in-depth discussion concerning the political pitfalls and snares of such an idea is beyond the scope of my contribution. What seems to me important to stress, for the argument I have been developing so far, is that these movements reclaim the necessity of thinking about politics but also about the economy and about health, from "the standpoint of place" while having, at the same time, an emancipatory vision: "anyone of any race, any religion, or origin is welcome, as long as they live well on the land (…) This sort of future culture is available to whoever makes the choice, regardless of background."[78]

[74] Holmgren 2011: 5 (my emphasis).

[75] Lockyer and Veteto 2013: 26. Lockyer and Veteto focus their analysis on the more recent American tradition of bioregionalism that emerged in the 1970s and expressed in the writings of such authors as Raymond Dasmann, Peter Berg, Gary Snyder, and Stephanie Mills. However, as pointed out by Aberley (who on this is following the analysis of Fitzpatrick Sale), American regionalists such as Howard Odum and Lewis Mumford can be considered as precursors of bioregionalist thinking, together with European regionalists such as Paul Vidal de la Blache and Patrick Geddes. See Aberley 1999. Geddes is, for example, a major influence in the bioregionalist approach developed by the "territorialist" Italian thinker Alberto Magnaghi. See Magnaghi 2014.

[76] Lockyer and Veteto 2013: 32.

[77] Holmgren 2011: 211.

[78] Using the words of Gary Snyder, an influential deep ecology American thinker and poet, quoted by Lockyer and Veteto: 34.

The idea of choosing to be committed to a place (and live *well* on it), while being engaged in nurturing a global network ensuring certain forms of circulation between "localities" (so as to share knowledge and experience and to support struggles for global justice),[79] is seen, by these authors, as a possible libertarian reaction against the standardization of space and time and the ever-increasing dominance of commensuration through the generalized marketization of all aspects of life induced by global capitalism. It also represents a challenge to the politically reactionary anti-modernist understanding of the place, as the homeland determining the entirety of one's identity, past, present, and future, in an immutable natural order of things.

In my understanding, in order to fully appreciate the originality of the political vision of "reinhabiting," as well as the specific political challenges that this idea raises, it is necessary to develop a (theoretical, analytical, and empirical) sensibility to the "emplaced" dimension of social processes, to which the emergence of what is described in literature as the phenomenological "sense of place" is, I argue, connected.[80] In particular, I contend that it is crucial to develop a (theoretical, analytical, and empirical) sensibility to *emplaced modes of valuation* orienting how human beings engage with their environments.

In the work I am currently developing with Gildas Renou, we define as a "mode of valuation" a culturally recognizable assemblage of sensorial and affective dispositions, cognitive capacities, and practical operations that are mobilized by agents, in the course of action, in order to judge what should count as valuable in a given situation and to evaluate, or test, this judgment.[81] Let me give an example in order to clarify this point, taken from my own fieldwork. The direct observation of a variety of permaculture initiatives (training courses, urban and rural projects, local, national, and international meetings), in a plurality of localities, is a necessary complement to the analysis of permaculture discourses and writings. Through observation, it is possible to examine how permacultural principles, tools, and instruments are appropriated in practice, put to work, and sometimes transformed in order to respond to the specific "normative expectations" of local actors.[82] Multi-site ethnography is necessary to understand how the bioregionalist imaginary of the permaculture movement is actually taking shape, or not, in the various localities in which permaculture initiatives are spreading. However, beside the differences one can observe in situ, I contend that there is a common feature, accounting for the movement to be recognizable as having an identity, which is the importance attrib-

[79] Quoting Gieryn (2000:472): "'Locality' is as much phenomenological as spatial, achieved against the ground of globalization or nationalization."

[80] For an extensive review of the literature oriented toward understanding the relevance of the distinction between space and place in sociology, see the article by Gieryn (2000).

[81] We define as *valuation* the identification/selection by an agent of what should count as valuable in a given situation, in a way that can be communicated and understood by others. This means that what counts as valuable relies on a widely shared cultural understanding of what is worth and what "matters." We define as *evaluation* the judgment of a situation according to a certain criterion of valuation, by means of convenient and appropriated activities. See Centemeri and Renou (2015).

[82] On the concept of "normative expectations," see Dodier and Barbot (2016).

uted to teaching and training activities in order to transfer not simply techniques but, more broadly, a specific evaluative framework.

The example I am going to discuss is related to my participant observation of a seminar on "designing a food forest," organized by the Italian association of permaculture (Milan, July 2015).[83] The teacher (a former president of the Italian association of permaculture and an experienced permaculture teacher) introduced the course telling the participants that in order to address the issue of "how to design a food forest," it is necessary, first, to address the question of "why we want to design a food forest." He then invited the participants to do the following exercise: to list the "good reasons" for designing a food forest. Students were divided in small groups and each drew up a list of good reasons. A spokesperson for each group presented their list to the plenary assembly of participants. A common list of "good reasons" to design a food forest was compiled at the end of the exercise by the teacher, starting from what participants said during this specific course but also from what previous students in previous, similar courses had said. The list of good reasons for creating a food forest thus compiled spanned from "the production of oxygen for the planet," "the preservation of traditional species," or "the recovering of environmentally damaged areas and the preservation and protection of soil" to "becoming more resilient," "having fun and let animals have fun too," "letting something spontaneous happen," "enjoying sounds," "sharing the experience of the forest with friends," etc. I propose to understand these "good reasons" to design a food forest as reasons according to which the food forest can be considered as something good, something that matters, and *valuable*. They are reasons to value the forest, and they point to practical ways in which to *valorize* the forest. The verb "to valorize" is used to indicate that a material reality is transformed or simply arranged according to a certain definition of value, in order to stabilize as much as possible a certain value experience. Only once the "good reasons" to value are clearly stated, can it be possible to start to actually draw a design of the food forest, which is meant to incorporate these evaluative judgments so as to valorize the environment accordingly. Quoting Mollison and Holmgren: "Net yield is only *one value to consider*. In commercial agriculture all value is converted into money (…). In subsistence agriculture *human needs* determine the value of yields and since our needs are diverse, so the yields should be diverse."[84]

What I am interested in, in this example, is the fact that, in Permaculture Design Courses, there is not just a transmission of technical knowledge on agroforestry and eco-design. Permaculture courses are meant to communicate certain criteria (and not others) defining *what should count as valuable and worthy in our ways of engaging with the environment*. These criteria are what we call, with Renou, "modes of valuation." They go with the teaching and learning of the knowledge, skills, capacities, and sensibilities needed to evaluate (and valorize) according to these

[83] The concept of "food forest" is based on the idea of designing a garden and emulating a real forest functioning in order to create a low-maintenance "forest garden," filled with the food plants and trees that one wants or is able to grow. See Jacke and Toensmeier 2005.

[84] Mollison and Holmgren 1978: 7 (my emphasis).

very diverse definitions of the valuable. It is important to notice that I am using here the term "evaluation" to point to both formalized procedures to assess the value according to conventional categories and to practices through which value is appraised by way of shared or customary local criteria. In both cases, evaluation does not always imply quantification.[85]

When speaking about emplaced modes of valuing the (socio-ecological) environment, we point to a form of knowledge and appreciation of the relationship of human beings to their environments that rests on the interrelationship of body-mind-environment and that takes place primarily at the esthetic level.[86] This same relationship, however, can also be understood and appreciated through distancing, when "the valuable" is independent from the emplaced experience. In this case, we speak about *modes of valuation through distancing*, and the key operation that has to be performed in order to evaluate and valorize is that of establishing equivalences, often but not necessarily by way of quantification. This implies the devising of tools and techniques of commensuration.[87] In emplaced modes of valuation, what counts as valuable in the relationship with the environment is, instead, some kind of sensorial and affective good, and evaluation rests on a judgment based on an embodied felt proximity to this (felt) good. In order to share this mode of (e) valuation with others, sensory skills and sensibilities must be developed and trained.

Going back to our list of different reasons to value a "food forest," "sharing the experience of the forest with friends" and "enjoying sounds" are reasons to value a food forest where what counts as valuable is based on the body experiencing the affective and esthetic influences of a specific environment and situation and appreciating this environment and situation for these particular sensorial and affective qualities. The valuation according to which what is valuable is "spontaneity" and "unexpectedness" in the interactions and collaborations between species in the ecosystem also relies on a form of knowledge dependent on the interrelationship of body-mind-environment and accounting for a type of observation that can detect such things as "spontaneity" in a specific environment. If we consider the valuation expressed through defining as a good reason to create a food forest the "production of oxygen for the planet" or the "preservation of traditional species," what is

[85] I adopt an approach to evaluation as a fundamental dimension of human action and not simply as a specialized field of activity. For an understanding of evaluative activities as fundamental to human existence, see Canguilhem (2006).

[86] Esthetic is used here in the sense of the level of *aesthesis*, the perception, and the corporeal sensibility. The concept of "being emplaced" originates in the vocabulary of the "phenomenology of place and space" developed by authors such as Edward Casey, Doreen Massey, Tim Ingold, and Sarah Pink. See in particular Pink 2009. We reinterpret this tradition through the lens of the "mesological" approach developed in France by Augustin Berque (2000). A major influence in our work is the "pragmatic sociology of proximity" (*sociologie pragmatique du proche*) developed by Laurent Thévenot and Marc Breviglieri. See Thévenot (2007).

[87] Commensuration takes place in the real world, and socio-technical "investments in forms" are devised in order to actually transform things that are different (in the personal experience that we have of them) into things that can be represented as the same (in value). These "investments" are meant to endure through time and space. Commensuration can imply, but not necessarily, the use of numbers. See on this Thévenot 1984, 2009.

valuable is defined in terms of an abstract representation of the environment, of its qualities, and of its beings. No interrelationship of body-mind-environment is necessary here in order to be able to judge what is valuable.

We contend that it is possible to identify a variety of emplaced modes of valuation, but this normative richness (or plurality) has been rather neglected by social sciences. The *ease* felt through the familiarity with "dwelled-in" places, people we are close to (emotionally), and appropriate tools and objects is a mode of emplaced valuation quite different from the *excitement* felt through the exploration of the newness or the *resonance* felt through attuning to an ambience.[88] The task of exploring these modes of emplaced valuation confronts us with methodological challenges, but it is crucial in order to understand what is at stake, practically and politically, in the process of reinhabiting, which is so crucial to the vision promoted by permaculture.

In my understanding, the capacity to care could be more precisely intended as related to the development of the awareness of the various emplaced modes of valuing we can experience in our relationship with the socio-ecological environment. But this awareness needs to be developed together with the consciousness of the plurality of detached modes of valuation – or of modes of establishing an equivalence – on which rest the possible diverse understandings of what justice means.[89] Only if both these capacities of emplaced (e)valuation and detached (e) valuation of our relationship with the environment have been developed and trained, does it become possible to design permaculturally healthy human settlements. They are healthy because they are designed according to principles in which earth care is not considered separately from "people care" and "fair share."

The permacultural design tries to valorize, in the socio-ecological organization of human communities (from the house to the bioregion), a wide variety of emplaced and detached goods, relying on the relationship human beings build and shape with their socio-ecological environments. The purpose is that of creating the condition both for human well-being and for the health of the biotic community. However, how concretely these plural ways of valuing can coexist harmoniously, and the role played by different local cultures in sustaining or hampering the conditions for this coexistence, is a question open to further investigation, through the in-depth analysis of local permaculture initiatives. This also implies investigating the extent to which innovations are able to circulate in the permaculture movements, given their solid local anchoring.

In conclusion, I would like to point to some problematic aspects raised by the ambitious value pluralism that informs permaculture's ethical framework. In the creation of a modern society based on a modern understanding of justice and a modern public sphere, those modes of valuation through distancing, in which "the valuable" is supposed to have universal validity (with science playing the role of

[88] See Breviglieri (2012), Auray and Vetel (2013), and Thibaud (2011).

[89] See, for example, the different understandings of justice that can guide an NGO's action of post-disaster neighborhood revitalization, the case study on post-Katrina New Orleans in Allen (2013).

establishing the terms of this universality), have been endowed with an unconditional privilege of legitimacy. This has implied the progressive eclipse of the acknowledgment of the role that emplaced modes of valuation play in social life as well as the progressive standardization of the wide variety of *local modes of valuation* that have always been crucial to the bioregional organization of socio-ecological communities (Doidy 2008). When they haven't been seen as the source of various dangers or impediments to the expression of "modern values," emplaced modes of valuation have been confined to the domain of an indistinct "informal," "tacit," and "embodied" dimension of social life.

As the permaculture approach to the design of human settlements tries to emphasize, emplaced modes of valuation are of crucial importance in the healthy socio-ecological organization of human communities. At the same time, they are disruptive of a certain modern understanding of the relationship between society and nature, which goes hand in hand with a certain modern understanding of the relationship between knowledge and experience and between reason and emotion. Permaculture thinking tries to overcome these dichotomies, one of them being particularly difficult to deal with: that opposing emancipation to spirituality.[90]

This disruption is necessary to imagine a post-capitalist future, but it can be turned into the reactionary destruction of fundamental acquisitions of modernity, in terms of social justice and individual freedoms. At the same time, it is fundamental to be aware of the radical and insurmountable tensions existing between emplacement and detachment.[91] Only with this awareness is it possible to fully measure the originality and the potentialities of the permaculture movement.

References

Aberley, D. 1999. Interpreting bioregionalism: A story from many voices. In *Bioregionalism*, ed. V. McGinnis, 13–42. London: Routledge.

Allen, B. 2013. Justice as measure of nongovernmental organization success in post disaster community assistance. *Science, Technology, & Human Values* 38 (2): 224–249.

Anker, P. 2010. *From Bauhaus to Ecohouse: A history of ecological design*. Baton Rouge: Louisiana State University Press.

[90] I cannot develop this point. I would like nevertheless to point out that emplaced modes of valuation are crucial in certain forms of basic spiritual experience, as Mollison implicitly acknowledges when he compares gardening to a form of meditation. As explained by Hache (2015), the difficulty, in the modern understanding of political engagement, not to oppose spirituality and emancipation, explains the skepticism toward some expressions of ecofeminism, such as that of Starhawk (who is a permaculturist). The importance to be attributed to the spiritual dimension, however, is quite controversial in the permaculture movement itself. See on this point Dawborn and Smith 2011: 40.

[91] In this sense, an in-depth scrutiny of the different kinds of goods that emplacement and detachment can provide the human being with is of paramount importance in social sciences. The research program of "sociology of regimes of engagement" developed by Laurent Thévenot is one of the rare examples of a serious investigation of this issue in social sciences.

Auray, N., and B. Vétel. 2013. L'exploration comme modalité d'ouverture attentionnelle. Design et regulation d'un jeu *freemium. Réseaux* 6. 182: 153–186.
Berque, A. 2000. *Ecoumène. Introduction à l'étude des milieux humains*. Paris: Belin.
Breviglieri, M. 2012. L'espace habité que réclame l'assurance intime de pouvoir: Un essai d'approfondissement sociologique de l'anthropologie capacitaire de Paul Ricoeur. *Études Ricoeuriennes/Ricoeur Studies* 3 (1): 34–52.
Calvario, R., and I. Otero. 2015. Back-to-the-landers. In *Degrowth: A vocabulary for a new era*, ed. G. D'Alisa, F. Demaria, and G. Kallis, 143–145. Oxon/New York: Routledge.
Canguilhem, G. 2006. *Le Normal et le pathologique*. Paris: PUF.
Casey, E. 1997. *The fate of place. A philosophical history*. Berkeley/Los Angeles: University of California Press.
Centemeri, L. and Renou, G. 2015. The contribution of a pragmatic sociology of valuation to the study of the social movement for sustainable degrowth. Paper presented at the *European Pragmatism Conference*. Paris, September 9th–11th.
Chatterton, P. 2010. So what does it mean to be anti-capitalist? Conversations with activists from urban social centres. *Urban Studies* 47 (6): 1205–1224.
Coste, J. 2014. Les maladies dominantes au xxe siècle. In *Histoire de la pensée médicale contemporaine – évolutions, découvertes, controverses*, ed. B. Fantini and L.L. Lambrichs, 259–278. Paris: Seuil.
Dawborn, K., and C. Smith, eds. 2011. *Permaculture pioneers. Stories from the new frontier*. Hepburn, Victoria (AU): Melliodora Publishing.
De Laet, M., and A. Mol. 2000. The Zimbabwe bush pump: Mechanics of a fluid technology. *Social Studies of Science* 30 (2): 225–263.
Dodier, N., and J. Barbot. 2016. La force des dispositifs. *Annales. Histoire, Sciences Sociales* 71, 2: 421–450.
Doidy, E. 2008. Cultiver l'enracinement. Réappropriations militantes de l'attachement chez les éleveurs jurassiens. *Politix* 83 (3): 155–177.
———. 2012. Reconstruction personnelle et critique sociale. Enjeux politiques de la réinsertion des anciens combattants par l'agriculture. *Lien Social et Politiques* 67: 33–50.
Ferguson, R.S., and S.T. Lovell. 2014. Permaculture for agroecology: Design, movement, practice, and worldview. A review. *Agronomy for Sustainable Development*. 34 (2): 251–274.
Gaille, M. 2011. L'idée de 'bioéthique globale' : un combat à reprendre ? Le travail de la philosophie et l'histoire tronquée de l'éthique médicale. *Cahiers philosophiques* 2 (125): 131–136.
Gieryin, Th.F. 2000. A space for place in sociology. *Annual Review of Sociology* 26: 463–496.
Hache, E. 2015. Préface. Where the future is. In *Rêver l'obscur. Femmes, magie et politique*, ed. Starhawk. Paris: Cambourakis.
Haenfler, R., B. Johnson, and E. Jones. 2012. Lifestyle movements: Exploring the intersection of lifestyle and social movements. *Social Movements Studies* 11 (1): 1–20.
Hill, S.B. 2011. Forward. In *Permaculture. Principles and pathways beyond sustainability*, ed. D. Holmgren. East Meon, Hampshire: Permanent Publications.
Holmgren, D. 2011. *Permaculture. Principles and Pathways Beyond Sustainability*. East Meon, Hampshire: Permanent Publications. First edition, 2002, Hepburn (Victoria, Australia): Holmgren Design Services.
Hopkins, R. 2011. *The transition companion: Making your community more resilient in uncertain times*. Totnes: Green Books.
Jacke, D., and E. Toensmeier. 2005. *Edible Forest gardens Vol. 1. Ecological vision and theory for temperate climate permaculture*. White River Junction: Chelsea Green Publishing.
Lockyer, J., and J.R. Veteto, eds. 2013. *Environmental anthropology engaging Ecotopia. Bioregionalism, permaculture, and Ecovillagies*. Berghahn: New York/Oxford.
Magnaghi, A. 2014. *La biorégion urbaine, petit traité sur le territoire bien commun*. Paris: Association culturelle Eterotopia France.
Maris, V. 2010. *Philosophie de la biodiversité. Petite éthique pour une nature en peril*. Paris: Buchet-Chastel.

McGinnis, V., ed. 1999. *Bioregionalism*. London: Routledge.
Michael, P., and W. Meachan. 2001. Permaculture as a design modality for healing and regeneration: Design with a deeper agenda. *The Design Journal: An International Journal for All Aspects of Design*. 4 (2): 42–49.
Mollison, B. 1988. *Permaculture – A Designer's Manual*. Tyalgum: Tagari Publications.
Mollison, B., and D. Holmgren. 1978. *Permaculture one: A perennial agricultural system for human settlements*. Tyalgum: Tagari Press.
Mulligan, M., and S.B. Hill. 2001. *Ecological pioneers: A social history of Australian ecological thought and action*. Cambridge: Cambridge University Press.
Petit, V. 2015. L'eco-design: design de l'environnement ou design du milieu? *Sciences du design* 2/21: 31–39.
Pink, S. 2009. *Doing sensory ethnography*. London: Sage.
Rosental, C. 2013. Toward a sociology of public demonstrations. *Sociological Theory* 31 (4): 343–365.
Saligram, N., and R. Nugent. 2016. Chronic diseases – The social justice issue of our time. *Lancet* 387 (10022): 943.
Schneider-Mayerson, M. 2015. *Peak oil. Apocalyptic environmentalism and libertarian political culture*. Chicago/London: The University of Chicago Press.
Smith, C. 2002. Learning for the environment: An examination of personal empowerment through learning permaculture. *Post-Script*. 3 (1): 12–25.
———. 2011. Introduction. In *Permaculture pioneers. Stories from the new frontier*, ed. K. Dawborn and C. Smith, X–XX. Hepburn, Victoria: Melliodora Publishing.
Starhwak. 2002. *Webs of power. Notes from the global uprising*. Gabriola Island: New Society Publishers.
Suh, J. 2014a. Towards sustainable agricultural stewardship: Evolution and future directions of the permaculture concept. *Environmental Values* 23: 75–98.
———. 2014b. Theory and reality of integrated rice–duck farming in Asian developing countries: A systematic review and SWOT analysis. *Agricultural Systems* 125: 74–81.
Tassin, J. 2011. Quand l'agro-écologie se propose d'imiter la nature. *Courrier de l'environnement de l'INRA* 61: 45–53.
Taylor, P.J. 1988. Technocratic optimism, H. T. Odum, and the partial transformation of ecological metaphor after world war II. *Journal of the History of Biology* 21 (2): 213–244.
Thévenot, L. 1984. Rules and implements: Investment in forms. *Social Science Information* 23 (1): 1–45.
———. 2007. The plurality of cognitive formats and engagements: Moving between the familiar and the public. *European Journal of Social Theory* 10 (3): 413–427.
———. 2009. Governing life by standards: A view from engagements. *Social Studies of Science* 39 (5): 793–813.
Thibaud, J.-P. 2011. A sonic paradigm of urban ambiances. *Journal of Sonic Studies* 1 (1). http://journal.sonicstudies.org/vol01/nr01/a02.
Tuan, Y.-F. 1977. *Space and place: The perspective of experience*. Minneapolis: University of Minnesota Press.
Turner, F. 2006. *From counterculture to Cyberculture. Stewart Brand, the Whole Earth network and the rise of digital utopianism*. Chicago/London: Chicago University Press.
Wall, D. 1998. *The no-nonsense guide to green politics*. Oxford: New Internationalist.
Yates, L. 2015. Rethinking prefiguration: Alternatives, micropolitics and goals in social movements. *Social Movement Studies* 14 (1): 1–21.

Chapter 14
"Affordances": A Notion that Reveals How Artificial It Is to Separate the Body and Its Environment

Anne-Lyse Chabert

Abstract In the world of medical treatment, the human body has gradually become so dissociated from the environment in which it lives and functions that doctors have come to consider the body as the sole and ultimate focus of intervention; it has become normal to isolate the body from its environment. Recently, however, attitudes have shifted somewhat, in an attempt to rehabilitate the contextual role of the environment as the agent of determinative value for the individual living within it. We take an even more radical view, i.e., not only is the environment a fundamental element to be taken into account when we study the individual associated with it, but in fact the environment is essential in shaping his or her existence. In other words, the body and the environment are extensions, or continuations, of one another. Our conceptual framework, in the context of this study, is the notion of "affordances," which link the body and the environment via an "invitation to act." As we outline this concept, we will strive to make it as resonant as possible by describing, and then generalizing, the example of people living with a disability. It will become clear that a precise reassessment and rearrangement of the entire environment are required in order that the individual's quality of life can improve.

Keywords Medical treatment · Human body · Environment · Disability · Rearrangement

14.1 Introduction

The body and the environment are often so sharply considered as dissociated that they are sometimes completely foreign to one another. We argue in favor of rethinking the issue. This does not mean rethinking each of the two entities separately (even by revisiting the nexus of their articulation) but rather rethinking the way they

---

A.-L. Chabert (✉)
Université de Paris, SPHERE, UMR 7219, CNRS-Université de Paris, Paris, France
e-mail: chabert.annelyse@gmail.com

F. Bretelle-Establet et al. (eds.), *Making Sense of Health, Disease, and the Environment in Cross-Cultural History: The Arabic-Islamic World, China, Europe, and North America*, Boston Studies in the Philosophy and History of Science 333, https://doi.org/10.1007/978-3-030-19082-8_14

are essentially interwoven, reviving the notion that the individual and his environment form a construct, a whole, chiefly so as to avoid going astray into therapeutic gestures that presuppose a harmonious relationship. This contribution will defend that thesis.

We propose to use the conceptual framework of *affordances* forged by James J. Gibson in 1977 to reveal the adjustment that must be made between the body and the environment in which it is inserted by means of "an invitation to act."[1] How can affordances become obvious when they remain silent as long as there is no dissonance between the body and the environment?

In this contribution we will work on a situation where a differential as to the quality of this link makes it feasible to grasp more clearly the environment's role and its impact on the individual's well-being, sorely tried by having to contend with a disability. To begin with, we will examine a specific example where the well-being of blind children and adolescents is enhanced when an adapted activity is made available to them. Starting from that concrete situation, we will try to generalize and see how the fact of being disabled makes the link described by the concept of affordances more visible.

14.2 The Framework of Affordances

As Marion Luyat and Tony Regia-Corte state in their article:

> When we move about freely within our environment, our posture and our locomotion adapt very flexibly (and unconsciously) to the terrain we cover. Even our minutest gestures are made without mishap and in a manner that is totally in phase with our aims. We do not attempt to reach unreachable objects, we do not attempt to sit on furniture that is too high or too low, nor do we attempt to walk on surfaces, such as water, where it is not feasible to walk.[2]

Human beings endowed with an ordinary perceptive and motor system constantly adapt their movements to the world around them by means of the tools at their disposal, be they organic or artificial and external. The sequence of their motor reactions is relatively fluid and does not seem to encounter any notable obstacles. This ability is actually the result of a lengthy apprenticeship on the part of "normal" individuals who have long matched their movements to their environment, generally without even realizing it.

Gibson, with his concept of affordances, sheds light on this harmony when he describes the way individuals have the possibility of attaching themselves technically to their external environment by selecting and processing information they receive from it; this is a perceptive operation. Their environment enables them or forbids them to carry out a number of actions.

[1] Gibson 1977

[2] Luyat and Regia-Corte 2009, p. 298.

The originality of this notion lies in the fact that it links an individual's action to their perception of their environment. Perception guides their action by enabling them to evaluate the potentialities that the situation authorizes them to use depending on their needs and the resources they can deploy. Thus individuals select from their environment the information that is relevant to the task they need to carry out, assigning different values within what is an undifferentiated physical environment to begin with. In so doing, and depending on the information they process, individuals structure and polarize the situation in which they are immersed.

Gestalt theory contributed to this line of thinking centered on the environment.[3] Gestalt theorists maintain that every single thing an individual perceives informs him about what can or cannot be done with that thing. Kurt Koffka, a colleague of Gibson at Smith College (1928–1941), points out that the German term "Aufforderungscharakter" illustrated the same idea that Gibson developed in his concept of affordance. "Each thing says what it is ... a fruit says 'Eat me' ; water says 'Drink me' ; thunder says 'Fear me.'"[4] Thus, a stairway can invite a passerby to climb it to the extent that their physical and psychological dispositions allow this action. Gestalt theory distinguishes from the outset between a behavioral environment and one that is simply geographical.

The neologism was coined from the word to afford, i.e., to offer. As Gibson writes, "the affordances of the environment are what it offers the animal, what it provides or furnishes, either for good or ill."[5] Thus the outside world cannot be grasped only by means of classical physical units of measurement. The environment must necessarily be considered in relation to individuals who take into account the capacities they can deploy within it, depending on the physiological and psychological parameters which constitute it. The concept of affordances is linked essentially to the animal-environment system, in line with an ecological theory which slightly predates the formation of the term and constitutes its mental framework (see the contribution by Perrine Michon in this same volume).

Since then the concept of affordances has been substantially reworked. Researchers seek to clarify a concept which does not seem to lend itself to one single definition. One understanding of it considers an affordance as a set of invariable properties of the environment, regardless of whether it is perceived or not,[6] while another considers it as a property of the animal-environment relationship.[7]

Thus the environment, taken in Jakob von Uexküll's sense of "Umgebung," is one vast field of opportunities, as Luyat and Regia-Corte write.[8] An individual perceives not only objects but a set of functional relationships which they must continually readjust to suit their own capacities and limits before they can use any

[3] Lewin 1926

[4] Koffka 1936 p. 7; Gibson 1986, p. 138.

[5] Gibson 1986, p. 127.

[6] Gibson 1979; Turvey 1992

[7] Stoffregen 2003

[8] Luyat and Regia-Corte 2009, p. 307.

of the objects effectively when they need to. For instance, an individual does not directly perceive the object "chair" but instead the relationship they project between the object and the use that their body enables them to make of it. In order to use it, their legs must be a certain length, and their body must adopt a certain posture – these are among the necessary faculties of which the individual must dispose. A chair may be used in a non-classical way, not in the ordinary way in which it is spontaneously used and was intended to be used. Conversely, several other objects (a bench, a stool, a rock) can also enable an individual to sit down after a relatively easy morphological adjustment. The notion "chair" can also take on other aspects depending on the user's criteria, expectations, and needs; the user can sit on surfaces that appear to be very different from one another. Thus an affordance does not stem from a specific object; it stems from a relationship projected by the user between their body and the environment they wish to be involved in.

Nor are affordances a fixed, unchanging set of any sort, or they may only seem to be so because of the habitual use we make of the objects surrounding us; they have to do with a set of actions which the potential actor, faced with a situation, constantly reevaluates. Here we see how closely the notion of affordance links action and perception; they become two quasi-indissociable entities, action-guiding perception and vice versa. The "acting body" is at the heart of such a concept; it remains very close to Merleau-Ponty's thinking about the body which, in turn, was strongly influenced by Gestalt theory.

Like Gibson, Merleau-Ponty stresses the primacy of the subject's lived experience in an environment from which he can never be abstracted altogether (see M. Gaille's contribution in this volume). As Merleau-Ponty says, perception relates to a very paradoxical activity by virtue of its very link with the perspective of action:

> Perception is thus a paradox, and the thing perceived is itself paradoxical. It exists only inasmuch as someone can perceive it. Not even for one moment can I imagine an object in itself. Thus there is no such thing as an object in itself. All there can ever be is an individual who projects what his body can do with the world that surrounds it. Perception is thus an operation whose aim is to explore and evaluate the possibilities that the environment offers to an individual in the effective realization of a performance which the individual will have decided upon according to his criteria for choosing and the feasibilities at his disposal.[9]

William H. Warren was the first to apply the notion of affordances.[10] He selects two groups of men and specifies their physical characteristics; in one group the men are relatively short (163.7 cm or about 5 ft 5 in.) and in the other, the men are taller (189.8 cm or about 6 ft 3 in.). Each subject is confronted with several stairways with steps of varying heights. The subject is then asked to judge but without making an actual physical gesture, whether he believes he can climb the steps on his own two feet without the help of his hands. The highest step the subject believes he can climb is, not surprisingly, lower for the short men (67.13 cm) than for the tall men (81.32 cm). However the ratio between the height of the step and the length of the subject's legs in each group is equivalent to a critical threshold of 0.88 cm. In short,

[9] Merleau-Ponty 1996

[10] Warren 1984

a stairway is no longer considered "climbable" when the ratio between the height of the steps and the length of the legs exceeds 0.88, which corresponds to the motor activity described biomechanically. Accordingly, Warren proposes an initial mathematical formalization of affordance by calculating ((*pi*)), the ratio that describes the interface between the environment and the individual's metric in accordance with the discipline of biomechanics.

Other concrete applications of the conceptual field of affordances have also been studied since Gibson and are listed in the article by Marion Luyat and Tony Regia-Corte,[11] for instance, the "sitability" of surfaces,[12] the "pass-throughability" of openings,[13] the "posturability" of a surface,[14] the "grabability" of an object,[15] the "crossability" of ditches,[16] and the possibility of fitting underneath a barrier.[17]

The word affordance refers to the suffix-ability; it does not attach to a specific object but to a use that fulfils certain functions. The interface and the tool are not elements whose appearance as a phenomenon is univocal within the same object; on the contrary, they can be the focus of several possible perceptions or usages for the same specific object. An affordance refers to an action harbored in the functional utilization of a tool with concrete and potentially proteiform manifestations.

Donald Norman, both an engineer and psychologist, used the term in 1988 in the context of interactions between humans and machines. The concept of affordances refers to the individual's possibilities of taking action and her direct perception of them. In *The Design of Everyday Things*,[18] he fleshes out the notion of affordances from the standpoint of the design of an object and the environmental interactions that it permits. Affordance does not only depend on an individual's physical capacities but also on her goals, her values, and all of her contextual background. Norman maintains that an object's affordance "spontaneously" suggests to the viewer the way she is going to interact with the object and the way she can use it to achieve her ends. The notion relies on a relational concept, far from being inherent in a subject or in an object. Norman's approach is the same as Gibson's except that with Norman, it also has a purpose which depends on practical applications. Norman asks how an object can be designed so as to be as effective as possible for its user.

An object is not used in the same way nor therefore perceived the same way depending on the individual's resources. Different users will use different grids of interpretation with a view to the action they plan to take by means of the tools they dispose of. A pair of scissors is an object; let's see how it illustrates the key points that Norman brings out.

---

[11] Luyat and Regia-Corte 2009, p. 310.

[12] Mark 1987; Mark and Vogele 1987

[13] Warren and Whang 1987

[14] Regia-Corte and Luyat 2004; Regia-Corte et al. 2004

[15] Solomon and Turvey 1988; Carello et al. 1989

[16] Burton 1994; Burton and McGowan 1997

[17] Van der Meer 1997

[18] Norman 1988

The way a tool is used depends on the purpose and on the user's dexterity, be he a child, a hair dresser, or a sculptor. It is because an ordinary individual is endowed with two hands, each with four fingers and an opposable thumb, that he will be able to use the tool the way he plans to. The scope of his perception tends toward the types of action he can envisage; depending on the goal he chooses, he will have different expectations about the way the tool he uses should be designed. Nor will he develop the same capacities for using it – e.g., is he using scissors to cut paper? Or to cut something thicker, like leather? From one gesture to another, he will be using different degrees of motor strength or of the precision the user must control in order to avoid doing damage (a hairdresser, for instance, has only a very small margin of error). In this context, a tool can be considered an extension of the body and at the same time an element of separation from the body since the body alone without the tool is not always in a position to take the action he envisaged.

Thus the concept of affordances has the merit of bearing on the entire environment/body field of discussion. It goes further than the idea of a preeminent adjustment between the two members of that duo since it makes it possible to take into account the equilibrium of the two parts of a whole; this in turn encourages functionality and dialogue between them. But in what situations is this concept expressed visibly? Is it not in a situation where the individual's well-being is seriously damaged? Once the dialogue has been court-circuited, how can it be recreated? How can a set of possibilities guaranteeing a sufficient range of opportunities for action be reestablished so that the environment ceases to be neutral or even hostile and becomes hospitable and inviting?

14.3 A Blind Football 5-a-Side Team: An Example of Adjusting Affordances

To make clear the importance of the environment, we will describe a specific case in which blind adolescents regained the possibility of playing soccer ("football 5-a-side") (Cabrol 2014). To begin with, we must clearly identify what prevents a blind or partially sighted person from feeling at ease and able to play soccer. How can the rules of affordances be adapted when they have been pre-established and firmly anchored in the rules of a time-honored game? What are the points to focus on in order to recreate an effective equilibrium so as to enable people with specific physical requirements to play? Answers will emerge from the description of a project to set up a football 5-a-side team in Mali; the project began in the late 2011 and was supported by an association, Libre Vue. The main stages of the project are represented in the timeline below (Fig. 14.1).

This example shows how the feasibility of playing a team sport is above all the proof of a result that has been achieved; the visible result stems from a chain of actions and a series of steps. Each step has to be workable before the next one can be carried out. What preceded these efforts was far less visible. We have to realize

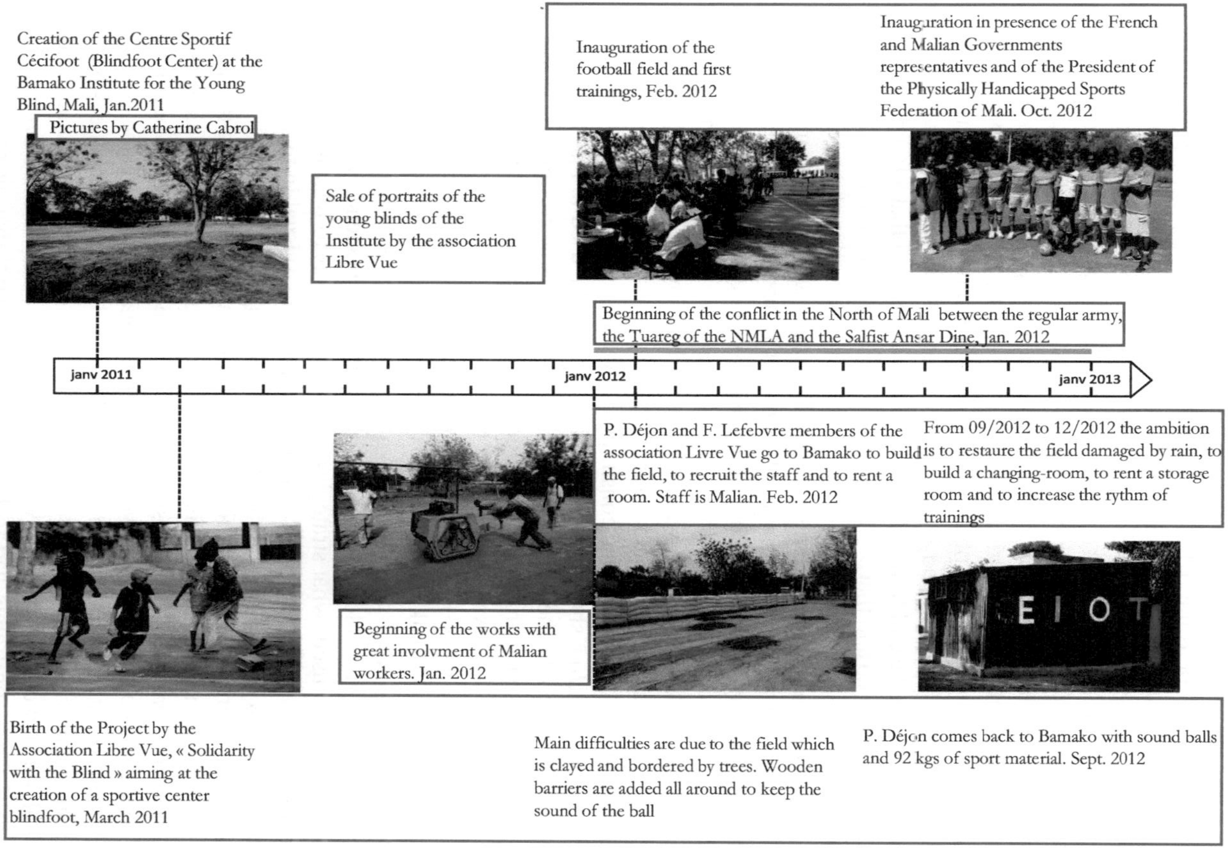

Fig. 14.1 The main stages of the project to set up a football 5-a-side team in Mali (2011–2013)

that each step in the process could have become an obstacle to each of the following ones if its purpose had not been achieved.

Catherine Cabrol, a professional photographer, instigated this project. Once she arrived in Mali to photograph blind or partially sighted children who lived in an National Institute for Young Blind People, she wished to do something to improve their quality of life by developing a suitable project. Soccer was a very popular sport in Mali, especially since 2002, when the 23rd African Cup of Nations was organized in Mali, so it was decided that soccer would be at the heart of the project. But how could a blind person play the game, even an adapted version of it? That is the first question an ordinary spectator (in the etymological sense of "he who observes") asks. How could the rules of ordinary soccer be modified so that blind people could engage in it even though their physiological state ordinarily excluded them?

After much discussion, and with the agreement of the educators who were in daily contact with the children at the institute, it was decided that a football 5-a-side team would be set up and that all the steps needed to do that would be planned. How should the work be organized? Starting from the ordinary rules of soccer, how could this project be firmly anchored in reality? What stages would that require?

Efforts would be needed on several technical fronts. First, creating a playing field, then building a clubhouse, providing appropriate equipment, and helping the young players to achieve a sensorial adaptation suited to the physical possibilities. Then, regular training sessions would be needed, to enable the young players to develop capacities that they could barely begin to perceive.

Given these specifications and constraints, how should the means available at the time be adapted? How should the available resources be used to make the project succeed? How should they be revealed and selected? How to organize a sort of "do-it-yourself" space in which to achieve each stage, which had to function effectively before the next stage could be broached? In terms of "affordance," what concerns us are the ways of getting around convention so as to readapt the ball itself. It was necessary to "think outside the box" rather than heeding statistical averages such as are found in all activities. How could a hitherto standard object be used differently so that it would be relevant to the specific competencies of each player? What tactics were used; what strategies put into play so as to reestablish a form of communication that dysfunctions as soon as a disability (visual in this instance) is factored in? How is the project to be pursued as time goes by?

When we attempt to identify the contextual constraints surrounding this project, we realize how many there are; this makes it harder to meet the specifications. The first constraint was Mali's overall situation at the time the project was beginning to take shape; the country did not have the means to organize such a game.

The terrain set aside for it by the local authorities did not lend itself easily to the creation of a playing field. Because of its heavy clay and its geo-climatic situation, easily flooded, work was often slowed by unpredictable weather conditions.

All of these constraints had to be taken into account; they were presented as relating to the setting in which the blind or partially sighted persons were used to living but which could be overcome by using a whole set of material, geographical, financial, and other resources. Those resources would make the project accessible to

Malian players of football 5-a-side. The key was to adopt a do-it-yourself outlook such as Claude Lévi-Strauss speaks of in *The Savage Mind* (Lévi-Strauss 1966), i.e., by coherently and judiciously assembling various means selected from among those available. This was the only way the desired goal could be reached – by organizing resources in such a way as to make them interdependent, each one augmenting the usefulness of the others, and ultimately making the project realizable.

The main elements that made it possible to put the situation to use were originally the organizers' motivation and inventiveness, but all of the participants likewise possessed those same qualities, and even more solidarity than was first expected emerged as time went by. People who lived near the National Institute for Young Blind People contributed an appreciable amount of unpaid assistance thanks to solutions that were invented on the spot. For instance, fences were made by stuffing straw into sacks used to carry millet; the auditory cues used to inform the players about their position were tin cans half-filled with pebbles. The project was legitimated by the local authorities and by media exposure, bringing the Bamako team's international recognition.

The rules by which football 5-a-side is played were adapted from the rules of the classic game. The adapted game is structured as follows. The field measures 40 meters by 20 meters. Playing time consists of two halftimes of 25 min each. Since information cannot be accessed or used visually, the game is oriented in such a way as to "translate" visual information into "auditory information." The equipment is adapted to that end. Inside the ball are a number of small bells. Audible beeps are fitted to the goal posts so that the players can situate them, and all around the playing field, there is a cord equipped with small bells. The goals are the same size as in handball and are either inflatable or equipped with protections so that the players will not bump into the edges. Each team consists of four players in the field and one goalkeeper, who is not blind. Similarly, the coach and the guide are sighted individuals. The goalkeeper is in charge of the defensive zone, the coach is in charge of the middle of the field and the attack, and the guide is placed behind the opposing team's goal.

When the ball is in play, each player must announce that he aims to move toward the one who has the ball: he must shout "voy" (Spanish for "I'm coming"). If not, he is penalized by a free kick or a yellow card. Much use is made of voices during the match as the players guide each other and reposition themselves. Yellow cards are counted as individual or collective errors. If a team has three collective errors, it is sanctioned by free kicks from a distance of 9 meters. A player who has accumulated five cards is excluded from the game. The player who replaces him must wait for 2 min of numerical inferiority before going onto the field. To equalize all degrees of partial vision or blindness, all players except the goalkeeper (who has normal eyesight) wear eyeshades or scarves tied over their eyes.

Once the rules were agreed upon, the project was carried out on a field measuring 1500 square meters (50 × 30), with a wooden fence along each length to keep the ball in play. When the field was constructed, the dimensions were deliberately enlarged compared to the normal official dimensions (40 m × 20).

Within that space, allotted by the UMAV (Malian Union of Blind People), six trees had to be uprooted, and then tons of earth (36 trucks, each with a capacity of 12 cubic meters) had to be added to offset the natural slope of the terrain. The photographs taken as the work proceeded for 21 days clearly show how closely involved the locally recruited Malian workers were and how vigorously they worked.

So that the players would not injure themselves by bumping into the fences, a system of protection had to be invented. The means available were modest. The organizers recuperated sacks usually used to carry wheat or millet. The seams were unstitched and then stitched again to the requisite dimensions (2 meters long by 1 m20 high), and the sacks were then stuffed with straw like makeshift mattresses. These made the conditions in which the game was played safe and acceptable for all.

A clubhouse was built, Malian style, using second-hand sheet metal and adding rustproof paint. Since the clubhouse was near the playing field, the protective mattresses were stored in it after each training session. The walls also served as a bulletin board.

Protective clothing had to be provided, along with gear of all sorts: socks and shoes in various sizes, jerseys, shorts, shin guards, gloves for the goalkeepers, whistles for the referees, megaphones for the coaches, blindfolds for the players, and, above all, balls equipped with bells inside. All of these items were imported directly from France through the specific financing of the project.

One of the vital challenges arising from this creativity was to continue the momentum by renewing or repairing whatever material objects were of direct or indirect interest to the players, whether their own equipment, the clubhouse, or the protective mattresses. Provision also had to be made for restoring the playing field after each rainy season, when the grass grows quickly and the ground develops gullies. More truckloads of soil had to be brought and sifted, some of the wooden poles had to be tarred to ward off termites, and fences needed repainting to weatherproof them. To keep the project ongoing, three regular training sessions per week were begun at the start of the 2012 school year. The six-member permanent staff, three men and three women, were all Malian; three of them had official diplomas as sport teachers and monitors authorized to work with young people. They were on hand at every training session – 1 h on Wednesday, 2 on Thursday, and 3 on Saturday. Each session included warm-up, relaxation, practice in following a sound-equipped ball, passing, kicking, and dribbling. Matches were organized.

From this example, it is clear that if the environment is modified with clear goals in mind, it can enable individuals to benefit from an improved quality of life. Keeping this example in mind, can we now draw more general conclusions about the link between affordances and disability?

14.4 The "Space" of Affordances in the Context of a Disability

The entire "space" of affordances is short-circuited, to begin with, when a disability is present, as we have just seen; it becomes apparent that the individuals concerned can no longer communicate with the interface of their environment. They are obliged to rebuild other mediations that will enable them to reestablish a lasting dialogue with their environment. One of La Fontaine's fables, "The Fox and the Stork," will make our point clear:

One day Mr. Fox decided to fork out
And invite old Mrs. Stork out.
The dinner wasn't elaborate –
Being habitually mean,
He didn't go in for haute cuisine –
In fact it consisted of a shallow plate
Of thin gruel.
Within a minute
Our joker had lapped his plate clean;
Meanwhile his guest, fishing away with her beak,
Got not a morsel of it.[19]

When the Stork arrives at the Fox's house, she cannot eat what the Fox offers her because the environment she suddenly finds herself in suits only to the Fox when he wants to eat, but does not suit her: her beak is too long. The problem for the Stork is the plate, so, when she in turn plays host to the Fox but in her own environment, she adapts or replaces the plate by a vase. She serves the food:

In a long-necked narrow-mouthed urn.
The Stork, easily stooping,
Enjoyed her fill.
With her long bill;
His snout, though, being the wrong shape and size,
He had to return to his den.
Empty-bellied, tail dragging, ears drooping....[20]

The tables have been turned, and the disability has been reversed, just as in the previous example of football 5-a-side, where a sighted person, who has been blindfolded, becomes disabled among the blind players even if he has a natural disposition for playing soccer. We see that the choice of objects, judiciously revised to suit the possibilities of a user who is initially disabled, can enable him to recover a role as an active participant in the world we all share.

We note that this transition occurs through several stages or elements; to all of them, the cooperative dimension is always central. The first element concerns the ways the necessary information is conveyed: that stage is organized, in this case, on the basis of the actual means the individual disposes of in order to receive it and on

---

[19] La Fontaine 1979

[20] Ibid.

the basis of his inventiveness. When people are equipped with different tools, it is logical for them to find different ways to decipher reality and have an impact on it.

A practicable path is created, or recreated, toward an order of action that takes into account the psychophysical and psychological possibilities of the individual concerned. He or she attempts to rearrange the surrounding conditions so that they will "match" him or her better, but the quest is difficult because they dispose of fewer "possible" means than those available to an ordinary person in the previous conditions. So, perceptual and motor capacities have to be reallotted and developed.

In this reorganization process, the logic of "recycling" is key. The notion of a circular process is clearly contained in "cycle," referring to an ecosystem where all of the elements are interdependent. The individual creates a new way of apprehending reality, not by creating part of it ex nihilo but by recombining certain sketches or fragments already present in order to arrive at a more orderly situation on his own scale. In the football 5-a-side example we have seen, the notion of recycling is clearly present; the solution was not to import means that did not exist on the spot but to discover alternatives that could be put to different uses. Sacks for millet were redimensioned and stuffed to become protective mattresses, and a tin can was filled with pebbles to become a ball that players could locate because they could hear it.

In addition, these blind children needed to mobilize all of their perceptive capacities and their mobility. They learned to fine-tune their auditory sensitivity and proprioceptive awareness to replace sight and everything that sight implies.

Such readjustment is among the operations that can be described and grouped under the heading of exaptation, proposed by Stephen J. Gould and Elisabeth S. Vrba. It designates the internal reorganization of the possibilities contained within the individual's body and within the environment. Here, it is the environment that is adapted through reliance on available resources since, when a disability is involved, some of the resources that hitherto assured the coherence of an equilibrium achieved in the ordinary way are missing.[21]

14.5 Conclusion

Taking the environment into account is indispensable when restoring the individual's well-being; this article explores the possibility of modifying a situation marked by disability. We can also point out that in the readjustment process, the affordance itself disappears since a balance between the body and environment is established anew; equilibrium is well constructed new; dialogue becomes possible anew. It can only be artificial to dissociate the body and the environment for, in invisible ways, they maintain a skillful interaction when they form a functional whole. Unless this relationship is clearly understood, any attempt to think about health or treatment will inevitably lead to errors whose practical consequences, especially in the fields of health and treatment, are harmful to the individual. How can you treat a patient if

[21] Gould and Vrba 1982.

you confine him or her within their illness? If you only consider the deficient body and the environment in which it is immersed? As we can see from the example of young blind children who are enabled to take part in a game of soccer, the improvement of someone's well-being is not just a "local" matter; it is not based solely on a "localized" adjustment targeting only the individual's body without any reference to the environment. On the contrary, a situation where a disability is present shows us that the entire environment is concerned in the interaction between the individual and environment, so that the individual can accede to a better quality of life.

References

Burton, G. 1994. Crossing without vision of path gaps. *Journal of Motor Behavior* 26 (2): 147–161.

Burton, G., and J. McGowan. 1997. Contact and posture in nonvisual judgment of gap crossability. *Ecological Psychology* 9 (4): 323–354.

Cabrol, C. 2014. Rapport d'activités du projet "Solidarité Aveugle". In *Création d'un Centre Sportif Cécifoot à l'Institut des Jeunes Aveugles de Bamako, au Mali*. Fontenay-Sous-Bois: Association Libre Vue.

Carello, C., A. Grosofsky, F. Reichel, H.Y. Solomon, and M.T. Turvey. 1989. Visually perceiving what is reachable. *Ecological Psychology* 1 (1): 27–54.

Gibson, J. J. 1979. *The Ecological Approach to Visual Perception*. Psychology Press.

Gibson, J.J. 1977. "The theory of affordances", dans R. Shaw, J. Bransford (sous la direction de), *Perceiving, Acting, and Knowing. Toward an Ecological Psychology*, 67–82. Hillsdale, Lawrence Erlbaum Associates.

———. 1986. *The Ecological Approach to Visual Perception*. Hillsdale: Lawrence Erlbaum Associates.

Gould, S.J., and E.S. Vrba. 1982. Exaptation. A Missing Term in the Science of Form. *Paleobiology* 8 (1): 4–15.

Koffka, K. 1936. *Principles of Gestalt Psychology*. New York: Harcourt Brace and Company.

La Fontaine, J. 1979. *Selected Fables*, Penguin Classics (trans. James Michie).

Levi-Strauss, C. 1966. *The Savage Mind*. London: Weidenfeld and Nicolson.

Lewin, K. 1926. Untersuchungen zur Handlungs-und Affektpsychologie II. Vorsatz, Wille und Bedürfnis. *Psychologische Forschung* 7: 330–385.

Luyat, M., and T. Regia-Corte. 2009. Les affordances: de James Jerome Gibson aux formalisations récentes du concept. *L'année psychologique* 109 (2): 297–332.

Mark, L.S. 1987. Eyeheight-scaled information about affordances: a study of sitting and stair climbing. *Journal of Experimental Psychology: Human Perception and Performance* 13 (3): 361–370.

Mark, L.S., and D. Vogele. 1987. A biodynamic basis for perceived categories of action: a study of sitting and stair climbing. *Journal of Motor Behavior* 19 (3): 367–384.

Merleau-Ponty, M. 1996. *Le primat de la perception et ses conséquences philosophiques*. Verdier: Lagrasse.

Norman, D.A. 1988. *The Design of Everyday Things*. New York: New York Basic Books.

Regia-Corte, T., and M. Luyat. 2004. Dynamic constraints on haptic judgment of slanted surfaces. *Current Psychology Letters* 12 (1).

Regia-Corte, T., M. Luyat, J.-C. Darcheville, and Y. Miossec. 2004. La perception d'une affordance de 'posturabilité' par les systèmes perceptivo-moteurs visuel et haptique. *L'Année Psychologique* 104: 169–202.

Solomon, H.Y., and M.T. Turvey. 1988. Haptically perceiving the distances reachable with hand-held objects. *Journal of Experimental Psychology: Human Perception and Performance* 14 (3): 404–427.

Stoffregen, T.A. 2003. Affordances as properties of the animal-environment system. *Ecological Psychology* 15 (2): 115–134.

Turvey, M.T. 1992. Affordances and prospective control: An outline of the ontology. *Ecological Psychology* 4 (3): 173–187.

van der Meer, A.L.H. 1997. Visual guidance of passing under a barrier. *Early Development and Parenting* 6: 149–157.

Warren, W.H., Jr. 1984. Perceiving affordances: visual guidance of stair climbing. *Journal of Experimental Psychology. Human Perception and Performance* 10 (5): 683–703.

Warren, W.H.Jr., and S. Whang. 1987. Visual guidance of walking through apertures: body-scaled information for affordances. *Journal of Experimental Psychology: Human Perception and Performance* 13 (3): 371–383.

Chapter 15
Gestalt Therapy and Its Contribution to the Understanding of the Link Between Health and the Environment

Perrine Michon

Abstract Resolutely inscribed into a sensitive and dynamic approach that emphasizes feelings and the perception of the body, Gestalt therapy focuses on the processes of contact between the organism and the environment, during which the forms (*gestalten*) are shaped and reshaped. This approach departs from the usual description of a subject in his/her environment to shed light the processes of interrelation *between* the subject and his/her environment. It abandons the classical and static pattern of separation between the inside and the outside, between the self and the other to focus on the modalities of contact, to its dysfunctions, to the mechanisms of regulation that arise, and to the permanent adjustment between one person and his/her environment. This premise that organism and environment are inseparable is innovative because it challenges the individualist and solipsistic view of the self. The Gestaltist innovation lies less in the libertarian minds of its founding fathers, who favored the expression of the body and of the emotions in normative societies than in its invitation to convert our usual way of thinking. The perspective of an "organism-environment" field invites us to escape from dualism by taking into account, in one situation, all the different elements that assemble to coexist, to leave the binary dynamic ("either you, or me") to consider the dynamic of a co-construction ("and you, and me"). Therefore, this innovative paradigm has effects on the conception of health and/or care: health is an unstable equilibrium, and disease is not reduced to organic deterioration but defines a way of living, an integral part of the human existence.

Keywords Gestalt · Organism · Environment · Contact · Self · Health

P. Michon (✉)
Lab'Urba EA 3482, CNRS-Université Paris Est Créteil UPEC, Creteil, France
e-mail: p.michon@u-pec.fr

F. Bretelle-Establet et al. (eds.), *Making Sense of Health, Disease, and the Environment in Cross-Cultural History: The Arabic-Islamic World, China, Europe, and North America*, Boston Studies in the Philosophy and History of Science 333, https://doi.org/10.1007/978-3-030-19082-8_15

15.1 Introduction

Gestalt therapy offers a stimulating conception of the relationship between the individual and the environment by positing the inseparability of the organism and the environment as a fundamental principle: primarily, organism and environment are in permanent and unintentional contact. Whatever we do, whatever we want, we are always connected to our environment. The process of breathing illustrates this double, permanent, unintentional movement between the organism and the environment that causes the organism to be permanently influenced by the environment and the environment influenced by the mere presence of the organism. By breathing in air, I absorb the oxygen necessary for my biophysical function, and, in return, I modify this air by, on expiration, rejecting, CO_2. The founding text by F. Perls, R. Hefferline, and P. Goodman proposes an original way of understanding the individual, his/her health (in the broad sense – psychological and physical), and his/her relationship to the environment.[1] They abandon the static, binary model of separation between the internal and the external, between me and not-me, and therefore between myself and others, or between the individual and the environment, to propose a vision based on the inseparability of organism and environment, which is essentially concerned with the contact between these two instances rather than with one or the other. Gestalt therapy focuses on the relationship, on the process of organism-environment contact, rather than on the individual or on the environment, considered as distinct and differentiated entities.

Included in the field of therapies with a holistic approach that were progressively constructed in the course of the twentieth century, in Europe and in the United States, Gestalt therapy is part of a lineage and in a critical position in relation to psychoanalysis. As the Germanic half of its name indicates, Gestalt therapy is interested in the "form" that emerges, in the figure that takes shape.[2] Beyond the linguistic difficulty that is represented by the absence of a French or English translation[3] of the term "gestalt" – which may have contributed to "fascinate and to diffuse a mysterious halo that stirs up curiosity without clearly revealing its identity,"[4] the term

[1] Frederick S. Perls, Paul Goodman, and Ralph Hefferline (2001, first re ed. 1951).

[2] "Gestalt," which can be translated as "shape," is the substantive of the German verb "gestalten," which means "take shape, create itself, and organize itself" and contains this idea of movement and process, which the translation does not fully capture.

[3] The choice of the term to designate this line of thinking and its therapeutic practices was the subject of debate and hesitation within the "group of 7" who participated in the institution and development of Gestalt therapy in the United States. F. Perls, who was more in the tradition of W. Reich, favored "concentration therapy," while Laura Perls preferred the term "existential therapy." R. Hefferline suggested "integrative psychotherapy" or "experiential therapy." Finally, the expression "Gestalt therapy" was adopted, referring to Gestalt theory. The vagaries of this denomination show that Gestalt therapy is the result of a collective co-construction, and not of a single theoretician, and testify to the hybridization of different practices and the assimilation of various traditions.

[4] Chantal Masquelier-Savatier 2015, p. 3.

"gestalt" refers to the idea of a movement, of an unceasing process. This idea of movement, this ever-changing and constantly renewed encounter between the organism and the environment, is at the heart of the approach of Gestalt therapy and the foundation of its principal theory: the theory of the *self*. "Let us call the "self" the system of contacts at any moment. As such, the self is flexibly various, for it varies with the dominant organic needs and the pressing environmental stimuli; it is the system of responses; it diminishes in sleep when there is less need to respond."[5]

In its therapeutic approach, Gestalt therapy addresses the form that appears in the encounter between the patient and the therapist; it is interested in the form that emerges. The therapeutic objective will be to set in motion forms that are rigid or trapped in an endless cycle of repetition – called unfinished "gestalts" – and to re-establish the exchanges between the organism and the "environment or make them more fluid. In this approach, the environment is taken in a very broad sense, ranging from the air that one breathes, to the individuals with whom one interacts.

Beyond the therapeutic methodology and tools, the Gestalt proposition and position can also be taken and understood as an invitation to rethink the relationship of man and the environment and to invite another way of being in the world. The foremost postulate, at the heart of this approach, posits the inseparability of the organism and the environment: in doing so, it makes it possible to envisage the contact between man and the environment as a permanent, dynamic, and consubstantial process for the identity and especially for the construction of the individual. It is this permanent process of transformations, adaptations, modifications, and unceasing adjustments that shapes the individual. At the same time, this postulate renders as inoperative and irrelevant the externalization or objectivation of the environment as something external to oneself, objectivable, even reifiable, and hence quantifiable or controllable. *Ultimately*, this paradigm of organism/environment inseparability invites us to operate a triple de-centration that leads us to re-examine our relationship to the environment, to the other, and to oneself.

Thus we will see firstly (I) how the Gestalt approach is perhaps a solution to overcoming the contemporary environmental aporia and the ecological dead-end which modernity has reached by externalizing man from his environment, separating the subject from the object and making space a mere abstraction in which men or objects (especially urban or architectural) can settle and spread out indiscriminately. In a second phase (II), we will see how the originality of the Gestalt approach is to take an interest in the interval, in the interrelation, and in contact (and in its different modalities and ways this unfolds) – through the prism of the theory of "self." Then we will see how this way of observing exchanges and contacts between man and his environment induces a necessary change of view on three levels (III): in our way of perceiving the environment, the other, and oneself.

[5]Perls et al. 1951, 2001, p. 58.

15.2 What Is the Link to the Environment in the Postmodern World?

15.2.1 A Weak Link to the Environment?

In a book hailed as "truly original" and bold by the journal *Science* that appeared in 1996 (published in French in 2013, under the title *Comment la Terre s'est tue. Pour une écologie des sens*),[6] the American philosopher and ecologist David Abram[7] seeks to understand: "Trees rarely, if ever, speak to us; animals no longer approach us as emissaries from alien zones of intelligence; the sun and the moon no longer draw prayers from us but seem to arc blindly across the sky. How is it that these phenomena no longer address us, no longer compel our involvement or reciprocate our attention?"[8]

In this book that seeks to re-establish a lost symbiosis between our senses and the world, David Abram attempts to explain the process by which civilization has closed in on itself, isolated itself from the "Earth that breathes." It starts from the observation that we are today: "so estranged from the stars, so utterly cut off from the world of hawk and otter and stone,"[9] we have cut ourselves off from our sensual and sensory landmarks, from other animals, and from the living earth. His objective is to reactivate and re-entrench the human experience within a wider ecology.[10] The human mind is not a reality of another world that has come to lodge in our physiology. Instead it is instilled, aroused by the sensory field itself, driven by the tensions and the involvement between our body and the living earth: "The invisible shapes of smells, rhythms of cricketsong, and the movement of shadows all, in a sense, provide the subtle body of our thoughts. Our own reflections, we might say, are a part of the play of light and its reflections. "The inner—what is it, if not intensified sky?"[11]

Recognizing such links between the inner psychological world and the world we perceive around us allows us to begin to turn inward, to free the psyche from its imprisonment within a purely human sphere, to let our capacity to feel make a return to the sensuous world that includes us. Intelligence is not exclusive to us, it is a property of the earth; we are in it, immersed in its depths. And as a result, each place, each ecology, seems to possess its own intelligence, its own dialect of soil, of leaves, and of sky. "Each place its own mind, its own psyche. Oak, madrone, Douglas fir, red-tailed hawk, serpentine in the sandstone, a certain scale to the

[6] David Abram 1996 (2013, for the French edition).

[7] David Abram occupied the "Arne Naess Chair in Global justice and the Environment" at the University of Oslo in 2014.

[8] Abram 1996, p. 130.

[9] Abram 1996, p. 261.

[10] For David Abram, this break between us and the environment constitutes a break that is of the order of a brutal ecological transformation.

[11] Abram 1996, p. 262.

topography, drenching rains in the winter, fog off-shore in the summer, salmon surging in the streams—all these together make up a particular state of mind, a place-specific intelligence shared by all the humans that dwell therein, but also by the coyotes yapping in those valleys, by the bobcats and the ferns and the spiders, by all beings who live and make their way in that zone. Each place its own psyche. Each sky its own blue."[12]

The issue here is not to seek to identify the temporal origin of this ecological rupture or of this lack of dialogue between man and the environment. The works of certain authors, such as Augustin Berque, associate this rupture and this dissociation of subject and object with the entry into modernity, marked by the works of Descartes and Newton. The concept of the modern man has abstracted man from his environment, making him a subject, surrounded by objects that he can or cannot master or possess but which are not part of his subjective being. This cleavage, posited by Descartes in the famous "cogito ergo sum" and developed by physics, introduces a dissociation between man and his environment that endures today. The issue here is more to shed light on the Gestalt position and see how the approach proposed by this line of thinking and the therapeutic practices based on it allow us to envisage the link between man and his environment in a different way.

15.2.2 The Gestalt Postulate: The Inseparability of Organism and Environment

The postulate which lies at the base of the Gestalt therapy approach posits, from the outset, and without this being due either to intentionality or will, permanent, ever-present, and always foremost contact between the organism and the environment. This postulate perhaps opens a way to rearticulate man and his environment in a sensory and bodily dimension. The hypothesis that we posit here is that looking at man and health issues from a Gestalt view and position may help to overcome in part the environmental disassociation and ecological rupture pointed out by David Abram in his anthropological reading of human relationships to the Earth and open up valuable and fruitful avenues in considering and questioning the relationship between health and the environment.

Gestalt proposes a new relational intelligence with our environment by moving the focal point from the individual, or the environment, to *the contact* between the organism and the environment, which is analyzed and deciphered through the "contact cycle, also called in the theoretical work which founds the Gestalt approach[13]: "the theory of the self." This position that decenters the view of the subject him/herself, or of the environment itself taken as an object of study, makes it possible to envisage other relational modalities between the individual and the environment

[12] Abram 1996, p. 262.

[13] Fritz Perls, Paul Goodman, and Ralph Hefferline, 2001 (1st ed. 1951).

and invites greater interest to be taken in the relationship and the interactions that occur between them. The point of departure, which will enable us to highlight what is playing out in the relationship, is what is felt and the attention paid and given to sensory and corporeal information.

In a therapeutic approach, it is these interruptions or alterations in contact that are the focus of the therapist's attention: Gestalt therapy considers that pathology results from interruptions or dysfunctions in the contact between organism and environment and that it is on these interruptions, or inhibitions to the contact, that the therapist's attention and therapeutic work focus. This way of paying attention to the contact and considering the pathology as the fixist reproduction of the same type of behavior and not adjusted to the situation can help to clarify various behavioral disorders (from the confluence which arises from the difficulty in differentiating and in distinguishing oneself from one's environment to egotism which is characterized, on the contrary, by a disconnection from one's environment). The objective is then, in a process of co-construction between the gestalt therapist and the patient, to set in motion rigid forms of contact, so that this permanent process of contact and of adjustment between the organism and the environment can take place in the most fluid way possible, which is at the very heart of the individual's growth toward greater self-fulfillment and autonomy.

"Gestalt therapy aims to set rigid forms in motion in order to improve the flow of exchange between the organism and the environment. It does not concern itself with the psyche, as a pre-existing entity, but it is interested in the experience of contact, in the hypothesis that the sedimentation of experiences constitutes the psyche."[14] Gestalt therapy proposes moving from an intrapsychic to an interpsychic approach where the individual cannot be separated from his/her environment.

15.3 The Gestalt Position: Another Way of Being in the World

Gestalt therapy is a practice that was constructed in reaction to psychoanalysis, as developed by Freud. Created by Fritz Perls, a German psychiatrist and psychoanalyst, a disciple then a dissident of Freud, and his wife Laura Perls, a dancer and doctor of psychology, the practice of Gestalt therapy was actually born in the United States where the Perls emigrated in the 1940s after having spent some time in South Africa. This is where the so-called group of 7, a prolific group of thinkers, composed of intellectuals from different backgrounds, who were found around Fritz and Laura Perls, was formed. From this collective and co-constructed reflection, the practical and theoretical foundations of Gestalt therapy were born, which were brought together in 1951 in a collective work entitled *Gestalt Therapy: Excitement and Growth in the Human Personality*, co-written by F. Perls, R. Hefferline, and

[14] C. Masquelier-Savatier 2015, p. 3.

P. Goodman. The first Institute of Gestalt therapy was founded in 1952 in New York. Then Fritz Perls crossed the United States and developed his experimental approach on the west coast. This practice would recross the Atlantic in the 1970s to be part, in France, of the field of psychophysical and emotional therapies that called into question the approach and position of the analytical cure in psychoanalysis.

15.3.1 History and Foundations: An Integrative Approach

Gestalt therapy draws on various sources and positions itself in the prolongation of the different currents which emerged more or less concomitantly in the cultural ferment of the late nineteenth and early twentieth centuries in Europe. Individually, the founders of this practice came from the field of psychoanalysis which would remain a reference, or an important counter-reference in relation to which Gestalt positioned itself. At the same time, Gestalt theory (or psychology of form), which was very prevalent in Germany at the beginning of the twentieth century, following in particular the works of Franz Brentano and his pupil Christian von Ehrenfels, also constituted a strong frame of reference, which underlies the Gestalt approach and proposition. Similarly, as a continuation of this trend, Kurt Lewin, who also emigrated to the United States, conceptualized field theory that strongly influences Gestalt therapy and applied this principle to group dynamics. Among the great philosophical currents which imbued the founders of Gestalt therapy, phenomenology (Husserl, Heidegger), existentialism (Sartre, Buber, Binswanger), and pragmatism (James, Dewey) are part of the common cultural background of the founders of this practice and mark the Gestalt position and the emphasis on subjective experience and the multiplicity of ways of seeing the world.

15.3.2 The Rebellious Child of Psychoanalysis

The main disagreement with S. Freud lies in the conception of child development and the elements of growth. F. Perls compares the stages of a child's growth with the relationship to food and the steps required to assimilate a foodstuff.[15] For both Perls and Freud, introjection[16] is a necessary step: initially, the subject accepts the external world with which he/she identifies totally. The child introjects the external world, that is, he/she assimilates it entirely without questioning or rejecting it. For Perls, growth then necessitates the destruction of the introjects, swallowed or absorbed in childhood, whereas for Freud, it is necessary to incorporate the values of previous generations, the law and social rules, through the constitution notably of

[15] F. Perls 1942 (1978 for the French edition).

[16] The concept of introjection was introduced by S. Ferenczi, a Hungarian psychoanalyst, in *Introjection and Transfer* (1909).

a superego. For Perls, just as the body retains the food it needs and rejects unnecessary or unwanted material, so the subject is invited to develop his/her creative potential and "aggressivity" (in the sense of the Latin etymology of the term "ad-gredere") to "go toward" and "get in touch with" his/her environment and, as he/she grows – or progresses therapeutically – to harvest that which he/she feels he/she needs, to choose, to sort, and to reject, especially among the introjections inherited from his/her past history. While the Freudian model relies on the management of desire and of repression to explain the identification in the adult and the construction of the subject, the Persian model starts from the need of the individual and his/her orientation in the environment to satisfy it in a dynamic of healthy aggressivity.

Gestalt therapy was also inspired by the work of Wilhelm Reich, himself a disciple of Freud. W. Reich distinguished himself from his master by attributing the origin of neurosis to the disorders of sexual satisfaction engendered by social repression. He forged the concept of "muscular armor," as a result of bodily tensions induced by the repression of impulses. It is then necessary to relax this rigid constraint to release physical and emotional energy. This view of things induces a different therapeutic position to psychoanalysis since Reich leaves the analytical reserve to become involved in contact with the patient. Gestalt therapy aligns itself with the Reichian tradition by giving an important place to the body and the emotions. However, there are some differences: in Gestalt therapy, it is not a question of attacking the armor, but of assisting the transformation of a resistance that may have been sound, useful, and suitable at some point in time, in a given environment. On the other hand, emotional release is not sought in itself in the process of Gestalt therapy, but emotion is welcomed as part of the process and as a possible lever for transformation. Moreover, by supporting the unity of organism and environment Gestalt therapy stands out from the Reichian vision, which reinforces a kind of dualism between the internal and the external, between a naturally good man and a repressive and bad society. Finally, the intervention of the therapist in Gestalt is less marked than in the Reichian perspective: the therapeutic work in Gestalt is a co-construction: awakening awareness in and through a copresence.

15.3.3 From Gestalt Psychology to Gestalt Therapy

Through the German substantive chosen to name this new practice, Gestalt therapy marks its direct connection with Gestalt psychology, or "psychology of form," that was on the rise and very influential in Germanic countries of the late nineteenth century.[17]

[17] The practice of Gestalt theory (or psychology of form), which developed research on perceptual organization (notably by elaborating Wertheimer's laws), brings together the works of Wolfgang Khöler, Kurt Koffka, and Max Wertheimer who founded the Berlin school before being forced to flee and emigrated to the United States.

Gestalt psychology is a psychological, philosophical, and biological theory according to which the processes of perception and mental representation spontaneously deal with phenomena as structured sets, as forms, and not as a simple addition or juxtaposition of elements that the brain aggregates and assembles. Contrary to the associationist or elementary tendency, the psychology of form formulates the opposite hypothesis: the landscape is unified not by the result of elementary operations but by a prime principle which acts at the very moment of perception. This unitary principle is called "form" or "gestalt." Christian Von Ehrenfels, a pupil of Brentano, who theorized the notion of form in 1890 in a work *Über Gestaltqualitäten*, emphasizes the law of constancy of form and illustrates it with a musical example: when we remember a melody, we recall a global musical structure and not a succession of notes taken in isolation.

15.3.4 The Theory of the **Self**

Born in the intellectual effervescence and cultural intermingling of early twentieth-century Europe, colored by the migratory path of its founders, rebellious heiress of psychoanalysis, and incorporating the contributions of Gestalt psychology regarding perception and the field perspective notably, Gestalt therapy differentiated itself from this legacy and forged its own identity on the other side of the Atlantic, and in particular on the basis of the theory of self. Gestalt therapy's principal and innovative contribution lies in the theory of *self* – which is as quick to say as it is complex to feel and understand in all its depth. The theory of self, also called contact cycle or contact sequence, offers an original checklist for considering the interactions between the individual and his/her environment. The theory of self-emerged from the usual description of a subject in an environment to endeavor to shed light on what happens between the subject and its environment – without prejudging what belongs to the subject and what is part of the environment – during this contact process. The self is this system of contacts at all times, this encounter perpetually in progress, perpetually moving between the organism and the environment. It is therefore not an entity that can be determinable and localizable: it has no existence of its own on the spatial or temporal plane; it is "the contact in the process of happening."[18] The self therefore does not designate the subject or the person; it is the process in action between the organism and environment. The self is not a pre-existing reality that reveals itself on the occasion of contact; it is the very fact of contact.

[18] C. Masquelier-Savatier 2015.

15.3.5 *"It Is the Contact That Is the Primary Reality and the Simplest"*[19]

In the Gestalt proposition, this system of contact is foremost: every organism, by existing, is in contact with the environment. Whatever the modalities, there is always contact between the organism and the environment. The individual is always in contact whether he/she likes it or not. Contact is not intentional: every organism is necessarily in contact with its environment; it is a state of fact. In this sense, the contact is foremost, anobjectal, and precedes the relation. According to Jean-Marie Robine, contact is the primary experience: "contact does not designate the relationship yet, but simply what articulates the subject with that which is not-me, human, or objectual."[20] The Gestalt approach, focusing on the self, emphasizes this primordial experience: contact is the primary and inevitable experience as the organism is continually in contact with the environment. "Contact is the phenomenon that precedes any organization of experience: pre-Oedipal, pre-objectual, pre-conscious, pre-representational, pre-emotional, pre-psychic. Before being sedimented into a 'psyche,' the experience was contact, and will be thereafter, since it is in and through the contact that the psyche will be able to 'ex-s ist' and preside over subsequent experiences. The pre is genesis but it is also structure."[21]

An infant in its cradle is in contact with the environment from the first second of its life: the first cry at birth, seen by the medical profession as a sign of good health and that everything has gone well in the birth, can be considered as the first sign – ex utero – of this inseparability of organism and environment and this pre-existing contact, foremost and before any other experience and any other relationship. It is the air which, by infiltrating the lungs of the newborn child, makes it cry out from the moment it appears in its new living environment: visible, manifested, and forceful dependence between the newborn and its environment illustrates the system of continuous and continual exchange between the organism and environment that will endure at every moment of life.

15.3.6 *Contact Is What Allows Growth*

This process of contact is vital: it is what allows the organism to survive and grow:

"The description of psychological health and disease is a simple one. It is a matter of the identifications and alienations of the self: if a man identifies with his forming self, does not inhibit his own creative excitement and reaching toward the coming solution; and conversely, if he alienates what is not organically his own and therefore cannot be vitally interesting, but rather disrupts the figure/back-ground,

---

[19] F. Perls 1942, p. 49.

[20] Jean-Marie Robine 1998.

[21] J.-M. Robine (2010), p. 3.

then he is psychologically healthy, for he is exercising his best power and will do the best he can in the difficult circumstances of the world. But on the contrary, if he alienates himself and because of false identifications tries to conquer his own spontaneity, then he creates his life dull, confused, and painful. The system of identifications and alienations we shall call the 'ego'."[22]

For Perls, the primary and fundamental movement in the human being is a movement of contact, of going toward, of displacement of the organism toward the environment: this movement of encounter between the organism and the environment in which the organism will take and destroy something from the environment, which will transform it and which in turn will transform the environment, is primordial and more fundamental than the impulses of life and death and repression in the unconscious, put forward by the Freudian approach and psychoanalysis. This process or contact is continuously active and strongly mobilized, as the environment is perpetually changing. To adapt to these ceaseless environmental movements, the individual develops a certain number of adjustments, which are said to be either creative or conservative – without either of these names having in themselves a connotation of value or judgment. A conservative adjustment can be perfectly suited to certain circumstances. It becomes invalid if it persists in time or if it is systematized to all or part of the new situations encountered. Creative adjustment consists in inventing new forms of being in contact, in imagining new ways of being in the world. This balance between the adjustments is the essential function of the self: "the fluidity between conservation and creation is indispensable. A creative adjustment without conservative adjustment would probably result in forms of hysteria, mania or madness. A conservative adjustment alone could lead to depression."[23]

The founding fathers of Gestalt therapy's vision of health was directly colored by this central position accorded to the *self*, to this system of contacts and to its modalities of regulation. Thus, for Perls and Goodman, pathology results from interruptions, inhibitions, or other accidents of contact that prevent creative adjustment from occurring and allowing a fluid and healthy exchange between the organism and the environment. "The anxiety of neurosis would be the consequence of brakes or halts in the harmonious development of the rise of excitement and the neurotic characters would result from rigid structures hindering the flexibility of creative adaptation."[24] The secure framework of the therapy would provide the opportunity to look at these blockages and to experiment with new contact processes during the therapeutic alliance between the therapist and the patient.

[22] F. Perls et al. 1951, p. 235.

[23] C. Masquelier-Savatier 2015, p. 66.

[24] Ibid.

15.3.7 The Contact Boundary: The Metaphorical Place for These Exchanges Between Organism and Environment

This perpetual process of contact takes place at the contact boundary, a metaphorical place for these constant interactions between the organism and the environment. The self can even be conceived of as "the contact boundary at work": "The self is the contact boundary at work; its activity consists is forming in forming figures and grounds."[25] The contact boundary simultaneously connects and differentiates. This metaphorical place is perhaps the most subtle and difficult concept of Gestalt therapy. The contact boundary is a virtual, symbolic space in which the process of contact between the organism and the environment expresses itself. It does not exist alone or in itself, it co-constructs itself in the exchange between the organism and the environment. It is therefore a space and virtual time: it is the place and the moment when the organism contacts the environment and is contacted by it. It can thus be compared with the skin, the only sensory organ that involves the passive and the active: one cannot touch without being touched, whereas one can see without being seen or hear without being heard. Like the skin, the contact boundary is that non-materializable space, which is at the same time the boundary between me and non-me, and the zone of exchange between the internal and the external. The image of the skin well illustrates this double movement of oneself toward the world and the world toward oneself that the contact boundary entails. The touch that contacts and opens to the world, at the same time contains and separates from the world. The self, or contact system, is defined by Perls as "touch touching something."[26]

15.3.8 The Stages of the Contact Cycle

Gestalt therapy identifies four stages in a cycle of contact that can be described as complete or completed: precontact, contact, full contact, and post-contact, steps that are punctuated by the appearance of a need that emerges from our awareness, develops and finds satisfaction in the environment, and then fades away, giving way to a new need. The "need" encompasses everything that is set in motion (appetite, curiosity, drive, desire, fear).

Phase 1: Precontact and "Id Mode"
The precontact phase is that of the emergence of the need from corporeal sensations. In the permanent flux of my sensations, of my needs, a precise stimulus suddenly becomes the figure that solicits my interest; it then detaches itself from the

[25] F. Perls et al. 1951, p. 58.
[26] F. Perls et al. 1951, p. 217.

background, "like a theatre actor who moves center-stage."[27] A gestalt emerges from the undifferentiated background to gradually take on a form that can trigger a "going toward" the environment. The organism is pushed toward something not knowing in advance "neither what, nor that it is." There is not yet necessarily an awareness of the nature or the object of the need. The emergence of this sensation acts within us in the contact. Then the organism/individual will identify its need (with the possibility of making mistakes and causing an alteration in the contact: a stomach ache could be understood and translated as "I'm hungry" when it should rather be "I'm afraid" and lead to an orientation toward the refrigerator when it was rather to finish writing the overdue article). In this precontact phase, the "id mode," constituted by my sensation and physio-corporal manifestations, is strongly mobilized and provides valuable information for the orientation toward the environment that will be made during Phase 2, relying more on the "ego mode."

Phase 2: Contact and the "Ego Mode"

The emergence of the figure, once identified, raises the excitement necessary for the orientation of the organism in the environment. This is the moment of commitment. I choose to satisfy my need and I mobilize my energy to take action: within me there is a turning toward the environment, a "going toward" which is triggered at that moment to seek in the environment which will feed me. This is where Perls and the Gestaltists speak of healthy aggressivity: there is an aggression of the environment by the organism that identifies, seizes, and ingests what it needs in the environment. This exploration phase allows for the examination of environmental resources. It is about identifying and alienating (in the sense of discarding, giving up) all other possibilities in order to make a choice. Proceeding with identifications and alienations puts the "ego function" to work. If the figure of hunger proves correct, I will locate the food that will enable me to satisfy this need in my environment, (the apple I remember in my bag, while I alienate the idea of taking the apple I see in the hand of the person sitting next to me on the subway). The transition from precontact to contact requires the organism to turn toward the environment: it is necessary to decenter from here (sensations) to go there (toward the surroundings).

Phase 3: Full Contact (Middle in Mode)

Phase 3 is that of full contact: it is that of satisfying the need. In this phase, there is an abolition of the contact boundary. There is no longer any distinction between the figure and the background, and this is a moment of healthy, creative, and nourishing fusion between the organism and the environment. This phase of full contact is the one archetypically where the middle in mode occurs. The middle in mode is a mode that is both passive and active where the environment contributes to the organism and the organism contributes to the environment. Establishing a full contact requires a letting go to immerse yourself totally in a moment of symbiosis or encounter. While it is about interpersonal relationships, it is the moment of the "we." This moment is not necessarily pleasant: it may be a loving or affectionate we, but also a conflictual we.

[27] G. Masquelier 2004, p. 54.

If I choose to take the apple from the bottom of my bag, this is the moment for chewing: I attack the environment: I transform the apple by chewing it into pieces and ingesting it, and it transforms me in turn by supplying me with vitamins and minerals.

Phase 4: Post-contact (Personality Mode)
Phase 4, called post-contact, is that of withdrawal and assimilation of the newness: a differentiation between the organism and the environment takes place again, and a contact boundary reappears. The organism digests the experience to assimilate it. It is in this phase that the personality function is mainly nourished and built. The figure processed in the previous phase gradually dissolves away. In the contact cycle, we return to a "point zero" where there is no longer a figure or background and where the organism finds itself available once more for the emergence of a new sensation, of another form.

15.3.9 Gestalt Therapy: A Therapy of Contact

The object of Gestalt therapy is therefore the form taken by the encounter between the organism and its environment, the figure that emerges in the moment, and the modalities of this contact. What interests Gestalt therapy is not so much the completed form, the figure on the background, but the process of the figure/background relationship, the process by which forms are constructed and deconstructed, which speak of a way of being in the world. Unlike other approaches that try to explain the psyche in static terms, Gestalt therapy positions itself in a dynamic perspective. It is interested in process, the continuous adjustment between an individual and his/her environment. In the face of changes in the environment, the adjustment has to be renewed unceasingly. Thus, the gestalt therapist focuses on the process, rather than on the content, that is, the "how" rather than the "why." Sometimes called "contact therapy," Gestalt therapy is primarily interested in this contact, ever present, always active, according to various modalities, more or less intense, and more or less fluid. The gestalt therapist works on the quality of this contact, the form it takes, the adjustments at work, and the dysfunctions that can alter it.

15.4 An Approach that Proposes a Triple Refoundation of the Relationship…

The postulate of the Gestalt approach, which posits the inseparability of organism and environment and makes this relationship one of the foundations for the growth of the individual, invites us to rethink the relationship of man to the world and to envisage a third way between an approach based on the individual (which could be that of the social sciences) and an approach based on the environment (which could be that of the life sciences or even ecological/environmental approaches).

The Gestalt proposition, by inviting a focus on the contact cycle and positing the inseparability of organism and environment as postulate, suggests a paradigm shift in the acceptation of both the individual and the environment. It upsets habits and patterns of thought that have a dual approach to the individual or the environment by proposing a dynamic, not static, reading and bringing the focus, not onto one of the two objects in a binary way (the individual or the environment), but onto the relationship that unites them. In this respect, we can see bridges between this largely empirical and pragmatic approach and the conception of mesology as developed by Augustin Berque, who, by defining it as the study of milieus,[28] proposes focusing on and analyzing how an individual seizes his/her environment and is primarily interested in the "in as much as" of the relationship (a mountain will be perceived in different ways according to the technical and symbolic conditions of a given society: it could be perceived as a resource, a constraint, a place of poetry, of contemplation, of pictorial support). This is what A. Berque summarizes in the formula R = S/P where R is the reality grasped by a subject (S) as a Predicate (P).

15.4.1 ...to the Environment: The Impossible Environmental Externalization

The Gestalt proposition radically alters the relationship that can be built with the environment, if one adopts this postulate of the inseparability of organism and environment. The founding work that underlies the Gestalt approach starts with the notion of "field organism/environment." This notion defines the interaction of the organism and the environment at work in any vital function: "It makes no sense to speak, for instance, of an animal that breathes without considering air and oxygen as part of its definition, or to speak of eating without mentioning food, or of seeing without light, or locomotion without gravity and supporting ground, or of speech without communicants."[29] The encounter between the organism and the environment creates an organism/environment field that forms an inseparable whole.[30] The primary contact, neither voluntary nor intentional and therefore not a matter of choice, supposes correlatively that the environment is always there, and always the not-me, around me. Even if I am alone on a desert island, there is always an unavoidable contact between the island and me. Wherever I am, there is always the sun that heats my skin, the soil that supports me, and the air that I breathe and these

[28] Berque developed this approach in various works and in particular: *Ecoumène. Introduction à l'étude des milieux humains*, Belin, Paris, 2000.

[29] F. Perls et al. 1951, p. 50.

[30] The founders of Gestalt therapy were inspired by field theory, developed by Kurt Lewin.

interactions are the source of permanent, moving, fluctuating contacts. They possibly form a primary and fundamental modality of the human-environment relationship.[31]

Accepting this, it can be seen that the externalization or the objectivation of the environment as something external to oneself, objectivable, even reifiable, and therefore measurable or controllable, is incompatible with this paradigm. This apparently simple proposition implies a profound upheaval of certain models or categories of thoughts. The unitary design of the organism/environment field shakes up the traditional dualistic conception of an individual separated from the environment. It renders inoperative the acceptation which tends to see the individual as an isolated entity subject to external uncertainties from his/her environment, as much as it undermines the possibility for the individual to consider his/her environment as a reservoir, a stock of resources, and an entity of which he/she could become master and possessor.

15.4.2 …to the Other: Healthy Aggressivity vs. Violence

The paradigm of the field, created ipso facto by the encounter between organism and environment, posits as a correlative dimension concerning my relationship with others: from the moment I exist, I have an effect on the other. Whatever I do, whatever the position of my act may be on a gradient going from the most visible to the least visible, from the most intrusive to the least identifiable, I exercise an influence on the other and the other exerts an influence on me by its presence alone. Man influences his surroundings at the same time as he is influenced by them: both affecting and affected, he is necessarily in contact with the other. Moreover, the filiation with phenomenology which Gestalt therapy claims for itself posits that the reality of the other is intrinsically and irreproachably different from mine, apart from the minimal phenomenological reduction that can be achieved. The organism/environment field is a phenomenal field "irreducible to the objective data that constitute it and inseparable from the person who organizes it."[32] The field is thus constituted both by objectivable facts and also by elements concerning the subjectivity of the person which are not going to speak only from the person concerned by this contact experience that he/she is participating in co-constructing with his/her environment. This apprehension of reality underlines that reality is different for everyone.

---

[31] In the therapeutic approach, one often considers the environment as the other, and of course it is often a question of going and looking at the quality of the contact, the ability to be in contact with the other, and the possible alterations of this contact. However the meaning of the environment proposed by the Gestalt reading of our relationship to the world does not reduce the environment to the other and to the human and encompasses a much broader meaning of the environment: contact, the encounter between the organism and the environment covers a spectrum that goes from the biophysical environment to the other human.

[32] C. Masquelier, p. 37.

This awareness of the existence of the other in my being in the world makes it possible to introduce a fundamental distinction between aggressivity and violence in the Gestalt approach. Gestalt therapy rehabilitates aggressivity as a life force, distinguishing it from violence, which is conceived of as a destructive force. For Perls, whose first work dealt with aggressivity and this primary movement of going toward, which he considers to be inscribed at the very heart of the vital impulse and growth of the individual, aggressivity is a central element.[33] For him, living and growing in the environment require it to be attacked in order to be able to feed on it. He takes the metaphor of food to illustrate this vital impetus "to go toward": to feed oneself, man must attack the food, that is to say bite, chew, and destroy it in order to take the substances necessary for his growth. In this contact process, the individual and the environment are linked by a singular and reciprocal effect: the individual who crunches an apple will destroy this apple by biting, chewing, and digesting it. He/she will participate in the modification of this environment-apple which in return will participate in the modification of the individual by bringing him sugars, vitamins, and minerals. Aggressivity is therefore perceived in Gestalt as a necessity for taking, selecting, and feeding on what is good in the environment. It is an essential factor for growth. On the other hand, violence, in the Gestalt approach, is perceived above all as a break in contact with the other and also with oneself. Violence is therefore fundamentally destructive. The etymology of the two terms illustrates these two differentiated postures: etymologically, aggressivity comes from the Latin "ad-gredere" which means "to go toward," this movement to go toward being synonymous with a movement of contact with the environment, while the Latin "violare" implies an act of force on someone or something, an abuse of strength. In one case, it is a matter of going toward, in the other of acting by force. While in aggressivity there is awareness of the existence of the other and encountering this other, violence is accompanied by a rupture of the relationship, the other becoming the object to be destroyed. Aggressivity and violence are therefore two opposing forces, the former leading to full contact and the latter to the rupture of contact.

15.4.3 ...to Oneself: For a Decentralization of the Subject

Finally, the Gestalt position invites us to modify the view and the acceptation that one can have of the ego and of oneself. By positing as the starting point the inseparability of organism and environment, the pre-existence of contact even before any relation and the importance of the self, the Gestalt proposition invites us to decenter our view of the individual on the relation, from oneself to the self. It proposes a conversion of view which ultimately makes the individual a second subject in relation to the self and to this process of contact which is permanently at work and which becomes the fundamental focal point.

[33] Frederick Salomon Perls 1942.

15.4.4 Another Conception of the Subject: The Self Is Not Oneself

The self, as a process of contact, is not oneself, contrary to what the English term might suggest: the Gestalt proposition shifts the focus of the subject to the contact or relationship and makes the individual a secondary subject, the central and main subject being this contact, this ever-changing, ephemeral and changing relationship between the organism and the environment. Consequently, the Gestalt self is different from the notion of an identity ego; it is defined not as a pre-existing entity, but as a contact movement, a construct that results from a continual adjustment to the novelty proposed by the environment. Defined thus this concept differs from the notion of oneself in the different meanings that this term can cover. The Gestaltist self does not refer to the common sense of "oneself" in French: this reflective personal pronoun of the third person singularizes the individual in its uniqueness, in its singularity, and in its individuality. The self is also distinguished from the Freudian instance of the ego (present in the Freudian trilogy: that, me, superego). Finally, the Gestaltist self-differentiates itself from its English namesake "self" which in Anglo-Saxon psychology refers to the awareness that a subject has of him/herself, of his/her individuality. The self in Gestalt is not an entity. It has no existence of its own, spatial or anterior. It is revealed only on the occasion of contact. This vision calls into question the notion of original identity or being. In no case can the person be reduced to a thing, a distinct individuality that should be analyzed, dissected, recovered, treated, or repaired. The self is a variable and temporal process. Instead of focusing on the identity, on what remains the same in a person in different situations – what is meant by the notion of oneself, which tends to fix a person's outline and places more emphasis on a particular and singular identity – the Gestalt self, on the contrary, designates what changes and adjusts. This unitary conception of the organism/environment field has upset the traditional conception of the subject. The Gestalt vision emphasizes the singular way each one is of being in the world and of contacting this world, opening up a fruitful possibility of movement and adjustments at the contact boundary, in a continuous to-and-fro between the world inside and outside, between the me and not-me, and the nonego and not-I.

15.4.5 A Radical Shift in the Subject

Contrary to the psychoanalytical position, it is not the psyche, the subject in his/her inner character that is the object of the therapeutic work, but the modalities of this subject being in contact, and being in the world, the way which he/she contacts and he/she lets him/herself be contacted by his/her environment. This conception of the subject and its place in the environment upsets the traditional apprehension of the individual or of the subject: I leave a finite world where I seek an absolute of myself in order to enter into an acceptation of a world of beings in motion. The subject of study then becomes being in the world rather than being. It is on the occasion of this contact movement, which infers that I necessarily and systematically include the environment and the other, that I exist and that I appear to myself. I exist only because I appear to the other, in this movement of contact. The I, the ego, as an absolute entity, which one could circumscribe, limit in time and space does not exist. I exist only because I appear to the other, am included and include the world in a contact movement. This calls for a decentralization of oneself and an important revolution in our society of individuals where the process of individuation has been increasing.

At the individual level of the life of a subject, the Gestalt position also invites a complex movement of decentering: after having spent a significant and crucial part of his/her life working on the awareness and the construction of an accomplished self, a process that begins at an early age in the first weeks and months, when the newborn begins to perceive him/herself as a distinct, differentiated entity from his/her mother – a fundamental step in differentiation in the life of a newborn child – a "demolition of the ego" is to be carried out with the Gestalt proposition; the apprehension of one's being as a moving being, who contacts and is contacted continually by his/her environment, invites us to leave a finite world in which I seek an absolute figure of oneself, distinct, identifiable, and de-limitable in a given space and time, to enter into a definition of a world of being in movement in which I exist only because I appear to the other. My ego as such does not exist, my ego is not a geolocalizable entity, and it is in this process of contact that I exist, that I transform myself, and that I bring myself into the world.

15.5 Conclusion

The Gestalt proposition, based on the principle of the unity of the organism/environment field, is stimulating insofar as it challenges the individualistic and solipsistic view of the person. It is an invitation to convert to a way of thinking that is inclined to conceive of the individual as a distinct entity. The Gestalt revolution relies more on this invitation to transform a way of thinking than on the libertarian spirit of the founders, which encouraged the expression of the body and the emotions in a normative society. The individualistic view of man is rooted in a paradigm that encloses us in an infinite duel (body/spirit, reason/emotions, positivism/subjectivism, etc.). Field theory invites us to emerge from dualism by considering the whole of a situation whose different elements come together to co-exist. It allows us to emerge from the binary, dual dynamic ("you, or me"), to envisage the dynamic of a co-construction ("you and me") and take into account the mutual influence that colors the specificity of the instant.

This innovative paradigm has implications for the design of health and care. Health is an unstable balance. Disease is not simply an organic degradation but defines a modality of existence; it is a form of human existence in its own right. Paul Goodman refers to "organic self-regulation" as the natural dynamic fit between the organism and the environment. This holistic conception, in the sense of a body interacting with an environment, gives an optimistic view of the capacity of each person to take care of him/herself, that is, to regain balance. In this perspective, the success of a cure does not come from the caregiver alone, but often depends on the nature of the relationship between the caregiver and the cared for. The therapist contributes to this recovery by the care and the respect given to the living. He/she is only an agent of change, which largely escapes him/her.

References

David Abram. 1996. *The Spell of the Sensuous. Perception and Language in a More-Than-Human World*, New York, (*Comment la Terre s'est tue. Pour une écologie des sens*, Editions La Découverte, Paris, 2013, for the French edition).

Masquelier, Gonzague. 2004. *Vouloir sa vie. La Gestalt aujourd'hui*. Paris: Retz.

Masquelier-Savatier, Chantal. 2015. *La Gestalt-Thérapie*. Paris: QSJ, PUF.

Perls, Frederick S., Ralph E. Hefferline, and Paul Goodman. 1951. *Gestalt Therapy: Excitement and Growth in the Human Personality*. London: Souvenir Press.

Perls, Frederick S. 1942. *Ego, Hunger and Aggression: A Revision of Freud's Theory and Method*. (*Le Moi, la faim et l'agressivité*, Tchou, 1978 for the French edition).

Robine, Jean-Marie. 1998. Le contact, expérience première. In *Gestalt-thérapie, la construction du soi*. Paris: L'Harmattan.

Robine, Jean-Marie. 2010. Le contact, aux sources de l'expérience. In *Le changement social commence à deux*, vol. 25. Bordeaux: L'Exprimerie.

Index

A
Aarab, A., 62–75
Abbasid caliphate, 210, 228
Abram, D., 140, 141, 350, 351
Acclimatation
  acclimatization, 117 (*see also* Adapt; Climate; Travel)
Accommodate, 12, 126, 132
Adapt
  adaptability, 213
  adaptation, 233 (*see also* Adjustment)
Adequate, 63, 127, 128
Adjustment, 128, 133, 203, 334, 336, 338, 345, 349, 352, 357, 360, 364
  creative, 357 (*see also* Adapt)
Aesthetics, 5, 9, 18, 139, 140, 213, 244, 293, 295, 304, 305, 327
Affordances, 18, 19, 333–345
Air
  circulation of, 17, 187, 290, 291
Albert the Great (*Albertus Magnus*), 40, 47, 54
Albertus Magnus
  Albert the Great, 47
Alexander Neckam (*Alexander Nequam*), 32
Al-Ǧāḥiẓ, 209–233
Al-Manṣūrī fī al-ṭibb (translated in Latin Liber ad Almansorem), 223
Alphand, A., 293
Altitudes, 97, 151, 154, 170, 171
Ambiances, 106, 139
Ancient, 4, 5, 10, 14, 20, 21, 28, 31, 32, 45, 51, 54, 63, 79, 82, 89, 93, 155, 166–169, 172, 174, 175, 178, 182, 185, 188, 190, 191, 193–196, 198, 199, 201, 202, 227, 259, 261, 269, 281
Animal
  reservoir, 147, 153
Animism, 136, 251
Anxieties, 128, 129, 136, 301, 323, 357
Aquinas, T., 35, 47, 54
Arab-muslim civilization, 15, 169, 210, 211, 220
Aristotle (*Aristoteles*)
  Aristotelian, 40, 41, 43–45, 48, 51, 54, 213, 214
  Aristotelianism, 31, 48
Arnoldus of Saxony (*Arnoldus Saxo*), 47
Astrology, 30, 66, 222, 232
Astronomy, 14, 30, 66, 215, 228, 232
Atkinson, S., 139
Atmosphere(s), 12, 113, 128, 136, 139, 187, 298, 323
  See also Gegenwelt; Merkwelt; Milieu; Umwelt
Auctoritas (authority), 32
Augustine of Hippo (*Augustinus Hipponensis*), 33, 35
Averroes, 47
Avicenna, 31, 45, 54

F. Bretelle-Establet et al. (eds.), *Making Sense of Health, Disease, and the Environment in Cross-Cultural History: The Arabic-Islamic World, China, Europe, and North America*, Boston Studies in the Philosophy and History of Science 333, https://doi.org/10.1007/978-3-030-19082-8

B

Bacteriology
bacteriological, 2, 3
Baḫtīšū 'Ibn Ǧibrīl, 62, 70, 76
Banner people, 258, 259, 264, 267, 278
See also Beijing city; Qing
Bardo *(中阴身 Zhongyin shen)*, 266
Barnacle, G., 53
Barriers, 83, 84, 171, 227, 315, 337
Bartholomew the Englishman (*Bartholomeus Anglicus*), 32
Behavior
behavioral changes, 65
behavioral setting, 137, 138
catastrophic behavior, 128, 129, 132, 136
ordered behavior, 128, 129, 132, 136
Beijing city
Beijing citizens, 16, 259, 269, 274 (*see also* Banner people)
Bensaad, M., 14, 169, 210
Berque, A., 8, 19, 319, 327, 351, 361
See also Mesology
Bertillon, L.-A., 117, 118
Bioethics, 313
Biology, 12, 51, 112, 117, 119, 134, 136, 137, 148–150, 153, 156, 159, 318
Biopolitics, 300
Bioregionalism, 310, 324
Biosphere, 140, 314
Biotic community, 241, 244, 251, 310, 311, 313, 328
Blanc, N., 8, 17, 187, 202, 287–306
Body
bodily, 13, 46, 49, 69, 128, 141, 171, 175, 195, 197, 198, 201, 224, 351, 354
body-subject, 140
Boethius, 45, 48–50, 53
Border(s), 14, 21, 91, 170, 179, 184, 187, 188
Bory de Saint-Vincent, J. B., 113
Botanist(s), 111, 115, 211, 216–219, 232
Braunstein J.-F., 130, 131
Brazil, 13, 146, 148–155, 157, 158
Breaths, 21, 83, 85, 89, 90, 94, 361
Bretelle-Establet, F., 13, 14

C

Cabet, E., 289, 290
Cafuas, 13, 147, 150–155
See also Housing
Calendar
lunar, 269, 276, 278
solar, 221, 229, 276
Canguilhem, G., 7, 12, 20, 117, 126, 127, 130–137, 141, 142, 219
Captivity, 10, 63–65, 75
Carel, H., 139
Cartographers, 220
See also Geographers; Map
Catholicon, 51
Cause
occult, 54
Centemeri, L., 8, 17, 18, 140
Chagas, C., 13, 146, 148–151, 153, 155, 156, 158, 159
Chagas disease, 13
Chansigaud, V., 7, 11, 106–121
Charmaz, K., 139
Charte d'Aalborg, 301
Charte d'Athènes, 301
Chartres (medieval school of), 30, 38, 51
China, 4, 5, 10, 20, 21, 256, 260, 261, 267, 269, 271, 279, 280, 294
See also Chinese
Chinese medical thinkers
Chinese medical thinking, 166
Chinese medicine, 167–176
Circulation
air, 17, 187, 290, 291
Classifications, 33, 41, 43, 80, 86, 217
Cleanliness, 69, 289, 300
See also Hygiene
Climate
climatology, 114 (*see also* Acclimatation; Geography; Season)
Closed, 70, 83, 90, 96, 99, 171, 175, 202, 271, 350
See also Confined
Coffins, 16, 259–269, 272–276, 279–281
Cold Damage Treatise (Shanghan lun 傷寒論), 173, 198
Colony
agricultural colonies, 190
colonial, 7, 114, 135, 152, 155
colonization, 13, 154, 155, 157, 190, 245, 318
Communication, 83, 97, 98, 152, 157, 290, 301, 340
Comparatist study(ies), 5
Complexions, 50, 51, 53, 54, 91, 170, 172, 197, 201
Comte, A., 112, 142
Configuration
configurationist, 2
Confined, 14, 21, 69, 110, 187, 202, 321, 329

Constantine the African (*Constantinus Africanus*), 30
Contact
boundary, 358–360, 364
cycle, 351, 355, 358–361
Contagion, 1, 31, 148
Contamination, 1, 2, 146, 148, 149, 151, 152, 154, 157, 186
Contra Faustum (Against Faustum), 35, 36, 41, 51
Corbin, A., 3, 187, 202
Correlation
correlative thinking, 11, 80, 168, 169, 269 (*see also* Correspondences)
Correspondences, 11, 32, 72, 80, 81, 84, 85, 87, 97, 99, 106, 107, 168, 169, 242
See also Correlation
Corruptibility (of bodies), 41
Cosmos, 11
correlative cosmology, 168, 172, 173, 175, 202 (*see also* Microcosm)
cosmology, 11, 30, 76, 168, 172, 173, 175, 202, 269, 280
Countryside, 17, 113, 202, 289, 294, 299, 303, 304
See also Rural
Creation, 10, 18, 27–56, 95, 100, 108, 109, 112, 114, 117, 118, 120, 136, 159, 203, 252, 279, 289, 296, 298, 311, 312, 315, 328, 340, 357
Cruz, O., 148–150, 152, 153
Cultures, 5–7, 18, 28, 31, 55, 83, 93, 119, 155, 156, 168, 169, 172, 177, 190, 194, 203, 211, 252, 280, 294, 299, 314, 317, 321, 324, 328

D

Dannenberg A.L., 139
De anima (On Soul), 42, 45
De animalibus (History of Animals), 40
De animalibus (On Animals), 42
Death
conception, 256, 260, 265–268, 275, 276, 280
observances, 258–265, 280 (*see also* Funeral rituals; Graveyard; Netherworld)
De celo et mundo (Heavens and Earth), 42, 44
De consolatione Philosophiae (The Consolation of Philosophy), 56
De floribus rerum naturalium, 32
Deforestation, 189, 190
See also Forest; Land clearance
De homine (On Man), 54
Delbourgo, J., 20
de Lille, A., 48, 49, 51–53
Delimitation, 17, 299
de Monet de Lamarck, J.-B., 112, 113
De naturis rerum, 32
Dependence
interdependence, 106–110, 243, 247, 251, 314
De proprietatibus rerum, 32
Derivationes, 52
Design
design activism, 312–320
Despeux, C., 10, 83, 88, 89, 93, 94, 166, 169, 183
Deterville, J.F.P., 112, 113
De trinitate (On the Trinity), 45
De vegetabilibus (On Plants), 40, 42
Didascalicon, 41
Diets, 1, 31, 64, 170, 224, 231, 316, 320
Dionysius the Areopagite (*Dionysius Areopagita*), 47
Disability, 19, 135, 141, 334, 340, 342–344
Disease
abnormal, 176, 193, 201
chronic, 2, 156, 305
infectious, 148, 160, 305
tropical, 147
vector-borne, 146, 148, 160
Dissonance, 334
Do-it-yourself, 340, 341
Domestic
domestication, 115, 117
Draelants, I., 9, 27–56
Dragmaticon, 38
Drugs, 170, 173, 175, 215, 222, 231
Duhl, L., 298
Dux, G., 21

E

Earth
earth care, 321, 328
Ecology
ecological, 17, 18, 140, 239–248, 251, 300–305, 312–324, 335, 349–351, 361
ecological conscience, 240, 243, 244
ecological health, 239, 248
ecological transition, 309–329
ecopsychology, 15, 140, 248, 250, 251, 302 (*see also* First person ecology)
Economic pressures, 13, 158

Ecosystem
ecosystem health, 239
ecosystem-related diseases, 145
Ego, 41, 250, 251, 357, 359, 363–365
See also Self
Egypt, 212, 228–232
Ehrenfels, C. von, 353, 355
Eight trigrams, 80
Elemented (*elementata*), 46
Elements (*elementa*), 46
Emanations, 174, 180–187, 202
Embodied dimension, 138, 329
Emotions, 89–90, 139, 140, 168, 215, 230, 329, 354, 365
Emplaced dimension
emplaced modes of valuation, 311, 325, 327–329
Empowerment, 315
Encounter, 6, 14, 130–136, 147, 160, 176, 202, 248, 305, 334, 349, 355, 357, 359–362
Encyclopaedias (medieval), 28, 53
Endemic, 146–148, 150, 151, 155–158, 160, 230
Entomologist, 113, 152, 153, 157
Environment (s)
contextualization of environment, 157, 299
cultural environment, 147
environmental causality, 160
environmental esthetics, 243–248
environmental ethics, 240, 319
environmental factors, 2, 10, 73, 147, 174, 217, 240, 298
environmental health, 239, 248
environmental miasmas, 3 (*see also* Fengtu 風土 (lit. Wind-Earth); huanjing 環境 (environment); Iqlim (climate); Keshva (territory); shanshui 山水 (lit. Mountain-Water); shuitu 水土 (lit. Water-Earth))
environment's impact, 14
environment's role, 334
harmful environment, 191, 201
hostile environment, 139
living environment, 19, 129, 134, 141, 209, 218, 298, 305, 356
material environment, 154, 300
medical environment, 99
physical environment, 11, 289, 335, 362
socio-economic environment, 152
terrifying environment, 14, 179, 190
Epidemic(s)
epidemiological, 147, 159, 166, 201
Equilibrium, 81, 98, 159, 250, 338, 344
Ethology, 12, 117
Etymologiae (The Etymologies), 40, 48, 311, 354, 363
Exchange
zone of exchange, 358
Existentialism, 353
Exposome, 2

F

Feelings, 5, 96, 128, 129, 131, 133, 139, 142, 246, 250, 251, 259, 302, 338
Fengshui (風水), 90, 267
See also Geomancy
Fengtu 風土 (lit. Wind-Earth), 7
See also Environment
Fescharek, R., 139
First person ecology, 15, 239–253
See also Ecology
Five agents (*wuxing*五行), 80, 81, 90, 97, 98
Flourishing (human), 12, 126–142, 311, 312, 318, 319
Fogs, 20, 170, 196, 199, 200, 230, 293, 351
Fons Vitae, 46
Food
food forest, 326, 327
Force (*virtus, vis*)
specific force (*virtus specifica*), 31
virtue (*virtus*), 47, 53
Forest(s)
dense forests, 184, 187
food forest, 326, 327
forester, 15, 240, 242
wild forests, 203 (*see also* deforestation)
Form (*forma*), 37, 45, 49
Foucault, M., 142, 300
Fourier, C., 289
Fragrance, 14, 184, 185
Freud, S., 353
Frunzeanu, E., 9, 27–56
Fu, D., 6, 7
Fudo, 8
See also Augustin; Berque; Mesology; Milieu
Fuller, S., 139
Funeral rituals, 16
See also Death

G

Gachelin, G., 13, 145–160
Ǧāhiẓ, 10, 61–77, 209–235
Gaille, M., 1–21, 125–142, 313, 336
Galen
galenic, 51, 54, 221 (*see also* humoral)

Garden (s)
  garden cities, 293, 294, 298, 299
Generation (*generatio*), 27–55, 67, 70, 87, 108, 168, 172, 212, 214, 222, 252, 258, 267, 295, 296, 312, 323, 353
Genesis, 29, 30, 55, 356
Geographers, 8, 19, 113, 119, 211, 219–221, 232, 303
  See also Cartographers
Geography
  geographical, viii, 5, 9, 15, 16, 19, 53, 61, 64, 73, 76, 93, 111, 115, 132, 134, 137–139, 147, 153–155, 160, 209–211, 216, 217, 220, 224, 227–229, 233, 294, 335, 341
  geographical environment, 134
  geographical setting, 5, 137, 138
Geomancy, 90, 97, 264, 266–269, 274, 279, 280
  See also Fengshui (风水)
Gestalt
  Gestaltists, 359, 364 (*see also* Goodman, P.; Perls, F.)
  theory, 335, 336, 348, 353, 354
  therapy, 19, 347–366
Gibson, J.J., 18, 334–337
Glosae super Platonem, 51
Godin, J.-B., 291
Goldstein, K., 12, 126–137, 141, 142
Goodman, P., 348, 351, 353, 357, 365
  See also Gestalt; Perls, F.
Graveyards, 16, 264–281
  See also Death
Greek, 4, 6, 9, 14, 30, 31, 33, 43, 50, 54, 62–64, 211–213, 220, 225, 231, 244
Greenway, R., 302
Gropius, W., 295, 296
Gros, F., 136
Guangdong, 14, 167, 175–179, 183, 187–191, 194–196
Guangxi, 14, 167, 175, 176, 178, 179, 182, 183, 185, 188–196
Guérin-Méneville, F.-E., 113
Guerric of Saint-Quentin (*Gerricus de Sancto Quintino*), 52
*Gu*蠱 poisons, 176, 178, 193, 196
Gurwitsch, A., 132

H

Habit (*habitus*), 10, 27–56, 67, 93, 113, 118, 170, 172, 174, 215, 222, 233, 273, 314, 321, 361
  habitat, 69, 82, 106, 111, 113–116, 120, 147, 153, 154, 217, 218, 224
  habitatio, 8, 11, 111
  habitation, 12, 17, 111–113, 232 (*see also* Environment; Milie; Statio)
Haeckel, E., 118, 160
Hainan Island, 181, 183, 185, 188, 190
Halimi, P., 139
Hanson, M., 20, 166, 168, 169, 171, 172, 174–177, 193
Harbsmeier, C., 6, 82, 91
Harmony, 16, 55, 81, 84, 89, 90, 95, 97, 99, 243, 247, 250, 272, 294, 318, 334
Harmful
  environment, 86, 89, 96, 97, 195, 199, 203
Health
  healthiness, 13, 290, 291
  healthy, 3, 7, 8, 12, 13, 16, 18, 20, 89, 107, 126–131, 133, 135, 175, 188, 191, 202, 216, 232, 246, 247, 251, 253, 256, 280, 290, 292, 294, 296, 300, 319–324, 328, 329, 354, 357, 359, 362, 363
  unhealthy, 3, 7, 8, 12, 13, 16, 21, 166, 174, 186, 193, 199–202, 229, 280, 288, 292, 295
Heavens, 28, 32, 42, 44, 46, 53, 80, 82, 84–88, 91–95, 169–172, 174, 177, 186, 192, 258, 279
Hebrew, 211
Hefferline, R., 348, 351, 352
Heidegger, 137, 353
Hexaemeron, *see* Genesis
Hierarchy, 95, 222
Hilarius of Poitiers, 50
Hippocrate
  hippocratic, 212, 214, 221, 223–225
  neo-hippocratic, 160
Holism
  holistic, 12, 18, 19, 126–128, 130–136, 312, 313, 322, 348, 366
Holmgren, D., 18, 311–315, 317–319, 321–324, 326
Homily 16 on Ecclesiastes, 36, 37
Horti-therapeutic gardens, 306
Hot
  scorching hot, 196 (*see also* Climate; Season)
Housing, 13, 146, 147, 151–155, 170, 272, 273, 289, 291, 294, 296, 299, 303
Howard, E., 293
Huanjing 環境 (environment)
  See also Environment
Hugh of Pisa (*Hugo Pisanus*), 52

Hugh of Saint-Victor (*Hugo de Sancto Victore*), 37, 38, 40, 41, 44
Human beings
human body, 10, 11, 62, 69, 70, 80, 81, 83–85, 90, 92–95, 98, 106, 168, 169, 172, 202, 212, 215, 223, 256
human experience (of environment and landscape), 35, 137, 138, 140, 319, 350
human flourishing, 12, 13, 126–142, 311, 312, 318, 319
human mind, 245, 350
human nature, 12, 129–136, 220, 250, 302
Humid
humidity, 14, 54, 86, 89, 174, 183, 190, 195, 197, 198, 200, 202, 230
Humors
humoral doctrines, 14
humoral medical pathology, 14, 215, 221
humoral medicine, 213, 215
humoral theories, 221 (*see also* Galen)
Ḥunayn Ibn Isḥāq, 211, 222, 225
Husserl, 137, 353
Hygiene
hygienic modernity, 256, 280
hygienist, 17, 288–292
hygienist utopias, 17, 288–292

I
Ibn al-Ǧazzār, 228, 229
See also Ibn Riḍwān; Physicians; Physiognomony
Ibn Gabirol, 46
Ibn Riḍwān, 209–233
Illness (experience of), 142
Imagination
imaginary, 98, 100, 251, 316, 321, 325
Imbert, C., 134
Immigration, 190, 191, 201
Incorruptibility (of bodies), 41
Indian, 64, 70, 154–157, 211
Indigenous societies, 324
Individual, 1, 40, 73, 80, 108, 126, 156, 192, 212, 242, 259, 296, 310, 334, 348
Influences, 4, 28, 65, 83, 106, 148, 170, 210, 213–218, 220–223, 229, 230, 249, 258, 295, 314, 336, 348
Inner Classic of the Yellow Thearch (Huangdi Neijing 黄帝内經*)*
Inner Classic, 80, 84–87, 89, 91–94, 99, 168–173
Insalubrity, 193, 194, 289, 298
Inseparability, 348, 349, 351, 352, 356, 360, 361, 363
Insiders, 14, 177–179, 194, 201
Institut Pasteur, 149
Interpenetration, 98
Iqlīm (climate), 8, 15, 220, 221, 232
Iraq, 10, 62, 64, 220
Isidore of Seville (*Isidorus Hispalensis*), 40, 41, 48, 52

J
Jackson R.J., 139
Jiangnan, 167, 174–176, 194, 195, 198, 201, 202, 276, 277
Jin Gui 靳貴, 14, 191–193, 202
John of Damascus (*Johannes Damascenus*), 55
John of Genoa (*Johannes Balbus of Genoa*), 48, 51, 52
Ju, X., 8, 16, 277

K
Katouzian-Safadi, M., 10, 14, 62–75, 169, 209–233
Kešvar (territory), 8, 15, 220
Kitāb al-Ḥayawān (The book of animals), 62, 63, 66, 69, 75, 76, 217
Kumar, A., 4

L
Lamouchi-Chebbi, K., 10, 62–75
Land
city landscape, 256, 269, 279
clearance, 190
ethic, 241, 243, 248
of exile, 176
health, 16, 239–253
landscape, 7, 11, 17, 139, 146, 151–153, 159, 187–190, 194, 199, 202, 203, 245, 250, 256, 257, 260, 269, 270, 279, 300, 301, 312, 318, 355
Larrère, C., 15, 16, 239–253, 312
Lassance, 148, 151, 152, 154, 158, 159
Latin America, 13, 147, 153, 154
Lavoisier, A.L., 187, 298
Le Blanc, G., 135, 142
Le Corbusier, 20, 295–297
Leopold, A., 15, 16, 239–253
Leroi, A.-M., 4
Leroux, M., 140

Leung, K.C.A., 172, 174–177, 202, 203
Liberal arts (*artes liberales*), 55
Liber de natura rerum, 41, 53
Life
  lifestyle, 18, 67, 155, 158, 258, 259, 311, 316, 321
  worth living, 127, 129, 136
Limoges, C., 126, 132, 134
Lingnan (South of Ling Mountains), 14, 176–179, 183, 184, 186, 187, 189, 191–193, 195–197, 199–201, 203
Links, 2, 29, 80, 81, 83, 84, 92, 95–100, 110, 130, 149, 150, 168, 209, 214, 219, 222, 223, 226, 228, 244, 266, 289, 315, 334, 350
Linnaeus, C., 108
Live
  lived space, 126
  lived world, 136–141
  living beings, 11, 12, 14, 15, 17, 86, 106, 115, 117, 131, 133–135, 140, 209, 211, 214, 218, 219, 221, 232, 240, 243, 251, 292, 300, 302
Local gazetteers, 171, 177–180, 193, 194
Lutz, A., 148

M

Mağlis cenacles, discussion circles, 63
Malaria, 147–149, 151, 153–155, 160, 179, 182, 193, 222
Manchu Four Rites (满洲四礼集Manzhou Sili Ji), 258, 261, 262
Map, 53, 141, 167, 177, 180, 210, 220, 221, 270, 274, 277
  See also Cartographers
Marratto, S.L., 138
Matter (*materia*), 4, 10, 11, 14, 30, 31, 37, 41, 45, 46, 49, 51, 53, 67, 130, 178, 186, 192, 193, 196, 201, 213, 240, 243, 246, 251, 288, 295, 325, 326, 345, 356, 361, 363
Meaning, 4, 6, 7, 9, 11, 14, 28, 29, 34, 35, 47–49, 52, 100, 106–121, 131, 132, 134–139, 160, 177, 180, 187, 190, 226, 244, 248, 250, 253, 263, 264, 279, 300, 304, 305, 314, 362, 364
Measure
  measurement, 114, 129, 335
Medicine, 14
  medical literature, 81, 173–175, 200, 201, 213
  medical text, 11, 95, 155, 167, 168, 173, 174, 178, 179, 193, 194, 200, 220, 224
  medical treatise, 14, 166, 168, 169, 171–173, 177, 179, 194, 197 (*see also* al-Manṣūrī fī al-ṭibb; Chinese Galen; Hippocrate; Ḥunayn Ibn Isḥāq)
Medieval
  medical cartographies, 3, 220
  medical geographies, 3, 147
  medical statistics, 3
Mental health, 62, 135, 297, 302
  See also Mind
Merleau-Ponty, M., 12, 13, 126, 136–140, 142, 305, 336
Mesology, 12, 19, 106, 117, 119, 361
  See also Berque, A.
Meteora (On Meteors), 42
Miasma (s)
  theory, 2
  miasmatic, 226, 300
  environmental miasmas, 3 (*see also* Zhang)
Michon, P., 19, 335, 348–366
Microcosm, 53, 81, 83, 85
  See also Cosmos
Milieu, 5, 31, 63, 106, 126, 172, 288, 319, 361
  See also Environment; Fudo; Habitat; \"Umgebung\"; \"Umwelt\";
Mind, 5, 9, 11, 13, 16, 17, 35, 39, 52, 95, 129, 133, 142, 154, 166, 180, 203, 210, 215, 219, 241, 242, 245, 249, 251, 271, 280, 297, 302, 321, 322, 341, 342, 350
  See also Mental health
Minerals, 54, 92, 99, 360, 363
Miracles, 34, 35, 41, 51–53
Mists, 14, 20, 96, 183, 184, 187, 189, 196
Mobility, 44, 97, 98, 344
Modern, 2, 4, 5, 7, 8, 10, 12, 13, 17, 21, 32, 35, 120, 179, 201, 233, 256, 269, 271, 281, 294, 295, 298, 300, 301, 328, 329, 351
Mollison, B., 17, 18, 311–315, 318, 320–323, 326, 329
Montgomery, S.L., 4
Moods, 15, 96, 216, 219, 303
Moon, 11, 84, 86, 96, 221–223, 232, 350
Motion
  motion (*motus*), 34, 37, 44–46, 49
Mountain (s)
  mountainous, 14, 183
Muchielli, L., 134

N

National, 150, 245, 316, 325
Nationalism, 149
Naturalis historia (Natural History), 33, 108–110, 114–117
Nature
 conservation, 243, 245, 247, 253
 natural, 2, 28, 62, 81, 106–110, 119, 127, 151, 156, 159, 166, 216, 248, 267, 288, 311, 342, 365
 natural environments, 2–6, 8, 10, 11, 13–15, 17, 20, 21, 68, 82, 85, 92, 152, 166, 168, 169, 183, 186, 188, 190, 196, 200, 202, 203, 219, 267, 268, 280, 289, 311
 natural history, 33, 114–117
 naturalists, 11, 12, 44, 106, 107, 110–115, 120, 217, 219, 221
 natural law, 28, 52, 169
 natural milieu, 7, 9, 11, 138, 172
 natural philosophy, 9, 30, 33, 35, 43, 54, 117, 232 (*see also* Philosophia naturalis)
 natural sciences, 10, 30, 42, 43, 62, 116, 127, 216, 292
 natural surroundings, 166, 168, 169
 natured (*natura naturata*), 34, 46
 naturing (*natura naturans*), 34, 46, 52
 preservation, 311
 protection of nature, 120, 240, 252, 253
 universal nature (*natura universalis*), 47, 48
Netherworld, 263, 265–271, 279–281
 See also Death
Network, 12, 28, 88, 90, 126, 135, 136, 142, 201, 266, 292, 314, 318, 321, 323, 325
Neurology
 neurologist, 12, 126, 127
Neuropsychiatry, 126, 127
Nine sectors, 80, 89
 See also Chinese medicine
Norm
 normal, 11, 67, 128, 130–132, 134, 135, 138, 175, 183, 246, 334, 341
 normal states, 126, 130, 131, 134
 normalcy, 12, 126, 130, 132, 135
 normative, 88, 133–135, 298, 310, 311, 325, 328, 365
 normativity, 132, 319 (*see also* Canghuilhem G.; Goldstein K.)
*Notes from Yuewei Cottage (*阅微草堂笔记*, Yuewei Caotang Biji)*, 267, 271
Nourishing, 79, 84, 359

O

Observations, 2, 5, 8, 10, 11, 13, 15, 21, 62–66, 71, 74, 75, 93, 109, 128, 138, 146, 147, 149, 151, 152, 157, 159, 166, 174, 177, 185, 196, 233, 234, 302, 318, 319, 322, 325–327, 350
Open, 9, 21, 48, 70, 83, 90, 91, 96, 158, 170, 171, 175, 187, 189, 191, 193, 202, 248, 251, 263, 269, 271, 276, 289, 291, 300, 302, 323, 328, 351, 358
Operations (*operatio*), 36–38, 46, 50, 53, 55, 74, 108, 128, 138, 293, 325, 327, 334, 336, 344, 355
Oral
 material, 259
 orally, 10
 testimonies, 8, 63, 259
Organism(s), 11, 12, 19, 67, 107–110, 112–115, 118, 127–134, 136, 137, 169, 247, 248, 348, 349, 351, 352, 354–366
Outsiders, 14, 16, 177–179, 190–192, 196, 201
Owen, R., 289

P

Painter, J., 139
Parasite
 paleo-parasitology, 156, 157
 parasitic diseases, 3, 148, 149
 parasitology, 3, 13, 146, 148–150
Pathogenic environment, 151, 153, 158
Pathological state(s), 12, 84, 126–135, 139, 141
Pathology/pathological, 3, 14, 127, 129–132, 134, 146, 156, 215, 221, 352, 357
Péquignot, H., 131
Perceptions, 13, 18, 29, 30, 36, 48, 114, 120, 137, 138, 140, 160, 201, 202, 232, 247, 271, 288, 292, 301, 303, 304, 327, 335–338, 355
 See also Merleau-Ponty, M.; Phenomenology
Perfumes/perfumed, 69, 185, 186, 226
 See also Fragrance
Perls, F., 348, 349, 351–354, 356–358, 361, 363
 See also Gestalt; Perls, L.
Perls, L., 348, 352
 See also Gestalt; Perls, F.
Permaculture, 17, 18, 140
Permeable
 permeability, 98, 172, 202

Perrier, J.-F., 139
Perse
  persian, 65, 210, 211, 219, 227, 231, 233
Pestilences, 225–226, 228, 230
Pests, 10, 63, 66, 68, 69
Pharmacists, 211, 216–219
Pharmacopeia, 216, 232
Pharmacopoeia, 106
Phenomenology
  phenomenological thinking, 137 (*see also* Merleau-Ponty, M.)
Philosophia naturalis natural philosophy, 10, 30
Philosophy, 9, 11, 12, 30, 36, 39, 41, 47, 54, 55, 112, 126–142, 222, 292, 312
  natural philosophy, 9, 30, 33, 35, 43, 54, 117, 232 (*see also* Existentialism; Phenomenology)
Phoenix, 53, 93
Physical
  environment, 11, 147, 149, 249, 289, 310, 335
  health, 249, 298
Physicians, 11, 14, 43, 50, 62, 70, 76, 82, 86, 91, 94, 107, 111, 117, 127, 129, 138, 155, 156, 166, 170, 175, 176, 179, 194, 202, 209–233, 290
  See also Medicine
Physics (*Physica*), 10, 31, 39, 41–43, 45, 46, 50, 51, 62, 110, 135, 136, 351
Physiognomony, 215, 216
  See also Medicine
Plant(s), 8, 14, 15, 29, 30, 40, 46, 53, 54, 64, 92, 93, 96, 99, 106–115, 118, 152, 183–187, 195, 199, 200, 211, 216, 217, 221, 222, 232, 233, 247, 248, 267, 268, 276, 279, 280, 294, 298, 303, 304, 312–314, 320, 323
Plato, 39, 50, 51, 214
Pliny the Elder (*Plinius*), 33, 44
Polluted
  pollution, 120, 300, 304, 311
Poor housing, 151
  See also Cafuas
Porosity, 6, 8
Powers, 16, 29, 36, 38, 50, 63, 68, 83, 92, 97–99, 108, 109, 114, 115, 139, 140, 149, 169, 210, 217, 220, 228, 242, 301, 304, 319, 322, 357
Pragmatism, 353
Pre-Islamic, 15, 64, 220, 221
Prevalence (of disease), 174, 175, 195, 197, 198
Property(ies)
  specific property *proprietas*, 50
Psychoanalysis, 19, 302, 348, 352–355, 357
Public health, 2, 13, 17, 135, 147, 150, 247, 288, 293, 298–300, 321, 322
Putrid areas, 289
  See also Miasma

Q
Qi 氣, 83
Qing (1644-1911), 167, 180, 195, 199, 256, 258–262, 264–267, 269, 270, 272, 276, 277, 280

R
Raj, K., 20
Rāzī, 69, 209–233
Records of Dreams under Window of Leisure (闲窗录梦*, Xianchuang Lumeng)*, 259
*Red and White Joyous Event (*红白喜事 *Hongbai Xishi)*, 261, 265, 274
Regeneration, 323
Region
  regional variations, 173, 175, 197–199, 201 (*see also* Geography; Land)
Reich, W., 348, 354
Reinhabitation, 310, 324
Relational ethics, 337, 351
Resilience, 318
Resonance, 8, 21, 80, 84, 89, 169, 328
Resources, 5, 12, 19, 119, 157, 194, 242, 246, 303, 306, 311, 312, 323, 335, 337, 340, 344, 359, 361, 362
Richardson, B. W., 290, 292
Rivasseau Jonveaux, T., 139
Rivers, 14, 64, 92–94, 96, 111, 113, 157, 167, 173–176, 180, 183, 184, 188, 189, 191, 195–197, 200–202, 233, 252, 267–271, 276, 297
Roberts, L., 20
Robic, M.-C., 134
Rocks, 92, 93, 99, 151, 336
Romantic movement, 306
Roncayolo, 291
Rosenberg, C., 1
Roszak, T., 140
Roudinesco Elizabeth, 130, 131
Rousseau, J.-J., 15, 240, 248–251, 289, 304, 306
Rural, 15, 114, 151, 152, 160, 179, 291, 299, 301, 313, 325
  See also Coutryside

S

Saint-Hilaire, É. G., 114, 115, 117
Saint-Victor (abbey and canons of), 36
Salerno (medical school of), 30
A Sand County Almanac, 15, 240, 241
Sassanid, 64, 73
Schaffer, S., 20
Scholastics, 28, 41, 48, 50, 54
Science
  history of science, 4–6
  scientific, 2–5, 9–11, 13–15, 18, 30, 32, 43, 45, 62, 63, 110, 111, 114, 116, 119, 120, 137, 141, 149, 150, 153, 156, 158, 160, 211, 215, 220, 224, 240, 243, 246, 251, 306
  scientific environment, 148–150
  scientific milieu, 5
  scientists, 4, 69, 70, 110, 187, 209–233
Sciencia naturalis, see Philosophia naturalis
Searles, H., 302
Season(s), 10, 11, 14, 73, 85, 87–90, 146, 168, 169, 183, 184, 188, 212, 222, 224–226, 229, 231–233, 303, 342
  See also Climate; Fog; Hot; Humid; Mists; Rain; Wind
Self
  self-reliance, 18, 314, 320–322
  theory of self, 355
Selin, H., 83
Senses
  sensitive, 31, 42, 44, 46, 203, 232, 244, 246, 249–251, 301
  sensory, 13, 18, 140, 232, 327, 350–352, 358
  sensory experiences, 15, 43, 140, 232 (*see also* Gestalt; Phenomenology)
  sensual, 109, 192, 350
  sensuous, 140, 350
Shanshui 山水 (lit. Mountain-Water), 91, 96, 202
  See also Environment
*Shuitu*水土 (lit. Water-Earth), 200, 202
  See also Environment
Sin (original sin), 35, 41, 50, 51
Skin, 83, 84, 87, 88, 90, 91, 96, 98, 141, 169–171, 175, 191, 202, 215, 358, 361
Smells, 3, 8, 13, 185, 244, 273, 290, 350
Soil, 3, 67, 93, 95, 98, 170, 191, 216–218, 241, 300, 310, 311, 320–324, 326, 342, 350, 361
Soul
  vegetative soul (*anima*), 42, 46
Sounds, 11, 63, 80, 84, 89, 97, 99, 140, 168, 203, 326, 327, 342, 354
Space
  lived space, 126
Species, 34, 40, 44, 46, 49, 50, 52, 53, 64–69, 71, 72, 75, 106, 109, 110, 112, 114, 115, 117, 128, 140, 148, 153, 154, 157, 214, 217, 312, 314, 326, 327
Speculum doctrinale
  The Doctrinal Mirror, 9, 33, 34, 41–43, 45, 46, 52
Speculum historiale
  The Mirror of History, 33, 42
Speculum maius
  The Great Mirror, 32
Speculum naturale
  The Mirror of Nature, 32, 33, 36–38, 44
Spencer, H., 111
Spirits
  spirituality, 329
Spitzer, L., 139
Standardization, 114, 321, 325, 329
Star(s)
  pole star, 87, 88
State, 31, 73, 80, 110, 128, 148, 168, 215, 242, 266, 289, 321, 334, 351
Statio
  station, 111–116, 120, 152 (*see also* Habitat; Milieu; Environment)
Steep banks, 183
Sternberg, E.M., 139
Stones, 30, 53, 85, 91, 92, 96, 170, 294, 350
Stuffy, 185
Subject
  subjective (experience), 137, 246, 259, 351, 353 (*see also* Self; Phenomenology)
Subsistence, 310, 311, 313, 326
Substance (*substantia*), 40, 46, 47, 49, 50, 69, 71, 92, 106, 107, 323, 363
Suffocating vapors, 174, 195, 197
Summa theologiae, 35, 47, 55
Sun, 11, 84, 86, 96, 212, 215, 222, 232, 291, 295–297, 350, 361
Sunlight, 291, 296
Super Dionysium de divinis nominibus, 47
Super Genesis ad litteram (On Genesis ad litteram), 55
Superstition, 280, 281
Surrounding(s)
  nature, 292
  world, 11, 12, 128, 133, 232, 233
Symbolism, 16, 91, 99, 100
Systematic correspondences, 168, 169, 173
  See also Cosmos

T
Tailor, 12, 129, 130, 135, 141
Taoism, 82, 83, 87, 323
Techniques
  laboratory techniques, 3, 152
  technology, 8, 292–294, 304, 316, 317
Temperament, 50, 51, 62, 69, 228
  See also Galen; Humors
Temples, 16, 260, 264, 266, 267, 269, 271–277, 279–281
Terrains, 90, 114, 170, 295, 296, 334, 340, 342
Theology (medieval), 9, 30, 31, 33, 36, 47, 54
Thibaud, J.-P., 139, 140, 328
Thick vegetation, 184
Thomas, C., 140
Thomas of Cantimpré (*Thomas Cantimpratensis*), 32, 41, 53
Thoreau, 240, 241, 251
Timaeus, 50, 51
Toombs, S.K., 138, 139
Topography, 2, 8, 183–187, 190, 199, 267, 351
Town
  planners, 296 (*see also* Urban)
Toxic, 69, 71, 91, 92, 99, 152, 170, 184, 195, 199
Transformations, 13, 92, 94, 95, 98, 100, 114–116, 120, 148, 154, 159, 172, 217, 219, 256, 259, 265, 277, 292, 293, 314, 315, 317, 321, 349, 350, 354
Transmission, 1, 31, 147–151, 158, 326
Travel
  travelers, 16, 63, 184, 227, 228, 230
Trees, 13, 29, 34, 35, 74, 85, 89, 92, 93, 96, 97, 99, 107, 155, 158, 189, 200, 217, 248, 267, 268, 271, 279, 290, 294, 295, 303, 342, 350
Trypanosomes, 146, 148–151, 153, 154, 157
Tuan, Y.-F., 140, 311
Turkish, 211

U
Uexküll, J. Von, 133, 134, 137, 335
Ulrich, R.S., 139
Umayyad caliphate, 220
Umgebung, 134
  See also Milieu
Umwelt, 134, 136
  See also Milieu
Unification, 98
Unity (principle of the), 15, 21, 98, 220, 228, 247, 354, 365
Universities (medieval), 28, 30, 31, 315
Urban
  areas, 288, 289, 292, 293, 295, 297–300, 303, 306
  environment, 17, 20, 256, 288, 291, 298, 301
  hygiene, 290, 291, 293, 295, 296, 298–300
  hygiene policies, 299
  planning, 19, 139, 256, 288, 293–305
  spaces, 16, 256–281, 295, 296, 298, 299
  urbanism, 294
  urbanization, 2
  utopias, 17, 288, 292, 296, 300

V
Valuation, 253, 305, 311, 325–329
Values, 8, 34, 50, 131, 133, 137, 233, 246, 253, 259, 289, 290, 322, 323, 326–328, 335, 337, 353, 357
van Beverwyck, J., 107
Vapors, 183, 187, 189, 195, 225
Vegetation, 92, 93, 96, 112, 183–187, 190, 241, 293, 296, 298, 305
Vermin (vermis, insects), 40
  See also Pests
Vigarello, G., 3, 169, 187, 200, 202
Vincent of Beauvais, 32, 33, 35, 36, 38, 40–43, 45–48, 51
Violence, 100, 263, 306, 362, 363
Virtue
  virtus, 44, 47, 53
  vis, 118, 159
  vis celestis, 54
  vis insita, 54
  See also Force
Vitalism, 136
Vocabulary, 11, 12, 106, 108–114, 118–120, 258, 292, 298, 314, 327
von Herder, J.G., 107

W
Warren, W.H., 336
Water
  water body, 49
  stagnant water(s), 3, 11, 91, 96
Wear, A., 8, 202
Wellbeing, 12, 126–142, 289, 302, 303, 305, 321, 328
Wild
  forests, 203
  wilderness movement, 306
Wild, C.P., 2

William of Conches, 30, 33, 38, 39, 44–45, 51
Williams, A., 139
Wind, 11, 84, 89, 90, 93, 96, 168–170, 173, 186, 187, 192, 197, 198, 212, 225, 229, 233, 276
Withering, W., 111
Wolf, F., 67
Womb, 95, 96, 99
Woods, 40, 81, 155, 156, 168, 169, 268, 291, 293, 297
See also Forests
World Health Organization (WHO), 239
Wright, F.L., 297

Y
Yunnan, 14, 175–178, 180–186, 188, 190, 194, 203

Z
Zakarīyā Ibn Muḥammad al-Qazwīnī, 66
Zhang 瘴*(miasma)*, 175, 264
See also Miasma
Zhuzi Family Rites(朱子家礼Zhuzi Jiali), 258, 259, 261
Zooethics, 12, 116, 117
Zoologists, 8, 62–75, 211, 216–219, 232

MIX
Papier aus verantwortungsvollen Quellen
Paper from responsible sources
FSC® C105338

If you have any concerns about our products,
you can contact us on
ProductSafety@springernature.com

In case Publisher is established outside the EU,
the EU authorized representative is:
Springer Nature Customer Service Center GmbH
Europaplatz 3, 69115 Heidelberg, Germany

Printed by Libri Plureos GmbH
in Hamburg, Germany